Lecture Notes in Control and Information Sciences 434

For further volumes:
http://www.springer.com/series/642

S. Bhatnagar, H.L. Prasad, and L.A. Prashanth

Stochastic Recursive Algorithms for Optimization

Simultaneous Perturbation Methods

Authors
Prof. S. Bhatnagar
Department of Computer Science
and Automation
Indian Institute of Science
Bangalore
India

L.A. Prashanth
Department of Computer Science
and Automation
Indian Institute of Science
Bangalore
India

H.L. Prasad
Department of Computer Science
and Automation
Indian Institute of Science
Bangalore
India

ISSN 0170-8643 e-ISSN 1610-7411
ISBN 978-1-4471-4284-3 ISBN 978-1-4471-4285-0 (eBook)
DOI 10.1007/978-1-4471-4285-0
Springer London Heidelberg New York Dordrecht

Library of Congress Control Number: 2012941740

Printed on acid-free paper

Springer is part of Springer Science+Business Media (www.springer.com)

To SB's parents Dr. G. K. Bhatnagar and Mrs. S.K. Bhatnagar, his wife Aarti and daughter Shriya

To HLP's parents Dr. H. R. Laxminarayana Bhatta and Mrs. G. S. Shreemathi, brother-in-law Giridhar N. Bhat and sister Vathsala G. Bhat

To LAP's daughter Samudyata

Preface

The area of *stochastic approximation* has its roots in a paper published by Robbins and Monro in 1951, where the basic stochastic approximation algorithm was introduced. Ever since, it has been applied in a variety of applications cutting across several disciplines such as control and communication engineering, signal processing, robotics and machine learning.

Kiefer and Wolfowitz, in a paper in 1952 (nearly six decades ago) published the first stochastic approximation algorithm for optimization. The algorithm proposed by them was a gradient search algorithm that aimed at finding the maximum of a regression function and incorporated finite difference gradient estimates. It was later found that whereas the Kiefer-Wolfowitz algorithm is efficient in scenarios involving scalar parameters, this is not necessarily the case with vector parameters, particularly those for which the parameter dimension is high. The problem that arises is that the number of function measurements needed at each update epoch grows linearly with the parameter dimension. Many times, it is also possible that the objective function is not observable as such and one needs to resort to simulation. In such scenarios, with vector parameters, one requires a corresponding (linear in the parameter-dimension) number of system simulations. In the case of large or complex systems, this can result in a significant computational overhead.

Subsequently, in a paper published in 1992, Spall proposed a stochastic approximation scheme for optimization that does a random search in the parameter space and only requires two system simulations regardless of the parameter dimension. This algorithm that came to be known as *simultaneous perturbation stochastic approximation* or *SPSA* for short, has become very popular because of its high efficiency, computational simplicity and ease of implementation. Amongst other impressive works, Katkovnik and Kulchitsky, in a paper published in 1972, also proposed a random search scheme (the *smoothed functional (SF) algorithm*) that only requires one system simulation regardless of the parameter dimension. Subsequent work showed that a two-simulation counterpart of this scheme performs well in practice. Both the Katkovnik-Kulchitsky as well as the Spall approaches involve perturbing the parameter randomly by generating certain *i.i.d.* random variables.

The difference between these schemes lies in the distributions these perturbation random variables can possess and the forms of the gradient estimators.

Stochastic approximation algorithms for optimization can be viewed as counterparts of deterministic search schemes with noise. Whereas, the SPSA and SF algorithms are gradient-based algorithms, during the last decade or so, there have been papers published on Newton-based search schemes for stochastic optimization. In a paper in 2000, Spall proposed the first Newton-based algorithm that estimated both the gradient and the Hessian using a simultaneous perturbation approach incorporating *SPSA-type* estimates. Subsequently, in papers published in 2005 and 2007, Bhatnagar proposed more Newton-based algorithms that develop and incorporate both SPSA and SF type estimates of the gradient and Hessian. In this text, we commonly refer to all approaches for stochastic optimization that are based on randomly perturbing parameters in order to estimate the gradient/Hessian of a given objective function as *simultaneous perturbation methods*. Bhatnagar and coauthors have also developed and applied such approaches for constrained stochastic optimization, discrete parameter stochastic optimization and reinforcement learning – an area that deals with the adaptive control of stochastic systems under real or simulated outcomes. The authors of this book have also studied engineering applications of the simultaneous perturbation approaches for problems of performance optimization in domains such as communication networks, vehicular traffic control and service systems.

The main focus of this text is on simultaneous perturbation methods for stochastic optimization. This book is divided into six parts and contains a total of fourteen chapters and five appendices. Part I of the text essentially provides an introduction to optimization problems - both deterministic and stochastic, gives an overview of search algorithms and a basic treatment of the Robbins-Monro stochastic approximation algorithm as well as a general multi-timescale stochastic approximation scheme. Part II of the text deals with gradient search stochastic algorithms for optimization. In particular, the Kiefer-Wolfowitz, SPSA and SF algorithms are presented and discussed. Part III deals with Newton-based algorithms that are in particular presented for the long-run average cost objective. These algorithms are based on SPSA and SF based estimators for both the gradient and the Hessian. Part IV of the book deals with a few variations to the general scheme and applications of SPSA and SF based approaches there. In particular, we consider adaptations of simultaneous perturbation approaches for problems of discrete optimization, constrained optimization (under functional constraints) as well as reinforcement learning. The long-run average cost criterion will be considered here for the objective functions. Part V of the book deals with three important applications related to vehicular traffic control, service systems as well as communication networks. Finally, five short appendices at the end summarize some of the basic material as well as important results used in the text.

This book in many ways summarizes the various strands of research on simultaneous perturbation approaches that SB has been involved with during the course of the last fifteen years or so. Both HLP and LAP have also been working in this area for over five years now and have been actively involved in the various aspects of the research reported here. A large portion of this text (in particular, Parts III-V as well as portions of Part II) is based mainly on the authors' own contributions to this area. The text provides a compact coverage of the material in a way that both researchers and practitioners should find useful. The choice of topics is intended to cover a sufficient width while remaining tied to the common theme of simultaneous perturbation methods. While we have made attempts at conveying the main ideas behind the various schemes and algorithms as well as the convergence analyses, we have also included sufficient material on the engineering applications of these algorithms in order to highlight the usefulness of these methods in solving real-life engineering problems. As mentioned before, an entire part of the text, namely Part IV, comprising of three chapters is dedicated for this purpose. The text in a way provides a balanced coverage of material related to both theory and applications.

Acknowledgements

SB was first introduced to the area of stochastic approximation during his Ph.D work with Prof. Vivek Borkar and Prof. Vinod Sharma at the Indian Institute of Science. Subsequently, he began to look at simultaneous perturbation approaches while doing a post doctoral with Prof. Steve Marcus and Prof. Michael Fu at the Institute for Systems Research, University of Maryland, College Park. He has also benefitted significantly from reading the works of Prof. James Spall and through interactions with him. He would like to thank all his collaborators over the years. In particular, he would like to thank Prof. Vivek Borkar, Prof. Steve Marcus, Prof. Michael Fu, Prof. Richard Sutton, Prof. Csaba Szepesvari, Prof. Vinod Sharma, Prof. Karmeshu, Prof. M. Narasimha Murty, Prof. N. Hemachandra, Dr. Ambedkar Dukkipati and Dr. Mohammed Shahid Abdulla. He would like to thank Prof. Anurag Kumar and Prof. K. V. S. Hari for several helpful discussions on optimization approaches for certain problems in vehicular traffic control (during the course of a joint project), which is also the topic of Chapter 13 in this book. SB considers himself fortunate to have had the pleasure of guiding and teaching several bright students at IISc. He would like to acknowledge the work done by all the current and former students of the Stochastic Systems Laboratory. A large part of SB's research during the last ten years at IISc has been supported through projects from the Department of Science and Technology, Department of Information Technology, Texas Instruments, Satyam Computers, EMC and Wibhu Technologies. SB would also like to acknowledge the various institutions where he worked and visited during the last fifteen years where portions of the work reported here have been conducted: The Institute for Systems Research, University of Maryland, College Park; Vrije Universiteit, Amsterdam; Indian Institute of Technology, Delhi; The RLAI Laboratory, University of Alberta; and the Indian Institute of Science. A major part of the work reported here has been conducted at IISc itself. Finally, he would like to thank his parents Dr. G. K. Bhatnagar and Mrs. S. K. Bhatnagar for their support, help and guidance all through the years, his wife Aarti and daughter Shriya for their patience, understanding and support, and his brother Dr. Shashank for his guidance and teaching during SB's formative years.

HLP's interest in the area of control engineering and decision making, which was essentially sown in him by interactions with Prof. U. R. Prasad at I.I.Sc., Dr. K. N. Shubhanga and Jora M. Gonda at NITK, led him to the area of operations research followed by that of stochastic approximation. He derives inspirations from the works of Prof. Vivek Borkar, Prof. James Spall, Prof. Richard Sutton, Prof. Shalabh Bhatnagar and several eminent personalities in the field of stochastic approximation. He thanks Dr. Nirmit V. Desai at IBM Research, India, collaboration with whom, brought up several new stochastic approximation algorithms with practical applications to the area of service systems. He thanks Prof. Manjunath Krishnapur and Prof. P. S. Sastry for equipping him with mathematical rigour needed for stochastic approximation. He thanks I. R. Rao at NITK who has been constant source of motivation. He thanks his father Dr. H. R. Laxminarayana Bhatta, mother Mrs. G. S. Shreemathi, brother-in-law Giridhar N. Bhat and sister Vathsala G. Bhat, for their constant support and understanding.

LAP would like to thank his supervising professor SB for introducing stochastic optimization during his PhD work. The extensive interactions with SB served to sharpen the understanding of the subject. LAP would also like to thank HLP, collaboration with whom has been most valuable. LAP's project associateship with Department of Information Technology as well as his internship at IBM Research presented many opportunities for developing as well as applying simultaneous pertubation methods in various practical contexts and LAP would like to thank Prof. Anurag Kumar, Prof. K.V.S. Hari of ECE department, IISc and Nirmit Desai and Gargi Dasgupta of IBM Research for several useful interactions on the subject matter. Finally, LAP would like to thank his family members - particularly his parents, wife and his daughter, for their support in this endeavour.

Bangalore,
May 2012

S. Bhatnagar
H.L. Prasad
L.A. Prashanth

Contents

Part I Introduction to Stochastic Recursive Algorithms

1 Introduction 3
1.1 Introduction 3
1.2 Overview of the Remaining Chapters 7
1.3 Concluding Remarks 11
References 11

2 Deterministic Algorithms for Local Search 13
2.1 Introduction 13
2.2 Deterministic Algorithms for Local Search 14
References 15

3 Stochastic Approximation Algorithms 17
3.1 Introduction 17
3.2 The Robbins-Monro Algorithm 18
3.2.1 Convergence of the Robbins-Monro Algorithm 19
3.3 Multi-timescale Stochastic Approximation 23
3.3.1 Convergence of the Multi-timescale Algorithm 24
3.4 Concluding Remarks 26
References 27

Part II Gradient Estimation Schemes

4 Kiefer-Wolfowitz Algorithm 31
4.1 Introduction 31
4.2 The Basic Algorithm 31
4.2.1 Extension to Multi-dimensional Parameter 35
4.3 Variants of the Kiefer-Wolfowitz Algorithm 36
4.3.1 Fixed Perturbation Parameter 36
4.3.2 One-Sided Variants 37

4.4 Concluding Remarks 38
References 38

5 Gradient Schemes with Simultaneous Perturbation Stochastic Approximation 41
5.1 Introduction 41
5.2 The Basic SPSA Algorithm 41
5.2.1 Gradient Estimate Using Simultaneous Perturbation 42
5.2.2 The Algorithm 43
5.2.3 Convergence Analysis 44
5.3 Variants of the Basic SPSA Algorithm 47
5.3.1 One-Measurement SPSA Algorithm 47
5.3.2 One-Sided SPSA Algorithm 49
5.3.3 Fixed Perturbation Parameter 49
5.4 General Remarks on SPSA Algorithms 51
5.5 SPSA Algorithms with Deterministic Perturbations 52
5.5.1 Properties of Deterministic Perturbation Sequences 52
5.5.2 Hadamard Matrix Based Construction 54
5.5.3 Two-Sided SPSA with Hadamard Matrix Perturbations 56
5.5.4 One-Sided SPSA with Hadamard Matrix Perturbations 62
5.5.5 One-Measurement SPSA with Hadamard Matrix Perturbations 63
5.6 SPSA Algorithms for Long-Run Average Cost Objective 65
5.6.1 The Framework 65
5.6.2 The Two-Simulation SPSA Algorithm 65
5.6.3 Assumptions 66
5.6.4 Convergence Analysis 68
5.6.5 Projected SPSA Algorithm 73
5.7 Concluding Remarks 75
References 75

6 Smoothed Functional Gradient Schemes 77
6.1 Introduction 77
6.2 Gaussian Based SF Algorithm 79
6.2.1 Gradient Estimation via Smoothing 79
6.2.2 The Basic Gaussian SF Algorithm 81
6.2.3 Convergence Analysis of Gaussian SF Algorithm 82
6.2.4 Two-Measurement Gaussian SF Algorithm 88
6.3 General Conditions for a Candidate Smoothing Function 91
6.4 Cauchy Variant of the SF Algorithm 92
6.4.1 Gradient Estimate 92
6.4.2 Cauchy SF Algorithm 94
6.5 SF Algorithms for the Long-Run Average Cost Objective 94
6.5.1 The G-SF1 Algorithm 95
6.5.2 The G-SF2 Algorithm 99

6.5.3 Projected SF Algorithms ... 100
6.6 Concluding Remarks ... 101
References ... 101

Part III Hessian Estimation Schemes

7 Newton-Based Simultaneous Perturbation Stochastic Approximation 105
7.1 Introduction ... 105
7.2 The Framework ... 106
7.3 Newton SPSA Algorithms ... 106
7.3.1 Four-Simulation Newton SPSA (N-SPSA4) ... 107
7.3.2 Three-Simulation Newton SPSA (N-SPSA3) ... 109
7.3.3 Two-Simulation Newton SPSA (N-SPSA2) ... 110
7.3.4 One-Simulation Newton SPSA (N-SPSA1) ... 111
7.4 Woodbury's Identity Based Newton SPSA Algorithms ... 113
7.5 Convergence Analysis ... 114
7.5.1 Assumptions ... 114
7.5.2 Convergence Analysis of N-SPSA4 ... 117
7.5.3 Convergence Analysis of N-SPSA3 ... 125
7.5.4 Convergence Analysis of N-SPSA2 ... 126
7.5.5 Convergence Analysis of N-SPSA1 ... 128
7.5.6 Convergence Analysis of W-SPSA Algorithms ... 130
7.6 Concluding Remarks ... 130
References ... 131

8 Newton-Based Smoothed Functional Algorithms ... 133
8.1 Introduction ... 133
8.2 The Hessian Estimates ... 134
8.2.1 One-Simulation Hessian SF Estimate ... 134
8.2.2 Two-Simulation Hessian SF Estimate ... 136
8.3 The Newton SF Algorithms ... 137
8.3.1 The One-Simulation Newton SF Algorithm (N-SF1) ... 137
8.3.2 The Two-Simulation Newton SF Algorithm (N-SF2) ... 138
8.4 Convergence Analysis of Newton SF Algorithms ... 139
8.4.1 Convergence of N-SF1 ... 139
8.4.2 Convergence of N-SF2 ... 146
8.5 Concluding Remarks ... 147
References ... 147

Part IV Variations to the Basic Scheme

9 Discrete Parameter Optimization ... 151
9.1 Introduction ... 151
9.2 The Framework ... 152
9.2.1 The Deterministic Projection Operator ... 153
9.2.2 The Random Projection Operator ... 154

9.2.3 A Generalized Projection Operator 155
9.2.4 Regular Projection Operator to $\bar{C}$ 157
9.2.5 Basic Results for the Generalized Projection Operator Case . 157
9.3 The Algorithms 160
9.3.1 The SPSA Algorithm 160
9.3.2 The SFA Algorithm 161
9.3.3 Convergence Analysis 162
9.4 Concluding Remarks 164
References 166

10 Algorithms for Constrained Optimization 167
10.1 Introduction 167
10.2 The Framework 168
10.3 Algorithms 171
10.3.1 Constrained Gradient-Based SPSA Algorithm (CG-SPSA) . 172
10.3.2 Constrained Newton-Based SPSA Algorithm (CN-SPSA) .. 173
10.3.3 Constrained Gradient-Based SF Algorithm (CG-SF) 175
10.3.4 Constrained Newton-Based SF Algorithm (CN-SF) 176
10.4 A Sketch of the Convergence 178
10.5 Concluding Remarks 185
References 186

11 Reinforcement Learning 187
11.1 Introduction 187
11.2 Markov Decision Processes 188
11.3 Numerical Procedures for MDPs 191
11.3.1 Numerical Procedures for Discounted Cost MDPs 192
11.3.2 Numerical Procedures for Long-Run Average Cost MDPs .. 193
11.4 Reinforcement Learning Algorithms for Look-up Table Case 194
11.4.1 An Actor-Critic Algorithm for Infinite Horizon Discounted Cost MDPs 195
11.4.2 The Q-Learning Algorithm and a Simultaneous Perturbation Variant for Infinite Horizon Discounted Cost MDPs 198
11.4.3 Actor-Critic Algorithms for Long-Run Average Cost MDPs 202
11.5 Reinforcement Learning Algorithms with Function Approximation . 206
11.5.1 Temporal Difference (TD) Learning with Discounted Cost .. 206
11.5.2 An Actor-Critic Algorithm with a Temporal Difference Critic for Discounted Cost MDPs 210
11.5.3 Function Approximation Based Q-learning Algorithm and a Simultaneous Perturbation Variant for Infinite Horizon Discounted Cost MDPs 213
11.6 Concluding Remarks 218
References 218

Part V Applications

12 Service Systems 225
12.1 Introduction 225
12.2 Service System Framework 226
12.3 Problem Formulation 228
12.4 Solution Methodology 232
12.5 First Order Methods 234
12.5.1 SASOC-SPSA 234
12.5.2 SASOC-SF-N 235
12.5.3 SASOC-SF-C 236
12.6 Second Order Methods 236
12.6.1 SASOC-H 237
12.6.2 SASOC-W 237
12.7 Notes on Convergence 238
12.8 Summary of Experiments 239
12.9 Concluding Remarks 240
References 240

13 Road Traffic Control 243
13.1 Introduction 243
13.2 Q-Learning for Traffic Light Control 245
13.2.1 Traffic Control Problem as an MDP 245
13.2.2 The TLC Algorithm 246
13.2.3 Summary of Experimental Results 248
13.3 Threshold Tuning Using SPSA 248
13.3.1 The Threshold Tuning Algorithm 249
13.3.2 Traffic Light Control with Threshold Tuning 250
13.3.3 Summary of Experimental Results 254
13.4 Concluding Remarks 255
References 255

14 Communication Networks 257
14.1 Introduction 257
14.2 The Random Early Detection (RED) Scheme for the Internet 258
14.2.1 Introduction to RED Flow Control 258
14.2.2 The Framework 259
14.2.3 The B-RED and P-RED Stochastic Approximation Algorithms 263
14.2.4 Summary of Experimental Results 266
14.3 Optimal Policies for the Retransmission Probabilities in Slotted Aloha 267
14.3.1 Introduction to the Slotted Aloha Multiaccess Communication Protocol 267
14.3.2 The SDE Framework 268

14.3.3 The Algorithm 270
14.3.4 Summary of Experimental Results 272
14.4 Dynamic Multi-layered Pricing Schemes for the Internet 272
14.4.1 Introduction to Dynamic Pricing Schemes 273
14.4.2 The Pricing Framework 273
14.4.3 The Price Feed-Back Policies and the Algorithms 275
14.4.4 Summary of Experimental Results 277
14.5 Concluding Remarks 278
References 279

Part VI Appendix

A Convergence Notions for a Sequence of Random Vectors 283
Reference 285

B Martingales 287
References 289

C Ordinary Differential Equations 291
References 294

D The Borkar-Meyn Theorem for Stability and Convergence of Stochastic Approximation 295
References 296

E The Kushner-Clark Theorem for Convergence of Projected Stochastic Approximation 297
References 300

Index 301

Acronyms

Lists of abbreviations used in the ensuing text, are as follows.

SPSA	Simultaneous Perturbation Stochastic Approximation
SF	Smoothed Functional
SFA	Smoothed Functional Approximation
MDP	Markov Decision Process
SDP	Stationary Deterministic Policy
SRP	Stationary Randomized Policy
RL	Reinforcement Learning
AC	Actor-Critic
SLA	Service Level Agreement
R-M	Robbins-Monro
K-W	Kiefer-Wolfowitz
FDSA	Finite Difference Stochastic Approximation
IPA	Infinitesimal Perturbation Analysis
p.d.f.	Probability Density Function
i.i.d.	Independent and Identically Distributed
a.s.	Almost Surely
w.p.1	With Probability One
ODE	Ordinary Differential Equation
SDE	Stochastic Differential Equation
OCBA	Optimal computing budget allocation
R&S	Ranking and Selection
MCP	Multiple Comparison Procedures
TLC	Traffic Light Control
QoS	Quality of Service
RED	Random Early Detection
CSMA	Carrier Sense Multiple Access
CSMA/CD	CSMA with Collision Detection
TP	Tirupati Pricing
PMP	Paris Metro Pricing

Part I
Introduction to Stochastic Recursive Algorithms

Stochastic recursive algorithms are one of the most important tools for problems of stochastic optimization. In recent times, an important class of such algorithms that are based on the simultaneous perturbation technique has become popular because of their superior computational time performance in converging to an optimum point. This has resulted in a flurry of research activity on stochastic algorithms that involve simultaneous perturbation.

This part of the book consists of three chapters. Chapter 1 gives an introduction to stochastic optimization problems and provides a motivation of where such problems arise and why they are important. It also provides an overview of the remaining chapters.

Chapter 2 discusses some of the well-known deterministic algorithms for optimization. Stochastic recursive algorithms turn out to be the stochastic analogs of these algorithms.

The basic stochastic recursive algorithm is the Robbins and Monro scheme. It is found to be applicable in a wide variety of settings, in particular, stochastic optimization. In Chapter 3, we discuss in detail the Robbins-Monro algorithm and analyze its convergence. The Robbins-Monro scheme (so named after its inventors, Robbins and Monro) is normally applicable when the objective function is an expectation of a noisy cost objective. Many times, one is faced with a problem of optimizing a long-run average cost objective in order to, say, optimize a steady-state system performance. Multi-timescale stochastic approximation plays an important role in such scenarios. We also present in Chapter 3, a general two-timescale stochastic recursive scheme and present its convergence analysis under general conditions.

Chapter 1
Introduction

1.1 Introduction

Optimization methods play an important role in many disciplines such as signal processing, communication networks, neural networks, economics, operations research, manufacturing systems, vehicular traffic control, service systems and several others. For instance, in a general communication network, a goal could be to optimally allocate link bandwidth amongst competing traffic flows. Similarly, an important problem in the setting of traffic signal control is to dynamically find the optimal order to switch traffic lights at signal junctions and the amount of time that a lane signal should be green when inputs such as the number of vehicles waiting at other lanes are provided. In the case of a manufacturing plant, an important problem is to decide the optimal order in which to allocate machine capacity for manufacturing various products on any day given the demand patterns for various products. These are only a few specific instances of innumerable problems across various disciplines that fall within the broad category of optimization problems. A usual way to model these problems analytically is by defining an objective or a cost function whose optimum constitutes the desired solution. For instance, in the case of the traffic signal control problem, a cost function could be the sum of queue lengths of vehicles waiting across all lanes at a red signal intersection. Thus, an optimal signal switching order would ensure that the sum of the queue lengths of waiting vehicles is minimized and thereby traffic flows are maximized. In general, a cost function is designed to penalize the less desirable outcomes. However, in principle, there can be several cost functions that have the same (or common) desired outcome as their optimum point. Suitably designing a cost objective in order to obtain the desired outcome in a reasonable amount of time when following a computational procedure could be a domain-specific problem. For instance, in the context of the traffic signal control problem mentioned above, another cost objective with the same optimum could be the sum of squared queue lengths of waiting vehicles instead of the sum of queue lengths. Optimization problems can be deterministic or stochastic, as well as they can be static or dynamic. We discuss this issue in more detail below.

S. Bhatnagar et al.: Stochastic Recursive Algorithms for Optimization, LNCIS 434, pp. 3–12.
springerlink.com

A general optimization problem that we shall be concerned about for the most part in this book has the following form:

$$\text{Find } \theta^* \text{ that solves } \min_{\theta \in C} J(\theta), \tag{1.1}$$

where $J : \mathscr{R}^N \to \mathscr{R}$ is called the objective function, θ is a tunable N-dimensional parameter and $C \subset \mathscr{R}^N$ is the set in which θ takes values. If one has complete information about the function J and its first and higher order derivatives, and about the set C, then (1.1) is a deterministic optimization problem. If on the other hand, J is obtained as $J(\theta) = E_\xi[h(\theta,\xi)]$, where $E_\xi[\cdot]$ is the expected value over noisy observations or samples $h(\theta,\xi)$ of the cost function with *random noise* ξ, and one is allowed to observe only these samples (without really knowing J), then one is in the realm of stochastic optimization. Such problems are more challenging because of the added complexity of not knowing the cost objective $J(\cdot)$ precisely and to find the optimum parameter only on the basis of the aforementioned noisy observations.

As we shall subsequently see, many times one resorts to search algorithms in order to find an optimum point, i.e., a solution to (1.1). In stochastic optimization algorithms, it is not uncommon to make a random choice in the search direction – in fact most of our treatment will be centered around such algorithms. Thus, a second distinction between deterministic and stochastic optimization problems lies in the way in which search progresses - a random search algorithm invariably results in the optimization setting being stochastic as well.

Suppose now that the objective function J has a *multi-stage* character, i.e., is of the form $J(\theta) = \sum_{i=1}^{N} E[h_i(X_i)]$, where N denotes the number of stages and X_i is the state of an underlying process in stage i, $i = 1,\ldots,N$. The state captures the most important attributes of the system that are relevant for the optimization problem. Further, h_i denotes a stage and state-dependent cost function. Let $\theta \stackrel{\triangle}{=} (\theta_1,\ldots,\theta_N)^T$ denote a vector of parameters θ_j, $j = 1,\ldots,N$ and let X_i depend on θ. The idea here is that optimization can be done one stage at a time over N stages after observing the state X_i in each stage i. Here, the value θ_i of the parameter in stage i has a bearing on the cost of all subsequent stages $i+1,\ldots,N$. This in short is the problem of dynamic optimization. Approaches such as dynamic programming are often used to solve dynamic optimization problems. Other manifestations of dynamic optimization, say over an infinite number of stages or in continuous time also exist. In relation to the above (multi-stage) problem, in static optimization, one would typically perform a single-shot optimization where the parameters $\theta_1,\ldots,\theta_N$ would be optimized all at once in the first stage itself. Broadly speaking while in a dynamic optimization problem with multiple stages, one makes decisions instantly as states are revealed, in static optimization, there is no explicit notion of time or perhaps even state as all decisions can be made at once.

An important class of multi-stage problems are those with an infinite number of stages and where the objective function is a long-run average over single-stage cost functions. More precisely, the objective function in this case has the form

$$J(\theta) = \lim_{N\to\infty} \frac{1}{N} E\left[\sum_{i=1}^{N} h_i(X_i)\right], \tag{1.2}$$

where X_i as before is the state in stage i that we assume depends on the parameter θ. An objective as (1.2) would in most cases not be analytically known. A usual search procedure to find the optimum parameter in such problems would run into the difficulty of having to estimate the cost over an infinitely long trajectory before updating the parameter estimate, thereby making the entire procedure very tedious.

Another important class of optimization problems is that of constrained optimization. Here, the idea is to optimize a given objective or cost function subject to constraints on the values of additional cost functions. Thus consider the following variation to the basic problem (1.1).

$$\text{Find } \theta^* \text{ for which } J(\theta^*) = \min_{\theta\in C}\{J(\theta) \mid G_i(\theta) \leq \alpha_i,\ i = 1,\ldots,p\}. \tag{1.3}$$

Here, $G_i(\cdot)$ and α_i, $i = 1,\ldots,p$ are certain additional cost functions and constants, respectively, that constitute the functional constraints. In the context of the traffic signal control problem where the objective function to be minimized is the sum of queue lengths on the various lanes, constraints could be put for the traffic on the side roads so that the main road traffic gets higher priority. For instance, a constraint there could specify that the traffic signal for a side road lane can be switched to green only provided the number of vehicles waiting on such a lane exceeds ten. Similarly, in a communication network, the objective could be to maximize the average throughput. A constraint there could specify that the average delay must be below a threshold. Another constraint could similarly be on the probability of packet loss during transmission being below a small constant, say 0.01.

While for the most part, we shall be concerned with optimization problems of the form (1.1), we shall subsequently also consider constrained optimization problems of the type (1.3). The objective function (and also the constraint functions in the case of (1.3)) will be considered to be certain long-run average cost functions.

We shall present various stochastic recursive search algorithms for these problems. Many of the stochastic search algorithms for optimization can be viewed as stochastic (i.e., with noise) counterparts of corresponding deterministic search algorithms such as gradient and Newton methods. In the setting of stochastic optimization, where the form of the objective function as well as its derivatives is unknown, one needs to resort to estimation of quantities such as the gradient and Hessian from noisy function measurements or else through simulation. A finite difference estimate of the gradient as proposed by Kiefer and Wolfowitz [18] requires a number of function measurements or simulations that is linear in the number of parameter components. A similar estimate of the Hessian [14] requires a number of function measurements that is quadratic in the number of measurements or simulations. When the parameter dimension is large, algorithms with gradient/Hessian estimators as above would be computationally inefficient because such algorithms would update once only after all the required function measurements have been made or simulations conducted. It is here that *simultaneous perturbation methods* play a significant role. In a paper published in 1992, Spall presented the Simultaneous Perturbation Stochastic Approximation (SPSA) algorithm that estimated the gradient of the objective function using exactly two function measurements (or simulations) made from perturbed values of the parameter, where each component of the parameter is perturbed along random directions using independent random variates most commonly distributed according to the Bernoulli distribution. A second well-known simultaneous perturbation technique that in fact came before SPSA was the smoothed functional (SF) scheme [17]. The idea in this scheme is some what similar to SPSA, however, the form of the gradient estimator is considerably different as perturbations that are distributed as per the Gaussian, Cauchy or uniform distributions can be used. A basic format for the simultaneous perturbation technique is described in Fig. 1.1.

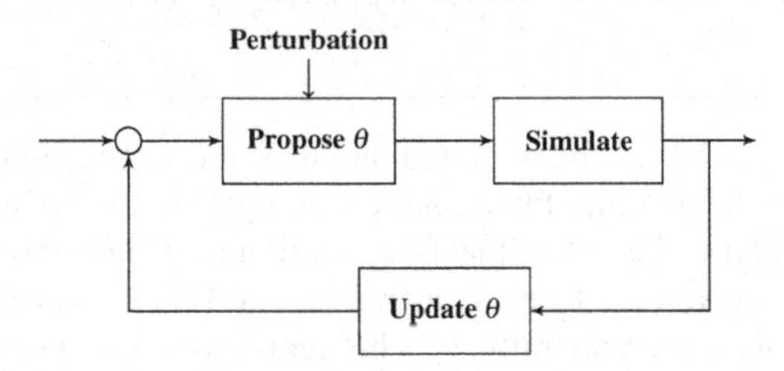

Fig. 1.1 Overall flow of a basic simultaneous perturbation algorithm.

During the course of the last ten to fifteen years, there has been a spurt of activity in developing Newton-based simultaneous perturbation methods. In [27] and [3], Newton-based analogs of the SPSA method were proposed. Further, in [4], Newton-based analogs of the SF algorithm have been proposed. We may mention here that in this text, by *simultaneous perturbation methods*, we refer to the entire family of algorithms that are based on either gradient or gradient and Hessian estimates that are obtained using some form of simultaneous random perturbations. While for the most part, we shall be concerned with static optimization problems, we shall also consider later, the problem of dynamic stochastic control or of decision making under uncertainty over a sequence of time instants. This problem will subsequently be cast as one of dynamic parameter optimization. We shall also present towards the end, applications of the proposed methods and algorithms to service systems, road traffic control and communication networks. A common unifying thread in most of the material presented in this text is of simultaneous perturbation methods.

1.2 Overview of the Remaining Chapters

We now provide a brief overview of the remainder of this book. In Chapter 2, we briefly discuss well-known local search algorithms. These have been described mainly for the case of deterministic optimization. However, we also discuss briefly the case of stochastic optimization as well. The algorithms for stochastic optimization that we present in later chapters will be based on these algorithms.

The fundamental stochastic algorithm due to Robbins and Monro [22] is almost six decades old. It estimates the zeros of a given objective function from noisy cost samples. Most stochastic search algorithms can be viewed as variants of this algorithm. In Chapter 3, we describe the R-M algorithm. We also present in this chapter, a general multi-timescale stochastic approximation algorithm that can be viewed as a variant of the R-M algorithm. Multi-timescale stochastic approximation algorithms play a significant role in the case of problems where the computational procedure would typically involve two nested loops where an outer loop update can happen only upon convergence of the inner loop procedure. A specific instance is the case when the objective function is a long-run average cost of the form (1.2). Such an objective function is useful in scenarios where one is interested in optimizing steady-state system performance measures, such as minimizing long-run average delays in a vehicular traffic network or the steady-state loss probability in packet transmissions in a communication network. A regular computational procedure in this case would perform the outer loop (parameter) update only after convergence of the inner loop procedure (viz., after obtaining the long-run average cost corresponding to a given parameter update). The same effect can be obtained with the use of coupled simultaneous stochastic updates that are however governed with diminishing step-size schedules that have different rates of convergence - the faster update governed with a slowly diminishing schedule and vice versa. Borkar [12, 13] has given a general analysis of these algorithms. We discuss the convergence of both the R-M and the multi-timescale algorithms.

Amongst the first stochastic gradient search algorithms based on estimating the gradient of the objective function using noisy cost samples is the Kiefer-Wolfowitz (K-W) algorithm [18] due to Kiefer and Wolfowitz. We review this algorithm in Chapter 4. While it was originally presented for the case of scalar parameters, in the case of vector-valued parameters, the K-W algorithm makes function measurements after perturbing at most one parameter component. Thus, K-W is not efficient under high-dimensional parameters since the number of function measurements or system simulations required to estimate the gradient grows linearly with the parameter dimension.

Spall invented the simultaneous perturbation stochastic approximation (SPSA) algorithm [23], [28] that requires only two function measurements at each instant regardless of the parameter dimension, by simultaneously perturbing all parameter components using a class of i.i.d. random variables. The most commonly used perturbations in this class are symmetric, ± 1-valued, Bernoulli-distributed random variables. A one-simulation version of this algorithm was subsequently presented in [24]. However, it was not found to be as effective as regular two-simulation SPSA. In [7], certain deterministic constructions for the perturbation random variables have been explored for both two-simulation and one-simulation SPSA. These have been found to yield better results as compared to their random perturbation counterparts. We review the SPSA algorithm and its variants in Chapter 5.

Katkovnik and Kulchitsky [17] presented a smoothed functional (SF) approach that is another technique to estimate the gradient of the objective function using random perturbations. This technique is some what different from SPSA. In particular, the properties required of the perturbation random variables here are seen to be most commonly satisfied by Gaussian and Cauchy distributed random variables. If one considers a convolution of the gradient of the objective function with a smoothing density function (such as that of Gaussian or Cauchy random variables), then through a suitable integration-by-parts argument, one can rewrite the same as a convolution of the gradient of the probability density function (p.d.f.) with the objective function itself. The derivative of the smoothing p.d.f. is seen to be a scaled version of the same p.d.f. This suggests that if the perturbations are generated using such p.d.fs, only one function measurement or system simulation is sufficient to estimate the gradient of the objective (in fact, the convolution of the gradient, that however converges to the gradient itself in the scaling limit of the perturbation parameter). A two-simulation variant of this algorithm that incorporates balanced estimates has been proposed in [29] and found to perform better than its one-simulation counterpart. We review developments in the gradient-based SF algorithms in Chapter 6.

Spall [27] presented simultaneous perturbation estimates for the Hessian that incorporate two independent perturbation sequences that are in the same class of sequences as used in the SPSA algorithm. The Hessian estimate there is based on four function measurements or system simulations, two of which are the same as those used for estimating the gradient of the objective. In [3], three other Hessian estimators were proposed. These are based on three, two and one system simulation(s), respectively. In Chapter 7, we review the simultaneous perturbation estimators of the Hessian. An issue with Newton-based algorithms that incorporate the Hessian is in estimating the inverse of the Hessian matrix at each update epoch. We also discuss in this chapter some of the recent approaches for inverting the Hessian matrix.

Bhatnagar [4] developed two SF estimators for the Hessian based on one and two system simulations, respectively, when Gaussian p.d.f. is used as the smoothing function. Using an integration-by-parts argument (cf. Chapter 6), twice, the Hessian estimate is seen to be obtained from a single system simulation itself. A two-sided balanced Hessian estimator is, however, seen to perform better than its one-sided counterpart. An interesting observation here is that both the gradient and the Hessian estimates are obtained using the same simulation(s). We review the SF estimators of the Hessian matrix in Chapter 8.

In Chapter 9, we consider the case when the optimization problem has a form similar to (1.1); however, the underlying set C is discrete-valued. Further, we shall let the objective function be a long-run average cost as with (1.2). In [11], two gradient search algorithms based on SPSA and SF have been proposed for this problem. A randomized projection approach was proposed there that is seen to help in adapting the continuous optimization algorithms to the discrete setting. We present another approach based on certain generalized projections that can be seen to be a mix of deterministic and randomized projection approaches, and result in the desired smoothing of the dynamics of the underlying process. Such a projection mechanism would also result in a lower computational complexity as opposed to a fully randomized projection scheme.

Next, in Chapter 10, we will be concerned with constrained optimization problems with similar objective as (1.3). We shall, in particular, be concerned here with the case when the objective has a long-run-average form similar to (1.2). Thus, in such cases, neither the objective nor the constraint region is known analytically to begin with. In [8], stochastic approximation algorithms based on SPSA and SF estimators for both the gradient and the Hessian have been presented. The general approach followed is based on forming the Lagrangian – the Lagrange multipliers are updated on a slower timescale than the parameter that, in turn, is updated on a slower scale in comparison to that on which data gets averaged. We will review these algorithms in Chapter 10.

Reinforcement learning (RL) algorithms [2] are geared towards solving stochastic control problems using only real or simulated data when the system model (in terms of the transition probabilities) is not known. Markov decision process (MDP) is a general framework for studying such problems. Classical approaches such as policy iteration and value iteration for solving MDP require knowledge of transition probabilities. Many RL algorithms are stochastic recursive procedures aimed at solving such problems when transition probabilities are unknown. Actor-critic (AC) algorithms are a class of RL algorithms that are based on policy iteration and involve two loops - the outer loop update does policy improvement while the inner loop procedure is concerned with policy evaluation. These algorithms thus incorporate two-timescale stochastic approximation. In [10, 1, 6], AC algorithms for various cost criteria such as infinite horizon discounted cost, long-run average cost as well as total expected finite horizon cost, that incorporate simultaneous perturbation gradient estimates have been proposed. We shall review the development of the infinite horizon algorithms in Chapter 11.

Chapter 12 considers the problem of optimizing staffing levels in service systems. The aim is to adapt the staffing levels as they are labor intensive and have a time varying workload. This problem is, however, nontrivial due to a large number of parameters and operational variations. Further, any staffing solution is constrained to maintain the system in steady-state and be compliant to aggregate SLA constraints. We formulate the problem using the constrained optimization framework where the objective is to minimize the labor cost in the long run average sense and the constraint functions are long run averages of the SLA and queue stability constraints. Using the ideas of the algorithms proposed in Chapter 10 for a generalized constrained optimization setting, we describe several simulation optimization methods that have been originally proposed in [19] for solving the labor cost optimization problem. The presented algorithms are based on SPSA and SF gradient/Hessian estimates. These algorithms have been seen in [19] to exhibit better overall performance vis-a-vis the state-of-the-art optimization tool-kit OptQuest, while being more than an order of magnitude faster than Optquest.

In Chapter 13, we consider the problem of finding optimal timings and the order in which to switch traffic lights given dynamically evolving traffic conditions. We describe here applications of the reinforcement learning and stochastic optimization approaches in order to maximize traffic flow through the adaptive control of traffic lights. We assume, however, as in the case of real-life situations that only rough estimates of the congestion levels are available, for instance, whether congestion is below a lower threshold, above an upper threshold or is in between the two. All our algorithms incorporate such threshold levels in the feedback policies and find optimal policies given a particular set of thresholds. For instance, in a recent work [21], we considered Q-learning-based traffic light control (TLC) where the features are obtained using such (aforementioned) thresholds. We also describe similar other algorithms based on simulation optimization methods. An important question then is to find optimal settings for the thresholds themselves. We address this question by incorporating simultaneous perturbation estimates to run in tandem with other algorithms. An important observation is that our algorithm shows significantly better empirical performance as compared to other related algorithms in the literature. Another interesting consequence of our approach is that when applied together with reinforcement learning algorithms, such methods result in obtaining an *optimal* set of features from a given parametrized feature class.

In Chapter 14, we select and discuss three important problems in communication networks, where simultaneous perturbation approaches have been found to be significantly useful. We first consider the problem of adaptively tuning the parameters in the case of random early detection (RED) adaptive queue management scheme proposed for TCP/IP networks. The original scheme proposed by Floyd [15] considers a fixed set of parameters regardless of the network and traffic conditions. We address this problem using techniques from constrained optimization [20] and apply simultaneous perturbation approaches that are found to exhibit excellent performance. Next, we consider the problem of tuning the retransmission probability parameter for the slotted Aloha multi-access communication system. The protocol as such specifies a fixed value for the same regardless of the number of users sending packets on the channel and the channel conditions. We propose a stochastic differential equation (SDE)-based formulation [16, 9] in order to find an optimal parameter trajectory over a finite time horizon. We also consider the problem of optimal pricing in the Internet. The idea here is that in order to provide a higher quality of service to a user who is willing to pay more, one needs to find optimal strategies for fixing prices of the various services offered. Our techniques [30] play a role here as well and are found to exhibit significantly better performance in comparison to other known methods.

Finally, in Appendices A-E, we present some of the basic material needed in the earlier chapters. In particular, we present (a) convergence notions for a sequence of random vectors, (b) results on martingales and their convergence, (c) ordinary differential equations, (d) the Borkar and Meyn stability result, and (e) the Kushner-Clark theorem for convergence of projected stochastic approximations. Some of the background material as well as the main results used in other chapters have also been summarized in these appendices.

1.3 Concluding Remarks

Stochastic approximation algorithms are one of the most important class of techniques for solving optimization problems involving uncertainty. Simultaneous perturbation approaches for optimization have evolved into a rich area by themselves from the viewpoint of both theory and numerous highly successful applications. Several estimators for the gradient and Hessian that involve simultaneous perturbation estimates have been developed in recent times that are seen to show excellent performance. SPSA and SF algorithms constitute powerful methods for stochastic optimization that have been found useful in many disciplines of science and engineering. The book reference of [28] provides an excellent account of SPSA. Surveys on the SPSA algorithm are available in [26], [25]. Also, [5] provides a more recent survey on simultaneous perturbation algorithms involving both SPSA and SF estimators. The current text is a significantly expanded version of [5].

References

1. Abdulla, M.S., Bhatnagar, S.: Reinforcement learning based algorithms for average cost Markov decision processes. Discrete Event Dynamic Systems 17(1), 23–52 (2007)
2. Bertsekas, D.P., Tsitsiklis, J.N.: Neuro-Dynamic Programming. Athena Scientific, Belmont (1996)
3. Bhatnagar, S.: Adaptive multivariate three-timescale stochastic approximation algorithms for simulation based optimization. ACM Transactions on Modeling and Computer Simulation 15(1), 74–107 (2005)
4. Bhatnagar, S.: Adaptive Newton-based smoothed functional algorithms for simulation optimization. ACM Transactions on Modeling and Computer Simulation 18(1), 2:1–2:35 (2007)
5. Bhatnagar, S.: Simultaneous perturbation and finite difference methods. Wiley Encyclopedia of Operations Research and Management Science 7, 4969–4991 (2011)
6. Bhatnagar, S., Abdulla, M.S.: Simulation-based optimization algorithms for finite horizon Markov decision processes. Simulation 84(12), 577–600 (2008)
7. Bhatnagar, S., Fu, M.C., Marcus, S.I., Wang, I.J.: Two-timescale simultaneous perturbation stochastic approximation using deterministic perturbation sequences. ACM Transactions on Modelling and Computer Simulation 13(2), 180–209 (2003)
8. Bhatnagar, S., Hemachandra, N., Mishra, V.: Stochastic approximation algorithms for constrained optimization via simulation. ACM Transactions on Modeling and Computer Simulation 21, 15:1–15:22 (2011)
9. Bhatnagar, S., Karmeshu, Mishra, V.: Optimal parameter trajectory estimation in parameterized sdes: an algorithmic procedure. ACM Transactions on Modeling and Computer Simulation (TOMACS) 19(2), 8 (2009)
10. Bhatnagar, S., Kumar, S.: A simultaneous perturbation stochastic approximation based actor-critic algorithm for Markov decision processes. IEEE Transactions on Automatic Control 49(4), 592–598 (2004)
11. Bhatnagar, S., Mishra, V., Hemachandra, N.: Stochastic algorithms for discrete parameter simulation optimization. IEEE Transactions on Automation Science and Engineering 9(4), 780–793 (2011)

12. Borkar, V.S.: Stochastic approximation with two timescales. Systems and Control Letters 29, 291–294 (1997)
13. Borkar, V.S.: Stochastic Approximation: A Dynamical Systems Viewpoint. Cambridge University Press and Hindustan Book Agency (Jointly Published), Cambridge and New Delhi (2008)
14. Fabian, V.: Stochastic approximation. In: Rustagi, J.J. (ed.) Optimizing Methods in Statistics, pp. 439–470. Academic Press, New York (1971)
15. Floyd, S., Jacobson, V.: Random early detection gateways for congestion avoidance. IEEE/ACM Transactions on Networking 1(4), 397–413 (1993)
16. Karmeshu, Bhatnagar, S., Mishra, V.: An optimized sde model for slotted aloha. IEEE Transactions on Communications 59(6), 1502–1508 (2011)
17. Katkovnik, V.Y., Kulchitsky, Y.: Convergence of a class of random search algorithms. Automation Remote Control 8, 1321–1326 (1972)
18. Kiefer, E., Wolfowitz, J.: Stochastic estimation of the maximum of a regression function. Ann. Math. Statist. 23, 462–466 (1952)
19. Prashanth, L.A., Prasad, H., Desai, N., Bhatnagar, S., Dasgupta, G.: Simultaneous perturbation methods for adaptive labor staffing in service systems. Tech. rep., Stochastic Systems Lab, IISc (2012), `http://stochastic.csa.iisc.ernet.in/www/research/files/IISc-CSA-SSL-TR-2011-4-rev2.pdf`
20. Patro, R.K., Bhatnagar, S.: A probabilistic constrained nonlinear optimization framework to optimize RED parameters. Performance Evaluation 66(2), 81–104 (2009)
21. Prashanth, L., Bhatnagar, S.: Reinforcement learning with function approximation for traffic signal control. IEEE Transactions on Intelligent Transportation Systems 12(2), 412–421 (2011)
22. Robbins, H., Monro, S.: A stochastic approximation method. Ann. Math. Statist. 22, 400–407 (1951)
23. Spall, J.C.: Multivariate stochastic approximation using a simultaneous perturbation gradient approximation. IEEE Trans. Auto. Cont. 37(3), 332–341 (1992)
24. Spall, J.C.: A one-measurement form of simultaneous perturbation stochastic approximation. Automatica 33, 109–112 (1997)
25. Spall, J.C.: An overview of the simultaneous perturbation method for efficient optimization. Johns Hopkins APL Technical Digest 19, 482–492 (1998)
26. Spall, J.C.: Stochastic optimization, stochastic approximation and simulated annealing. In: Webster, J.G. (ed.) Wiley Encyclopedia of Electrical and Electronics Engineering, vol. 20, pp. 529–542. John Wiley and Sons, New York (1999)
27. Spall, J.C.: Adaptive stochastic approximation by the simultaneous perturbation method. IEEE Trans. Autom. Contr. 45, 1839–1853 (2000)
28. Spall, J.C.: Introduction to Stochastic Search and Optimization. John Wiley and Sons, New York (2003)
29. Styblinski, M.A., Tang, T.S.: Experiments in nonconvex optimization: stochastic approximation with function smoothing and simulated annealing. Neural Networks 3, 467–483 (1990)
30. Vemu, K.R., Bhatnagar, S., Hemachandra, N.: Optimal multi-layered congestion based pricing schemes for enhanced qos. Computer Networks (2011), `http://dx.doi.org/10.1016/j.comnet.2011.12.004`

Chapter 2
Deterministic Algorithms for Local Search

2.1 Introduction

Search algorithms can be broadly classified into two major categories – global search and local search algorithms. Global search algorithms aim at finding the global minimum while local search algorithms are mainly concerned with finding a local minimum point. More formally, for the optimization problem (1.1), we say that $\theta^* \in C$ is a global minimum of the function J if $J(\theta^*) \leq J(\theta)$ $\forall \theta \in C$. On the other hand, we say that $\theta^* \in C$ is a local minimum of J if there exists an $\varepsilon > 0$ such that $J(\theta^*) \leq J(\theta)$ $\forall \theta \in C$ with $\| \theta - \theta^* \| < \varepsilon$. Many times, as we do, the norm $\| \cdot \|$ is chosen to be the Euclidean norm. A necessary condition for existence of local minima of a function J, assuming it is differentiable at all points within C, is that

$$\nabla J(\theta) = 0 \text{ for } \theta \in C^o,$$

where C^o is the interior of the set C. This condition may, however, not be satisfied if the local minimum is a boundary point of C. Similarly, a sufficient condition for a point $\theta \in C^o$ to be a local minimum point is

$$\nabla J(\theta) = 0 \text{ and } \nabla^2 J(\theta) \text{ is a positive definite matrix,}$$

assuming that the function J is twice differentiable.

A well-known example of a global search technique is simulated annealing [5, 4]. Even while it is desirable to converge to a global minimum, global search techniques are often known to be slow and impractical and many times one has to be content with local search methods. A typical local search algorithm (ignoring random noise effects for now) has the form [1], [2], [3]:

$$\theta(n+1) = \theta(n) - a(n)[D(\theta(n))]^{-1} \nabla J(\theta(n)), \tag{2.1}$$

S. Bhatnagar et al.: Stochastic Recursive Algorithms for Optimization, LNCIS 434, pp. 13–15.
springerlink.com

where $D(\theta(n))$ is a positive definite and symmetric $N \times N$ matrix and $\nabla J(\theta(n))$ is the gradient of $J(\theta)$ evaluated at $\theta = \theta(n)$. Also, $a(n), n \geq 0$ is a sequence of step-sizes that are positive and asymptotically diminishing to zero. Note that if $\nabla J(\theta(n))$ and $D(\theta(n))$ are analytically known quantities, then recursion (2.1) can proceed as is and we are in the domain of deterministic optimization. On the other hand, if $J(\theta(n))$ is of the form $J(\theta(n)) = E_\xi[h(\theta(n), \xi)]$ and we only have access to noisy cost samples $h(\theta(n), \xi)$, then quantities $\nabla J(\theta(n))$ and in many cases $D(\theta(n))$ need to be estimated. The algorithms for stochastic optimization are thus noisy or stochastic in nature because of the presence of noise in the cost samples. In addition, the estimators of $\nabla J(\theta(n))$ and $D(\theta(n))$ may introduce additional randomness as happens for instance in the SPSA and SF gradient and higher order algorithms, see Chapters 5–8. Thus the search direction could be random as well.

2.2 Deterministic Algorithms for Local Search

In order to bring out ideas clearly, we assume here that $\nabla J(\theta(n))$ and $[D(\theta(n))]^{-1}$ are analytically known quantities, i.e., we have a deterministic optimization framework with (2.1) as our search algorithm. This will, however, not be the case in the later sections where we shall primarily be concerned with the stochastic optimization setting.

Given $\theta(n) \in C$ such that $\nabla J(\theta(n)) \neq 0$, any $x(n) \in \mathbb{R}^N$ satisfying $x(n)^T \nabla J(\theta(n)) < 0$ is called a *descent* direction since the directional derivative $x(n)^T \nabla J(\theta(n))$ along the direction $x(n)$ is negative and thus by a Taylor's expansion one obtains

$$J(\theta(n) + a(n)x(n)) = J(\theta(n)) + a(n)x(n)^T \nabla J(\theta(n)) + o(a(n)). \tag{2.2}$$

Now since $x(n)^T \nabla J(\theta(n)) < 0$ and $a(n) > 0 \,\forall n$, it follows that $J(\theta(n) + a(n)x(n)) < J(\theta(n))$ for $a(n)$ sufficiently small. Now since $D(\theta(n))$ is a positive definite and symmetric matrix, so is $D(\theta(n))^{-1}$. When $x(n) = -D(\theta(n))^{-1} \nabla J(\theta(n))$, then from (2.2), we have

$$\left.\begin{aligned} J(\theta(n) - a(n)D(\theta(n))^{-1}\nabla J(\theta(n))) &= J(\theta(n)) \\ &\quad - a(n)\nabla J(\theta(n))^T D(\theta(n))^{-1}\nabla J(\theta(n)) + o(a(n)). \end{aligned}\right\} \tag{2.3}$$

Now since $D(\theta(n))^{-1}$ is positive definite and symmetric, it follows that

$$\nabla J(\theta(n))^T D(\theta(n))^{-1} \nabla J(\theta(n)) > 0 \text{ for all } \nabla J(\theta(n)) \neq 0.$$

Hence, $x(n) = -D(\theta(n))^{-1} \nabla J(\theta(n))$ is a descent direction as well. Algorithms that update along descent directions are also called *descent algorithms*.

The following well-known algorithms are special cases of (2.1):

1. **Gradient Algorithm :** This is the most commonly used descent algorithm. Here, $D(\theta(n)) = I$ (the N-dimensional identity matrix). This is also called the *steepest descent* algorithm since its updates are strictly along the direction of negative gradient.
2. **Jacobi Algorithm :** In this algorithm, $D(\theta(n))$ is set to be an $N \times N$-diagonal matrix with its ith diagonal element $\nabla^2_{i,i} J(\theta(n))$, which is also the ith diagonal element of the Hessian $\nabla^2 J(\theta(n))$. For $D(\theta(n))$ to be a positive definite matrix in this case, it is easy to see that all elements $\nabla^2_{i,i} J(\theta(n))$, $i = 1, \ldots, N$, should be positive.
3. **Newton Algorithm :** Here, $D(\theta(n))$ is chosen to be $\nabla^2 J(\theta(n))$, the Hessian of $J(\theta(n))$.

The $D(\theta(n))$ matrices in Jacobi and Newton algorithms, respectively, need not be positive definite (for all n), in general, as they vary with $\theta(n)$ and hence should be projected appropriately after each parameter update so as to ensure that the resulting matrices are positive definite [1, pp.88-98]. With proper scaling provided by the $D(\theta(n))$ matrix, the descent directions obtained using Jacobi and Newton algorithms are preferable to the one using gradient algorithm. However, obtaining estimates of the Hessian in addition to the gradient, in general, requires much more computational effort. In subsequent chapters, we will present several algorithms which, in principle, choose a descent direction similar to one of the above three types. However, all the algorithms discussed subsequently will be stochastic in nature involving random estimates of the descent direction. Consequently, the evolution of the optimization parameter updates $\theta(n)$ in those algorithms is also stochastic.

References

1. Bertsekas, D.P.: Nonlinear Programming. Athena Scientific, Belmont (1999)
2. Bertsekas, D.P., Tsitsiklis, J.N.: Parallel and Distributed Computation. Prentice Hall, New Jersey (1989)
3. Bhatnagar, S.: Adaptive multivariate three-timescale stochastic approximation algorithms for simulation based optimization. ACM Transactions on Modeling and Computer Simulation 15(1), 74–107 (2005)
4. Gelfand, S.B., Mitter, S.K.: Recursive stochastic algorithms for global optimization in $\mathscr{R}^{d_*}$. SIAM Journal on Control and Optimization 29(5), 999–1018 (1991)
5. Kirkpatrick, S., Gelatt, C.D., Vecchi, M.: Optimization by simulated annealing. Science 220, 621–680 (1983)

Chapter 3
Stochastic Approximation Algorithms

3.1 Introduction

The development in the area of stochastic algorithms (not necessarily for optimization) started in a seminal paper by Robbins and Monro [17]. They considered the problem of finding the zeros of a function $L : \mathbb{R}^N \to \mathbb{R}^N$ under noisy observations. The Robbins-Monro algorithm finds immense applications in various disciplines. For instance, in the case of the gradient search algorithm for the problem of finding a local minimum of the function $J : \mathbb{R}^N \to \mathbb{R}$, see Chapter 4, one can let $L(\theta) = \nabla J(\theta)$. Similarly, in scenarios where the aim is to find a fixed point of a function $F : \mathbb{R}^N \to \mathbb{R}^N$, one may choose $L(\theta) = F(\theta) - \theta$. Situations requiring fixed point computations arise often, for instance, in reinforcement learning, see Chapter 11, where one estimates the value of a given policy. The corresponding update is many times a fixed point recursion aimed at solving the Bellman equation for the given policy.

We first discuss in detail the R-M algorithm in Section 3.2. Next, we review the multi-timescale variant of the R-M algorithm in Section 3.3. Such algorithms are characterized by coupled stochastic recursions that are individually driven by different step-size schedules or timescales. The step-sizes typically converge to zero with different rates. An important application of multi-timescale stochastic approximation that we consider in this book is one of minimizing long-run average costs. In order to apply the regular R-M scheme in such cases, one requires estimates of the average cost corresponding to a given parameter update. One approach that is however computationally tedious is to sample long enough cost trajectories using Monte-Carlo simulation each time to estimate the average cost corresponding to a given parameter update. This difficulty is avoided through the use of multi-timescale stochastic approximation as the 'faster' recursion in this case can estimate the average cost corresponding to a given parameter update while the 'slower' recursion updates the parameter.

S. Bhatnagar et al.: Stochastic Recursive Algorithms for Optimization, LNCIS 434, pp. 17–28.
springerlink.com

3.2 The Robbins-Monro Algorithm

Let $\theta(n)$ denote the nth update of the parameter θ. Let the observed sample of $L(\theta(n))$ be $L(\theta(n)) + M(n+1)$ where $M(n+1)$ is a suitable noise term that we assume to be a martingale difference. The case when noise enters into the argument of the cost function, such as (say) $g(\theta(n), \xi(n))$, where $\xi(n), n \geq 0$ are some $\mathbb{R}^k$-valued independent and identically distributed (i.i.d.) random vectors and $g : \mathbb{R}^N \times \mathbb{R}^k \rightarrow \mathbb{R}^N$ can also be handled in our framework, since one can in such a case write

$$g(\theta(n), \xi(n)) = L(\theta(n)) + M(n+1), \tag{3.1}$$

where $L(\theta(n))$ can be set to be $L(\theta(n)) = E_{\xi}[g(\theta(n), \xi(n))]$. Here, $E_{\xi}[\cdot]$ denotes the expectation with respect to the common distribution of $\xi(n)$. Also, $M(n+1) = g(\theta(n), \xi(n)) - L(\theta(n))$, $n \geq 0$ can be seen to be a martingale difference sequence with respect to a suitable filtration. In the original R-M scheme [17], the noise random vectors $M(n+1)$ are considered i.i.d. and zero-mean. Note that the i.i.d. assumption there is across $M(n)$, not across individual components of $M(n)$. Equation (3.1) represents a popular generalization of the original R-M scheme with the additive noise generalized to a martingale difference instead of just i.i.d. noise.

The Robbins-Monro stochastic approximation algorithm is as follows: For $n \geq 0$,

$$\left.\begin{aligned} \theta(n+1) &= \theta(n) + a(n)g(\theta(n), \xi(n)) \\ &= \theta(n) + a(n)(L(\theta(n)) + M(n+1)) \end{aligned}\right\}, \tag{3.2}$$

where $a(n), n \geq 0$ is a sequence of positive real numbers called step-sizes.

Remark 3.1. To derive intuition regarding the above recursion, lets ignore the noise term $M(n+1)$ for a moment. Then, one can see that if the recursion (3.2) converges after some iterations (say N), then $\theta(n+1) = \theta(n) = \theta^*, \forall n \geq N$, where θ^* represents the converged parameter value. This when used in the above recursion (3.2), gives us $L(\theta^*) = 0$. The recursion (3.2) serves the purpose of computing a zero of the given function $L(\cdot)$. Of course, with the introduction of the noise term $M(n+1)$, more detailed analysis would be necessary along with certain restrictions on the step-sizes $a(n), n \geq 0$, which are discussed in the next section.

If $\theta(n)$ are constrained to take values within a prescribed set $C \subset \mathbb{R}^N$ (with C being a strict subset of $\mathbb{R}^N$), one will have to project after each iterate the value of $\theta(n+1)$ to the set C. The new value of $\theta(n+1)$ would then correspond to its projected value after the update. We discuss the convergence analysis of the algorithm in Section 3.2.1 primarily for the case when $C = \mathbb{R}^N$. It will, however, be assumed that the iterates $\theta(n)$ will stay uniformly bounded almost surely.

3.2.1 *Convergence of the Robbins-Monro Algorithm*

Convergence in the mean-square sense, of the R-M scheme with i.i.d. noise terms $M(n+1)$ is shown in [17]. As with [5] and [13], we show convergence in the almost sure sense, of the R-M scheme with the generalized martingale difference noise-term $M(n+1)$. In order to prove convergence of recursions such as (3.2), one needs to first ensure that the iterates in these recursions remain stable or uniformly bounded. If the iterates stay uniformly bounded, then convergence in almost sure sense would imply convergence in the mean-square sense as well (see Appendix A). The converse is however not true in general, i.e., if they converge in the mean square sense, then they need not converge almost surely, even when they are uniformly bounded.

Let $\mathscr{F}(n) = \sigma(\theta(m), M(m), m \leq n)$, $n \geq 0$ denote a sequence of increasing sigma fields. Our convergence analysis is based on the ordinary differential equation (ODE) approach, for instance, see [5, Chapter 2]. Consider the following ODE associated with (3.2):

$$\dot{\theta}(t) = L(\theta(t)). \tag{3.3}$$

We make the following assumptions:

Assumption 3.1. The map $L : \mathbb{R}^N \to \mathbb{R}^N$ is Lipschitz continuous.

Assumption 3.2. The step-sizes $a(n), n \geq 0$ satisfy the requirements

$$\sum_n a(n) = \infty, \quad \sum_n a(n)^2 < \infty. \tag{3.4}$$

Assumption 3.3. The sequence $(M(n), \mathscr{F}(n))$, $n \geq 0$ forms a martingale difference sequence. Further, $M(n)$, $n \geq 0$ are square integrable random variables satisfying

$$E[\|M(n+1)\|^2 \mid \mathscr{F}(n)] \leq K(1 + \|\theta(n)\|^2) \text{ a.s., } n \geq 0, \tag{3.5}$$

for a given constant $K > 0$.

Assumption 3.4. The iterates (3.2) remain almost surely bounded, i.e.,

$$\sup_n \|\theta(n)\| < \infty, \text{ a.s.} \tag{3.6}$$

Assumption 3.5. The ODE (3.3) has $H \subset C$ as its set of globally asymptotically stable equilibria.

Assumption 3.1 ensures that the ODE (3.3) is well posed. Assumption 3.2 is also a standard requirement. In particular, the first condition in (3.4) is required to ensure that the algorithm does not converge prematurely. The second condition there is required to reduce the effect of noise. Common examples of $\{a(n), n \geq 0\}$ that are seen to satisfy Assumption 3.2 include

- $a(n) = \frac{1}{n}$, $\forall n \geq 1$ and $a(0) = 1$,
- $a(n) = \frac{1}{n^{\alpha}}$, $\forall n \geq 1$ with $a(0) = 1$ and any $\alpha \in (0.5, 1)$,
- $a(n) = \frac{\ln n}{n}$, $\forall n \geq 2$ with $a(0) = a(1) = 1$,
- $a(n) = \frac{1}{n \ln n}$, $\forall n \geq 2$ with $a(0) = a(1) = 1$.

Assumption 3.3 is a general requirement [5] that is seen to be satisfied in many applications. For instance, it is seen to be easily satisfied by most reinforcement learning algorithms.

We now discuss in more detail Assumption 3.4 even though it is routinely assumed in many references. An easy way by which Assumption 3.4 can be satisfied is if the set C in which θ takes values is a bounded subset of $\mathbb{R}^N$ as in such a case (as mentioned previously), one would project the iterates after each update to the set C, thereby ensuring that the resulting parameters are both feasible (i.e., take values in the set where they are allowed to take values in) and remain bounded. In the case when C is unbounded (such as $C = \mathbb{R}^N$ as here) but one roughly knows the region of the space where the asymptotically stable equilibria lie, one could choose a large bounded set that contains the above region as the constraint set for the algorithm and use projection (as before) to ensure that the iterates remain bounded. This would also imply that the remainder of the space is not visited by the algorithm which may in fact be good since the algorithm in such a case would not waste its resources in exploring the region of the space that does not contain the equilibria. The projection technique is often used to ensure the stability of iterates. Other approaches to prove stability of the iterates (for instance when $C = \mathbb{R}^N$) include the stochastic Lyapunov technique [13] and the recently proposed approach in [6], [5] whereby one does a scaling of the original iteration (3.2) to approximate the same with a deterministic process in a manner similar to the construction of the fluid model of [9], [10]. This approach is remarkable in that using just an ordinary differential equation (ODE)-based analysis, one can prove both the stability and the convergence of the original random iterates. Another approach [8] is to define a bounded constraint region for the iterates, use projection as above, but gradually increase the size of the constraint region as iterations progress. Nevertheless, we will assume that $C = \mathbb{R}^N$ in this analysis and that the iterates stay bounded under Assumption 3.4.

Define a sequence of time points $t(n)$, $n \geq 0$ as follows: $t(0) = 0$ and for $n \geq 1$, $t(n) = \sum_{m=0}^{n-1} a(m)$. It follows from (3.4) that $t(n) \uparrow \infty$. The map $n \mapsto t(n)$ can be viewed as a map from the "algorithmic time" to the "real time". Define now a continuously interpolated trajectory $\bar{\theta}(t), t \geq 0$ (obtained from the algorithm's updates) as follows: Let $\bar{\theta}(t(n)) = \theta(n), n \geq 0$, with linear interpolation on the interval

$[t(n),t(n+1)]$. By Assumption 3.4, it follows that $\sup_{t\geq 0}\|\bar{\theta}(t)\| = \sup_n \|\theta(n)\| < \infty$ a.s. Let $\bar{T} > 0$ be a given real number. Define another sequence $\{T(n), n \geq 0\}$ as follows: $T(0) = t(0) = 0$ and for $n \geq 1$,

$$T(n) = \min\{t(m) \mid t(m) \geq T(n-1) + \bar{T}\}.$$

Let $I(n)$ denote the interval $[T(n), T(n+1))$. From its definition, there exists an increasing sub-sequence $\{m(n)\}$ of $\{n\}$ such that $T(n) = t(m(n))$, $n \geq 0$. Also, let $\theta^n(t), t \geq t(n)$ denote the trajectory of the following ODE starting at time $t(n)$ and under the initialization $\theta^n(t(n)) = \bar{\theta}(t(n)) = \theta(n)$:

$$\dot{\theta}^n(t) = L(\theta^n(t)),\ t \geq t(n). \tag{3.7}$$

Let $Z(n)$, $n \geq 0$ be defined according to

$$Z(n) = \sum_{m=0}^{n-1} a(m)M(m+1).$$

Lemma 3.1. *The sequence $(Z(n), \mathcal{F}(n))$, $n \geq 0$ is a zero-mean, square integrable, almost surely convergent martingale.*

Proof. It is easy to see that each $Z(n)$ is $\mathcal{F}(n)$-measurable and integrable. Further, $Z(n), n \geq 0$ are square integrable random variables since $M(n+1)$ are square integrable by Assumption 3.3. Consider now the process $\{B(n)\}$ defined by

$$\begin{aligned}
B(n) &= \sum_{m=0}^{n-1} E\left[\|Z(m+1) - Z(m)\|^2 \mid \mathcal{F}(m)\right], \\
&= \sum_{m=0}^{n-1} E\left[a(m)^2\|M(m+1)\|^2 \mid \mathcal{F}(m)\right], \\
&= \sum_{m=0}^{n-1} a(m)^2 E\left[\|M(m+1)\|^2 \mid \mathcal{F}(m)\right], \\
&\leq \sum_{m=0}^{n-1} a(m)^2(1 + \|\theta(n)\|^2),
\end{aligned}$$

by Assumption 3.3. Now, from Assumptions 3.2 and 3.4, it follows that

$$B(n) \to B_\infty < \infty \text{ a.s.}$$

The claim follows from the martingale convergence theorem (Theorem B.2). □

Proposition 3.2. *We have*

$$\lim_{n\to\infty} \sup_{t\in I(n)} \|\bar{\theta}(t) - \theta^n(t)\| = 0,\ a.s.$$

Proof. (*Sketch*) The proof for a similar result is given in detail in [5, Chapter 2, Lemma 1]. The proof follows by following a series of steps that involve bounding the various terms that upper-bound the norm difference between the algorithm's and the ODE's trajectories. The Lipschitz continuity of L ensures that the growth in the recursion is at most linear. That together with Assumptions 3.3 and 3.4 ensure that the iterates do not blow up. Moreover, the norm difference can then be bounded from an application of the Gronwall's inequality (Lemma C.1) and the upper bound is seen to vanish asymptotically as $n \to \infty$. We refer the reader to [5, Chapter 2, Lemma 1] for details. □

Note that by Assumption 3.5, H is the globally asymptotically stable attractor set for the ODE (3.3). Recall from Definition C.10 that given $\bar{T}, \Delta > 0$, we call a bounded, measurable $\theta(\cdot) : \mathbb{R}^+ \cup \{0\} \to \mathbb{R}^N$, a $(\bar{T}, \Delta)$-perturbation of (3.3) if there exist $0 = T(0) < T(1) < T(2) < \cdots < T(r) \uparrow \infty$ with $T(r+1) - T(r) \geq \bar{T}\ \forall r$ and solutions $\theta^r(y), y \in [T(r), T(r+1)]$ of (3.3) for $r \geq 0$, such that

$$\sup_{y \in [T(r), T(r+1)]} \|\theta^r(y) - \theta(y)\| < \Delta.$$

Theorem 3.3. *Under Assumptions 3.1 to 3.5, the iterates $\theta(n), n \geq 0$ obtained from the algorithm (3.2) converge almost surely to H.*

Proof. From Proposition 3.2, $\bar{\theta}(t)$ serves as a $(\bar{T}, \Delta)$-perturbation for the ODE (3.3). The claim follows by applying the Hirsch lemma (Lemma C.5), for every $\varepsilon > 0$. □

A detailed ODE argument showing convergence of the stochastic iterates to a *compact connected internally chain transitive invariant set* of the corresponding ODE has been shown in [1], [5]. In most applications, as we consider, the associated ODEs either have a unique stable equilibrium or else a set of asymptotically stable isolated equilibria. Thus, if $H = \{\theta^*\}$ is a singleton, i.e., contains a unique asymptotically stable equilibrium of (3.3), then by Theorem 3.3, $\theta(n) \to \theta^*$ a.s. as $n \to \infty$. In the case of multiple isolated equilibria, the algorithm shall converge to one amongst them depending on the noise and initial condition. (Here by isolated equilibria, we mean that one can construct certain sufficiently small open neighbourhoods such that exactly one equilibrium is contained within each neighbourhood.) Further, in case H does not contain isolated equilibria, Theorem 3.3 merely says that the recursion $\theta(n)$ asymptotically converges to H. Other ODE-based analyses of the stochastic recursion include [14], [12], [2] and [13].

We have considered till now the basic R-M scheme which is used to compute a zero of the given function $L(\cdot)$ under noisy observations. The case where there are coupled functions $L_1(\cdot,\cdot)$ and $L_2(\cdot,\cdot)$ with two sets of parameters operating at different timescales is considered in the next section.

3.3 Multi-timescale Stochastic Approximation

We consider the case of two timescales here, i.e., recursions involving two different step-size schedules. Similar ideas as described below carry over when the number of timescales is more than two. Let $\theta(n) \in \mathbb{R}^N$ and $\omega(n) \in \mathbb{R}^d$ be two sequences of parameters that are updated according to the following coupled stochastic approximation recursions: $\forall n \geq 0$,

$$\theta(n+1) = \theta(n) + a(n)\left(L_1(\theta(n), \omega(n)) + M^1(n+1)\right), \quad (3.8)$$

$$\omega(n+1) = \omega(n) + b(n)\left(L_2(\theta(n), \omega(n)) + M^2(n+1)\right), \quad (3.9)$$

where $M^1(n+1)$ and $M^2(n+1)$ are martingale difference noise terms (see Appendix B.2). The step-sizes $a(n), b(n), n \geq 0$ satisfy the following requirement:

Assumption 3.6. $a(n), b(n) > 0, \forall n \geq 0$, Further,

$$\sum_n a(n) = \sum_n b(n) = \infty, \ \sum_n \left(a(n)^2 + b(n)^2\right) < \infty, \text{ and,} \quad (3.10)$$

$$\lim_{n \to \infty} \frac{a(n)}{b(n)} = 0. \quad (3.11)$$

Remark 3.2. To understand the set of recursions (3.8) and (3.9), let us ignore the noise terms $M^1(n+1)$ and $M^2(n+1)$ for the moment and consider a case with $a(n) = \frac{1}{n}$ and $b(n) = \frac{1}{n^{0.6}}$, $n \geq 1$, which satisfies both equations (3.10) and (3.11). Under these simplifications, the following insight can be derived:

1. It follows that for a given $N \geq 0$,

$$t(N) \triangleq \sum_n^N a(n) < \sum_n^N b(n) \triangleq \tau(n),$$

   and higher the value of N, the further apart the above two summations are. In other words, the time line $\tau(n), n \geq 1$ with time steps $b(n)$ reaches infinity faster than the time line $t(n)$ with step-sizes $a(n)$. So, we say that the recursion of the ω parameter is on a "faster" timescale than the recursion of θ.
2. From equation (3.11), it follows that as we go further in the recursions, the updates to θ will be quasi-static compared to those for ω. Hence, the updates to ω would appear to be equilibrated for the current quasi-static θ. In other words, for a given θ, the updates to ω would appear to have converged to a point ω^* such that $L_2(\theta, \omega^*) = 0$ (assuming there is a unique such point corresponding to θ). Thus, one expects that the updates of ω would converge to a $\omega^* \triangleq \gamma(\theta)$. For

the updates to θ, the ω would, for all practical purposes, be ω^* itself. Hence, the updates to θ, ignoring the noise term, would appear to be

$$\theta(n+1) = \theta(n) + a(n)L_1(\theta(n), \gamma(\theta(n))).$$

Following the analysis of the R-M scheme, this recursion would converge to a point θ^* (assuming it is unique) where $L_1(\theta^*, \gamma(\theta^*)) = 0$. These concepts are formalized and discussed along with the necessary assumptions in Section 3.3.1.

Remark 3.3. Suppose both the updates of θ and ω were performed with the same step-size sequence, say $a(n), n \geq 0$, then both the recursions could be combined together and analyzed as one recursion of the basic R-M scheme. These updates would then together converge to a point θ^*, ω^* (assuming such a point is unique), where $L_1(\theta^*, \omega^*) = 0$ and $L_2(\theta^*, \omega^*) = 0$ simultaneously. This is in contrast to the case of two timescales where the solution would be θ^*, ω^* such that $\omega^* = \gamma(\theta^*)$ and $L_1(\theta^*, \gamma(\theta^*)) = 0$.

Remark 3.4. Like in the previous section, one can consider the case where the noise term enters the cost function itself. Thus, let the two recursions be

$$\theta(n+1) = \theta(n) + a(n)g_1(\theta(n), \omega(n), \xi^1(n)), \tag{3.12}$$

$$\omega(n+1) = \omega(n) + b(n)g_2(\theta(n), \omega(n), \xi^2(n)), \tag{3.13}$$

where $\xi^1(n), n \geq 0$ are i.i.d. random vectors and so are $\xi^2(n), n \geq 0$. Then one can rewrite

$$g_1(\theta(n), \omega(n), \xi^1(n)) = L_1(\theta(n), \omega(n)) + M^1(n+1),$$
$$g_2(\theta(n), \omega(n), \xi^2(n)) = L_2(\theta(n), \omega(n)) + M^2(n+1),$$

where $M^1(n+1), M^2(n+1), n \geq 0$ are suitable martingale difference sequences. In this manner (3.12) and (3.13) can be recast as the recursions (3.8) and (3.9), respectively.

Remark 3.5. The above discussion which is for two timescales, can be easily generalized to multiple timescales by starting the analysis from the "fastest" timescale to the "slowest" timescale.

3.3.1 Convergence of the Multi-timescale Algorithm

A general analysis of two-timescale algorithms is available in [4] as well as Chapter 6 of [5]. We present a sketch of the same here. We make the following assumptions:

Assumption 3.7. The functions $L_1 : \mathbb{R}^N \times \mathbb{R}^d \to \mathbb{R}^N$ and $L_2 : \mathbb{R}^N \times \mathbb{R}^d \to \mathbb{R}^d$ are both Lipschitz continuous.

Assumption 3.8. $M^1(n), M^2(n),\ n \geq 0$ are both martingale difference sequences with respect to the filtration $\mathscr{F}(n) = \sigma(\theta(j), \omega(j), M^1(j), M^2(j), j \leq n), n \geq 0$. Further,

$$E\left[\|M^i(n+1)\|^2 \mid \mathscr{F}(n)\right] \leq K\left(1 + \|\theta(n)\|^2 + \|\omega(n)\|^2\right),\ i = 1,2.$$

Assumption 3.9. The iterates are a.s. uniformly bounded, i.e.,

$$\sup_n (\|\theta(n)\| + \|\omega(n)\|) < \infty, \text{ w.p.1.}$$

Assumption 3.10. The ODE

$$\dot{\omega}(t) = L_2(\theta, \omega(t)), \tag{3.14}$$

has a globally asymptotically stable equilibrium $\gamma(\theta)$, uniformly in θ, where $\gamma : \mathbb{R}^N \to \mathbb{R}^N$ is a Lipschitz continuous map.

Assumption 3.11. The ODE

$$\dot{\theta}(t) = L_1(\theta(t), \gamma(\theta(t))), \tag{3.15}$$

has a globally asymptotically stable equilibrium $\theta^* \in \mathbb{R}^N$.

Assumptions 3.7-3.9 are seen to be similar to analogous assumptions for the R-M algorithm except for the requirement in (3.11) that suggests that $a(n)$ approaches zero at a rate faster than $b(n)$ does.

Let us define $t(n)$ in the same manner as before. Also, let $\tau(n), n \geq 0$ be defined according to $\tau(0) = 0$ and $\tau(n) = \sum_{m=0}^{n-1} b(m)$. Note that from the viewpoint of the (slower) timescale governed by $b(n), n \geq 0$, the recursion (3.8) can be rewritten as

$$\theta(n+1) = \theta(n) + b(n)\eta^1(n), \tag{3.16}$$

where $\eta^1(n) = \dfrac{a(n)}{b(n)}\left(L_1(\theta(n), \omega(n)) + M^1(n+1)\right) = o(1)$, since $a(n) = o(b(n))$ from (3.11). Now (3.16) is seen to track the ODE

$$\dot{\theta}(t) = 0. \tag{3.17}$$

Also, a similar analysis as described in Section 3.2.1 can be used to show that (3.9) asymptotically tracks the ODE

$$\dot{\omega}(t) = L_2(\theta(t), \omega(t)). \tag{3.18}$$

In other words, over intervals $\hat{I}(n) = [\hat{T}(n), \hat{T}(n+1)]$ of length approximately $\hat{T} > 0$, with $\hat{T}(n) = \tau(m(n)) \approx n\hat{T}$, the norm difference between the interpolated trajectories of the algorithm's parameter iterates and the trajectories of the ODEs (3.17)-(3.18) vanishes almost surely as $n \to \infty$ (cf. Proposition 3.2). Now, as a consequence of (3.17), the ODE (3.18) can be rewritten as

$$\dot{\omega}(t) = L_2(\theta, \omega(t)). \tag{3.19}$$

By Assumption 3.10, from an application of Hirsch's lemma (Lemma C.5), it follows that the recursion (3.9) for given θ asymptotically converges to $\gamma(\theta)$ almost surely, and in fact, $\|\omega(n) - \gamma(\theta(n))\| \to 0$ as $n \to \infty$ almost surely. A similar argument as Proposition 3.2 can now be applied to show that the norm difference between the trajectory obtained from the θ-recursion (3.8) when interpolated using the time instants $t(n)$ and that of the ODE (3.15) again vanishes asymptotically over intervals $I(n) = [T(n), T(n+1)]$, with $T(n), n \geq 0$ defined in a similar manner as Proposition 3.2. Now, by another application of the Hirsch lemma, it can be shown that $\theta(n) \to \theta^*$ as $n \to \infty$ almost surely. We thus have the following result (that is similar to Theorem 2 on pp.66 of [5]):

Theorem 3.4.

$$\lim_{n\to\infty} (\theta(n), \omega(n)) = (\theta^*, \gamma(\theta^*)) \ a.s.$$

3.4 Concluding Remarks

The R-M algorithm has been analyzed in detail in [3], [12], [13], [11], [7], [5] and several other books and papers. The ODE method is one of the techniques used to study its convergence. A second approach based entirely on probabilistic arguments is also popular in the literature. Because of its wide applicability, the R-M algorithm is still very popular even six decades after it was originally invented.

A general analysis of two-timescale stochastic approximation using the ODE approach is provided in [4], [5]. Multi-timescale algorithms are helpful in cases when in between two successive updates of the algorithm, one typically has to perform an inner-loop procedure recursively until it converges. Thus, one would in practice have to wait for a long time before updating the algorithm once. Using a multi-timescale algorithm as in (3.8)-(3.9), both recursions (for the inner and outer loops) can run together, and convergence to the desired point can be achieved. Key application areas where this procedure has been succesfully applied are simulation optimization and adaptive control that we study in later chapters. There is another reason why multi-timescale algorithms can be interesting. In [15], averaging of stochastic approximation iterates in the case of one-timescale algorithms such as (3.2) has been seen to improve the rate of convergence. The same procedure can be accomplished using a two-timescale algorithm such as (3.8)-(3.9) wherein the 'averaging' is performed along the faster timescale.

Multi-timescale algorithms are also useful in other situations. In [18], a smoothed version of SPSA is presented that is seen to improve performance. The idea that is similar to Polyak averaging and the resulting algorithm has a multi-timescale nature. In [16], [19], the step-sizes $a(n), n \geq 0$ are adaptively set according to the objective function value obtained. Since the update direction in SPSA is random, a move in the descent direction (in their scheme) is rewarded by a slightly higher step-size in the next update step while a move in the ascent direction attracts a penalty. Moreover, if the objective function value becomes worse, a certain blocking mechanism is enforced whereby starting from the previous estimate, a new gradient evaluation is made with a reduced step-size $a(n)$. The procedure of [16], [19] is performed for the smoothed version of SPSA making the overall scheme again of the multi-timescale type.

References

1. Benaim, M.: A dynamical systems approach to stochastic approximations. SIAM Journal on Control and Optimization 34(2), 437–472 (1996)
2. Benveniste, A., Métivier, M., Priouret, P.: Adaptive Algorithms and Stochastic Approximations. Springer, Berlin (1990)
3. Benveniste, A., Priouret, P., Métivier, M.: Adaptive algorithms and stochastic approximations. Springer-Verlag New York, Inc. (1990)
4. Borkar, V.S.: Stochastic approximation with two timescales. Systems and Control Letters 29, 291–294 (1997)
5. Borkar, V.S.: Stochastic Approximation: A Dynamical Systems Viewpoint. Cambridge University Press and Hindustan Book Agency (Jointly Published), Cambridge and New Delhi (2008)
6. Borkar, V.S., Meyn, S.P.: The O.D.E. method for convergence of stochastic approximation and reinforcement learning. SIAM Journal of Control and Optimization 38(2), 447–469 (2000)
7. Chen, H.: Stochastic approximation and its applications, vol. 64. Kluwer Academic Pub. (2002)
8. Chen, H.F., Duncan, T.E., Pasik-Duncan, B.: A Kiefer-Wolfowitz algorithm with randomized differences. IEEE Trans. Auto. Cont. 44(3), 442–453 (1999)
9. Dai, J.G.: On positive Harris recurrence for multiclass queueing networks: A unified approach via fluid limit models. Annals of Applied Probability 5, 49–77 (1995)
10. Dai, J.G., Meyn, S.P.: Stability and convergence of moments for multiclass queueing networks via fluid limit models. IEEE Transactions on Automatic Control 40, 1889–1904 (1995)
11. Duflo, M.: Random iterative models, vol. 34. Springer (1997)
12. Kushner, H.J., Clark, D.S.: Stochastic Approximation Methods for Constrained and Unconstrained Systems. Springer, New York (1978)
13. Kushner, H.J., Yin, G.G.: Stochastic Approximation Algorithms and Applications. Springer, New York (1997)

14. Ljung, L.: Analysis of recursive stochastic algorithms. IEEE Transactions on Automatic Control AC-22, 551–575 (1977)
15. Polyak, B.T., Juditsky, A.B.: Acceleration of stochastic approximation by averaging. SIAM J. Control and Optim. 30(4), 838–855 (1992)
16. Renotte, C., Wouwer, A.V., Remy, M.: Neural modeling and control of a heat exchanger based on SPSA techniques. In: Proceedings of the American Control Conference, Chicago, IL, pp. 3299–3303 (2000)
17. Robbins, H., Monro, S.: A stochastic approximation method. Ann. Math. Statist. 22, 400–407 (1951)
18. Spall, J.C., Cristion, J.A.: Nonlinear adaptive control using neural networks: estimation with a smoothed form of simultaneous perturbation gradient approximation. Statistica Sinica 4, 1–27 (1994)
19. Wouwer, A.V., Renotte, C., Remy, M.: Application of stochastic approximation techniques in neural modelling and control. International Journal of Systems Science 34, 851–863 (2003)

Part II
Gradient Estimation Schemes

Most of the important stochastic recursive algorithms that are based on some form of gradient estimation were studied in the previous century. These algorithms are geared towards solving an associated stochastic optimization problem. When the cost objective is a simple expectation over noisy observations or cost samples, the Robbins-Monro algorithm in conjunction with a suitable gradient estimator is applied. Under long-run average cost objectives, a multi-timescale stochastic algorithm with a gradient estimator is used. This part of the book comprises three chapters and deals with efficient gradient estimation approaches.

The earliest gradient estimation scheme is the Kiefer-Wolfowitz algorithm (presented originally in a paper in 1952 by Kiefer and Wolfowitz) that relies on generating a sufficient number of samples by perturbing each individual component of the parameter, one at a time. There are primarily two versions of this scheme. The first version involves generating $2N$ cost samples (each corresponding to a different perturbed parameter) while the second requires $(N+1)$ cost samples, where N is the parameter dimension. These schemes as well as some of their variants are reviewed in Chapter 4, for both cases of cost objectives (when they are simple expectations and also when they have a long-run average form).

Spall, in a paper in 1992, presented a remarkable gradient estimator that requires only two function evaluations regardless of the parameter dimension N. This estimator is based on simultaneously perturbing all parameter components using i.i.d. random variables satisfying certain properties that are most commonly satisfied by symmetric Bernoulli random variates. The Robbins-Monro algorithm in conjunction with this estimator has become famously known in the literature as the simultaneous perturbation stochastic approximation (SPSA) algorithm. In a later paper, Spall also presented a one-measurement gradient estimator using a similar perturbation methodology that however does not perform well. Bhatnagar, Fu, Marcus and Wang subsequently presented a simultaneous perturbation methodology that is based on deterministic (regular) perturbation sequences instead of random. A one-simulation variant of SPSA based on Hadamard matrix perturbations is seen to exhibit significantly better performance as compared to the one-simulation randomized difference algorithm of Spall. In Chapter 5, we discuss in detail the various versions of the SPSA scheme, both for cost objectives that are an expectation over noisy cost samples as well as those that are certain long-run averages. We also present the convergence analyses for the various cases.

Katkovnik and Kulchitsky presented in a paper in 1972, a scheme based on smoothing the gradient of the cost objective using one of the following probability density functions for convolution with the gradient: Gaussian, Cauchy or Uniform. It is observed using an integration-by-parts argument that the convolution of the smoothing density function with the objective gradient is the same as the convolution of the objective function itself with a scaled density function. A one-measurement estimator of the gradient is thus obtained. Two-measurement balanced versions of these estimators are seen to show better performance. We call the resulting estimates as the smoothed functional (SF) estimates. Chapter 6 discusses in detail the smoothed functional gradient estimators and the resulting algorithms along with their convergence analyses.

Chapter 4
Kiefer-Wolfowitz Algorithm

4.1 Introduction

In the Robbins-Monro algorithm (3.2), suppose that $g(\theta(n),\xi(n))$ is an observation or sample (with noise) of the negative of the gradient of a cost objective $J(\theta(n))$ evaluated at the nth iteration, i.e., $g(\theta(n),\xi(n))$ is a noisy observation of $L(\theta(n)) = -\nabla J(\theta(n))$. Here, $\xi(n), n \geq 0$ denotes the i.i.d. noise sequence. One can show, as we do below, under certain standard conditions that (3.2) converges to a local minimum of J. We are now in the domain of stochastic gradient algorithms, i.e., gradient algorithms (Chapter 2) with noise. We shall assume, in particular, that the objective function is a simple expectation over the noisy cost samples or $J(\theta) = E_{\xi}[h(\theta,\xi)]$. If $\nabla h(\theta,\xi)$ exists for any given noise sample ξ, then under certain regularity conditions, see for instance, [8], [9], [10], it may be possible to interchange the expectation and the gradient operators to obtain $\nabla J(\theta) = E[\nabla h(\theta,\xi)]$. In such a case, one may set $g(\theta(n),\xi(n)) = -\nabla h(\theta(n),\xi(n))$ in (3.1) and obtain asymptotic convergence to a local minimum of J. Infinitesimal Perturbation Analysis (IPA) and its variants are largely based on this idea, see [12], [7], [10], [8], [9]. When applicable, IPA shows excellent performance. In practice, however, one often does not have access to direct gradient measurements. It is also possible that while the function J is continuously differentiable with bounded higher order derivatives, the function h itself is not so. In such cases, one requires other gradient estimation techniques. The finite difference stochastic approximation (FDSA) [13], which is usually referred to as the Kiefer-Wolfowitz algorithm (after its inventors) is perhaps the earliest known algorithm that is used for estimating the gradient under noisy measurements. In Section 4.2, we describe the basic Kiefer-Wolfowitz scheme. Its variants are then explained in Section 4.3.

4.2 The Basic Algorithm

The original Kiefer-Wolfowitz algorithm [13] was proposed for the case where θ is a one-dimensional parameter taking values in a bounded interval $C_1 \subset \mathbb{R}$. We first

S. Bhatnagar et al.: Stochastic Recursive Algorithms for Optimization, LNCIS 434, pp. 31–39.
springerlink.com

discuss this one-dimensional case and subsequently its multi-dimensional parameter version. In the Kiefer-Wolfowitz algorithm, the underlying scheme is still (3.2) except that the single noisy measurement $g(\theta(n),\xi(n))$ in (3.2) is replaced by

$$\bar{g}(\theta(n),\xi^+(n),\xi^-(n)) = -\left(\frac{h(\theta(n)+\delta(n),\xi^+(n)) - h(\theta(n)-\delta(n),\xi^-(n))}{2\delta(n)}\right). \tag{4.1}$$

Here, $\xi^+(n)$, $\xi^-(n)$, $n \geq 0$ are $\mathbb{R}$-valued independent noise samples. Further, $h(\theta(n)+\delta(n),\xi^+(n))$ and $h(\theta(n)-\delta(n),\xi^-(n))$ are two independent noisy measurements of the objective with perturbed parameter values $\theta(n)+\delta(n)$ and $\theta(n)-\delta(n)$, respectively. It can be seen that if one filters out the noise, then for $\delta(n)$ sufficiently small, (4.1) will be a noisy approximation of $-\nabla J(\theta(n))$. In particular,

$$E[\bar{g}(\theta(n),\xi^+(n),\xi^-(n)) \mid \theta(n)] = -\nabla J(\theta(n)) + o(\delta(n)).$$

The K-W algorithm (4.2) proceeds along the negative gradient direction in order to find a local minimum.

$$\theta(n+1) = \theta(n) - a(n)\frac{h(\theta(n)+\delta(n),\xi^+(n)) - h(\theta(n)-\delta(n),\xi^-(n))}{2\delta(n)}, \tag{4.2}$$

$n \geq 0$. The scalar parameters $\delta(n), n \geq 0$ should be carefully chosen so that $\delta(n) \to 0$ (as $n \to \infty$) at a rate slow enough that the variance in the FDSA estimates does not blow up. We now present our assumptions.

Assumption 4.1. The map $J : \mathbb{R} \to \mathbb{R}$ is Lipschitz continuous and is twice differentiable with its second order derivative being bounded. Further, the function $L(\theta)$ defined by $L(\theta) = -\frac{dJ(\theta)}{d\theta}$, $\forall \theta \in \mathbb{R}$ and the map $h : \mathbb{R} \times \mathbb{R} \to \mathbb{R}$ are both Lipschitz continuous.

The above is mainly a technical requirement that ensures that the corresponding ODE is well posed and its trajectories bounded. Further, the smoothness requirements on $J(\theta)$ ensure via a Taylor series argument that the algorithm converges to a local minimum. Reference [13] has a more general setting where $J(\theta)$ need not be differentiable but should satisfy a set of regularity conditions [13, Conditions 1-3]. However, we limit our discussion to the case where J is differentiable. Though the result in [13] is for a generalized case where J need not be differentiable, the result shown is for convergence only in probability while here in our discussion, we show almost sure convergence.

Assumption 4.2. The step-sizes $a(n), \delta(n) > 0$, $\forall n$ and

$$a(n), \delta(n) \to 0 \text{ as } n \to 0, \quad \sum_n a(n) = \infty, \quad \sum_n \left(\frac{a(n)}{\delta(n)}\right)^2 < \infty. \tag{4.3}$$

Thus, $a(n)$ and $\delta(n)$ are both diminishing sequences of positive numbers with $\delta(n)$ going to zero slower than $a(n)$. The first condition above is analogous to a similar condition in (3.4). The last condition is a stronger requirement and ensures convergence of the resulting martingale noise sequence.

Assumption 4.3. $\xi^+(n)$, $\xi^-(n)$, $n \geq 0$ are independent random variables having a common distribution and with finite second moments.

Assumption 4.4. The iterates (4.2) remain almost surely bounded, i.e.,

$$\sup_n |\theta(n)| < \infty, \text{ a.s.} \tag{4.4}$$

Consider the ODE:

$$\dot{\theta}(t) = -\frac{dJ(\theta(t))}{dt}. \tag{4.5}$$

Let $S = \left\{\theta \,\middle|\, \frac{dJ(\theta)}{d\theta} = 0\right\}$ denote the set of all fixed points of (4.5).

Theorem 4.1. *Under Assumptions 4.1-4.4, the parameter updates (4.2) satisfy $\theta(n) \to S$ with probability one.*

Proof. Note that the algorithm (4.2) can be rewritten as follows:

$$\theta(n+1) = \theta(n) - a(n)\left(\frac{dJ(\theta(n))}{d\theta} + \beta(n) + \eta(n)\right), \tag{4.6}$$

where

$$\eta(n) = \frac{h(\theta(n)+\delta(n),\xi^+(n)) - h(\theta(n)-\delta(n),\xi^-(n))}{2\delta(n)} - \frac{J(\theta(n)+\delta(n)) - J(\theta(n)-\delta(n))}{2\delta(n)}, \text{ and,}$$

$$\beta(n) = \frac{J(\theta(n)+\delta(n)) - J(\theta(n)-\delta(n))}{2\delta(n)} - \nabla J(\theta(n)).$$

It is easy to see that $\eta(n), n \geq 0$, is a martingale difference sequence with respect to the filtration $\mathscr{F}(n) = \sigma(\theta(m), \xi^+(m), \xi^-(m), m \leq n), n \geq 0$. Further, $\hat{\xi}(m) = \sum_{n=0}^{m} a(n)\eta(n) (m \geq 0)$ forms a martingale with respect to the same filtration. We shall first show below that $\{\hat{\xi}(m)\}$ is an almost surely convergent martingale sequence. This result will follow from the martingale convergence theorem (Theorem B.2) if we can show that $\sum_{m=0}^{\infty} E\left[(\hat{\xi}(m+1) - \hat{\xi}(m))^2 \mid \mathscr{F}(m)\right] < \infty$ almost surely. Now note that

$$|h(\theta(n), \xi(n))| - |h(0,0)| \leq |h(\theta(n), \xi(n)) - h(0,0)| \leq \bar{K}(|\theta(n)| + |\xi(n)|).$$

In the above, $\bar{K} > 0$ denotes the Lipschitz constant of the function h. It follows that

$$|h(\theta(n), \xi(n))| \leq \bar{K}_1(1 + |\theta(n)| + |\xi(n)|),$$

where $\bar{K}_1 = \max(\bar{K}, |h(0,0)|)$. Similarly, since $J(\theta)$ is also Lipschitz continuous, it is easy to see that

$$|J(\theta(n)| \leq \bar{K}_2(1 + |\theta(n)|),$$

for some $\bar{K}_2 > 0$. Now,

$$\Sigma_{m=0}^{\infty} E\left[(\hat{\xi}(m+1) - \hat{\xi}(m))^2 \mid \mathscr{F}(m)\right] = \Sigma_{m=0}^{\infty} a(m)^2 E\left[\eta(m+1)^2 \mid \mathscr{F}(m)\right]$$

$$\leq \Sigma_{m=0}^{\infty} \frac{a(m)^2}{\delta(m)^2} E\left[h(\theta(m)+\delta(m), \xi^+(m))^2 + h(\theta(m)-\delta(m), \xi^-(m))^2 \mid \mathscr{F}(m)\right]$$
$$+ \Sigma_{m=0}^{\infty} \frac{a(m)^2}{\delta(m)^2} E\left[(J(\theta(m)+\delta(m))^2 + J(\theta(m)-\delta(m))^2 \mid \mathscr{F}(m)\right]$$

$$\leq 8\bar{K}_1 \Sigma_{m=0}^{\infty} \frac{a(m)^2}{\delta(m)^2} E\left[1 + (\theta(m))^2 + \delta(m)^2 + (\xi^+(m))^2 \mid \mathscr{F}(m)\right]$$
$$+ 8\bar{K}_2 \Sigma_{m=0}^{\infty} \frac{a(m)^2}{\delta(m)^2} E\left[1 + (\theta(m))^2 + \delta(m)^2 \mid \mathscr{F}(m)\right].$$

It follows now as a consequence of Assumptions 4.2-4.4, that

$$\sum_{m=0}^{\infty} E\left[(\hat{\xi}(m+1) - \hat{\xi}(m))^2 \mid \mathscr{F}(m)\right] < \infty \text{ a.s.}$$

Now, using Taylor series expansions of $J(\theta(n)+\delta(n))$ and $J(\theta(n)-\delta(n))$, respectively, around the point $\theta(n)$, i.e.,

$$J(\theta(n)+\delta(n)) = J(\theta(n)) + \delta(n)\frac{dJ(\theta(n))}{d\theta} + O(\delta(n)^2),$$
$$J(\theta(n)-\delta(n)) = J(\theta(n)) - \delta(n)\frac{dJ(\theta(n))}{d\theta} + O(\delta(n)^2),$$

we obtain,

$$\beta(n) = O(\delta(n)), \text{ i.e., } \beta(n) \to 0 \text{ as } n \to \infty.$$

Thus in lieu of equation (4.6), the iteration scheme in (4.2) can be seen to track the negative gradient of J but with diminishing noise. Thus, (4.6) can be viewed as a noisy Euler discretization of the ODE

$$\dot{\theta} = -\frac{dJ(\theta)}{d\theta}, \tag{4.7}$$

but with diminishing step increments. The result now follows by an application of the Hirsch lemma (Lemma C.5). □

Remark 4.1. Note that S corresponds to the set of all fixed points of the ODE (4.5) and not merely local minima. Points in S that are not local minima will however be unstable equilibria. In principle, the scheme can converge to an unstable equilibrium. By assuming noise to be sufficiently rich or by introducing additional noise [6, 17], one can ensure that the scheme does not get stuck in an unstable equilibrium. In most practical applications, however, stochastic approximation algorithms such as (4.2) are seen to converge to local minima.

4.2.1 Extension to Multi-dimensional Parameter

For $\theta \in \mathbb{R}^N$, a natural extension of the original scheme (4.2) is as given below:

$$\theta_i(n+1) = \theta_i(n) - a(n)\frac{h(\theta(n)+\delta(n)e_i, \xi_i^+(n)) - h(\theta(n)-\delta(n)e_i, \xi_i^-(n))}{2\delta(n)}, \tag{4.8}$$

for $i = 1,2,\ldots,N$, where $e_i = \left(0, \ldots, 0, 1, 0, \ldots, 0\right)^T \in \mathbb{R}^N$, with 1 at the ith location, is the unit vector along the ith coordinate direction in $\mathbb{R}^N$. Further, $\xi_i^+(n)$, $\xi_i^-(n)$, $i = 1,\ldots,N$ are the corresponding i.i.d. noise samples that are also independent of each other and have a common distribution with finite second moments. Also, $\theta_i(n) \in \mathbb{R}$ denotes the ith component of the parameter vector $\theta(n) \in \mathbb{R}^N$, at update instant n. Under similar assumptions as Assumptions 4.1-4.4, the convergence of the multi-dimensional K-W algorithm (4.8) can be shown and similar conclusions as in Theorem 4.1 can be drawn. We leave this as an exercise for the interested reader.

4.3 Variants of the Kiefer-Wolfowitz Algorithm

There are some variants of the Kiefer-Wolfowitz algorithm available in the literature where either different perturbation parameter schemes are explored or the selection of the perturbation noise is varied. In [16], [14], it is seen that the use of common random numbers, i.e., $\xi_i^+(n) = \xi_i^-(n)$ reduces the estimator variance. This may be possible in a few simulation-based settings with random variables that are obtained from the same pseudo-random sequence. However, in most practical settings, this is difficult to achieve even when simulation is used. We discuss below two popular variations, one where the perturbation parameter is held constant and another in which one-sided perturbations are employed.

4.3.1 Fixed Perturbation Parameter

Quite often it makes sense to simply set $\delta(n) \equiv \delta$ for a 'small' $\delta > 0$ as has been done in [3], [4] and [5] (see also [15, pp. 15] for a discussion along these lines). With a fixed perturbation parameter, δ, the iteration scheme in (4.2) can be re-written as

$$\theta(n+1) = \theta(n) - a(n)\frac{h(\theta(n)+\delta,\xi^+(n)) - h(\theta(n)-\delta,\xi^-(n))}{2\delta}. \tag{4.9}$$

Note that we consider $\theta(n)$ to be scalar-valued again for simplicity. The case of vector $\theta(n)$ can be handled as explained in Section 4.2.1. The analysis of recursion (4.9) can be shown under weaker requirements on the step-size sequence $a(n), n \geq 0$ than those in Assumption 4.2. The convergence result that one obtains in this case is also weaker than the one given in Theorem 4.1. Specifically, we replace Assumption 4.2 with the following:

Assumption 4.5. The step-sizes $a(n) > 0$, $\forall n$ and

$$\sum_n a(n) = \infty, \quad \sum_n a(n)^2 < \infty. \tag{4.10}$$

This is essentially the same requirement as (3.4). For $\varepsilon > 0$, let

$$S^\varepsilon = \{\theta \mid |\theta - \theta^*| < \varepsilon \text{ for some } \theta^* \in S\},$$

denote the set of points that are in an ε-neighborhood of the set S. We have the following result:

Theorem 4.2. *Under Assumptions 4.1, 4.3, 4.4 and 4.5, given $\varepsilon > 0$, there exists $\bar{\delta} > 0$ such that for every $\delta \in (0, \bar{\delta}]$, the iterates $\theta(n), n \geq 0$, governed according to (4.9) converge a.s. to S^{ε}.*

Proof. (*Sketch*) One can rewrite (4.9) as

$$\theta(n+1) = \theta(n) - a(n)\left(\frac{dJ(\theta(n))}{d\theta} + \beta(n) + \eta(n)\right), \tag{4.11}$$

where

$$\eta(n) = \frac{h(\theta(n)+\delta, \xi^+(n)) - h(\theta(n)-\delta, \xi^-(n))}{2\delta} - \frac{J(\theta(n)+\delta) - J(\theta(n)-\delta)}{2\delta}, \text{ and}$$

$$\beta(n) = \frac{J(\theta(n)+\delta) - J(\theta(n)-\delta)}{2\delta} - \frac{dJ(\theta(n))}{d\theta},$$

respectively. As in the proof of Theorem 4.1, one can see that $\hat{\xi}(m), m \geq 0$ defined according to $\hat{\xi}(m) = \sum_{n=0}^{m} a(n)\eta(n) (m \geq 0)$ forms a convergent martingale sequence. Further, using Taylor series expansions of $J(\theta(n)+\delta)$ and $J(\theta(n)-\delta)$ around $\theta(n)$, it is easy to see that $\beta(n) = O(\delta)$. The result again follows from the Hirsch lemma (Lemma C.5). □

4.3.2 One-Sided Variants

The gradient estimates in (4.2) are also called two-sided finite difference (or balanced) estimates while those that we describe below in (4.12) are called one-sided finite difference (or unbalanced) estimates. In one-sided FDSA, the scheme is as follows:

$$\theta(n+1) = \theta(n) - a(n)\frac{h(\theta(n)+\delta(n), \xi^+(n)) - h(\theta(n), \xi^-(n))}{\delta(n)}. \tag{4.12}$$

The same for N-dimensional parameter θ can be re-written as

$$\theta_i(n+1) = \theta_i(n) - a(n)\frac{h(\theta(n)+\delta(n)e_i, \xi_i^+(n)) - h(\theta(n), \xi_i^-(n))}{\delta(n)}, i = 1,2,\ldots,N. \tag{4.13}$$

Proofs of convergence of both of these recursions follow along the same lines as Theorem 4.1. One-sided variants bring in a computational advantage by requiring approximately half of the number of simulations compared to the original Kiefer-Wolfowitz scheme in the case of multi-dimensional parameter where the original algorithm (4.8) requires $2N$ function measurements in order to obtain one estimate of the gradient while a one-sided variant (4.13) requires $(N+1)$ function measurements to obtain a gradient estimate. In the setting of simulation optimization [7], [8], [11], [12], [5], [1], [2], where one does not have access to function measurements but needs to simulate the whole system, one requires in effect $2N$ (resp. $(N+1)$) parallel simulations of the entire system when using two-sided (resp. one-sided) estimates. These algorithms become computationally inefficient when N becomes large and therefore one requires more computationally efficient methods for gradient estimation.

4.4 Concluding Remarks

Building on the stochastic algorithms which seek to obtain a zero of a function with noisy measurements, this chapter introduced and discussed a class of stochastic algorithms performing gradient descent on a cost objective. The Kiefer-Wolfowitz algorithm [13] marks the beginning of the development of this class of stochastic-gradient algorithms. When applied to N-dimensional parameter settings (with $N > 1$), these algorithms require $2N$ or $N+1$ noisy function measurements depending on whether two-sided or one-sided estimates are used. When N is large, these algorithms can become computationally inefficient because of the need to generate so many noisy cost observations. We address this scalability issue in the next two chapters that deal with the simultaneous perturbation stochastic approximation (SPSA) and the smoothed functional (SF) gradient algorithms, respectively.

References

1. Bhatnagar, S.: Adaptive multivariate three-timescale stochastic approximation algorithms for simulation based optimization. ACM Transactions on Modeling and Computer Simulation 15(1), 74–107 (2005)
2. Bhatnagar, S.: Adaptive Newton-based smoothed functional algorithms for simulation optimization. ACM Transactions on Modeling and Computer Simulation 18(1), 2:1–2:35 (2007)
3. Bhatnagar, S., Borkar, V.S.: Multiscale stochastic approximation for parametric optimization of hidden Markov models. Prob. Engg. and Info. Sci. 11, 509–522 (1997)
4. Bhatnagar, S., Borkar, V.S.: A two time scale stochastic approximation scheme for simulation based parametric optimization. Prob. Engg. and Info. Sci. 12, 519–531 (1998)
5. Bhatnagar, S., Fu, M.C., Marcus, S.I., Bhatnagar, S.: Two timescale algorithms for simulation optimization of hidden Markov models. IIE Transactions 33(3), 245–258 (2001)
6. Brandiere, O.: Some pathological traps for stochastic approximation. SIAM J. Contr. and Optim. 36, 1293–1314 (1998)
7. Cassandras, C.G.: Discrete Event Systems: Modeling and Performance Analysis. Aksen Associates, Boston (1993)

8. Chong, E.K.P., Ramadge, P.J.: Optimization of queues using an infinitesimal perturbation analysis-based stochastic algorithm with general update times. SIAM J. Cont. and Optim. 31(3), 698–732 (1993)
9. Chong, E.K.P., Ramadge, P.J.: Stochastic optimization of regenerative systems using infinitesimal perturbation analysis. IEEE Trans. Auto. Cont. 39(7), 1400–1410 (1994)
10. Fu, M.C.: Convergence of a stochastic approximation algorithm for the $GI/G/1$ queue using infinitesimal perturbation analysis. J. Optim. Theo. Appl. 65, 149–160 (1990)
11. Fu, M.C., Hill, S.D.: Optimization of discrete event systems via simultaneous perturbation stochastic approximation. IIE Trans. 29(3), 233–243 (1997)
12. Ho, Y.C., Cao, X.R.: Perturbation Analysis of Discrete Event Dynamical Systems. Kluwer, Boston (1991)
13. Kiefer, E., Wolfowitz, J.: Stochastic estimation of the maximum of a regression function. Ann. Math. Statist. 23, 462–466 (1952)
14. Kleinman, N.L., Spall, J.C., Naiman, D.Q.: Simulation-based optimization with stochastic approximation using common random numbers. Management Science 45, 1570–1578 (1999)
15. Kushner, H.J., Yin, G.G.: Stochastic Approximation Algorithms and Applications. Springer, New York (1997)
16. L'Ecuyer, P., Glynn, P.W.: Stochastic optimization by simulation: convergence proofs for the GI/G/1 queue in steady-state. Management Science 40(11), 1562–1578 (1994)
17. Pemantle, R.: Nonconvergence to unstable points in urn models and stochastic approximations. Annals of Prob. 18, 698–712 (1990)

Chapter 5
Gradient Schemes with Simultaneous Perturbation Stochastic Approximation

5.1 Introduction

Spall [26], [29] invented a remarkable algorithm that has become popular by the name simultaneous perturbation stochastic approximation (SPSA). It is remarkable in that it requires only two function measurements for a parameter of any dimension (i.e., any $N \geq 1$) and exhibits fast convergence (that is normally faster than the Kiefer-Wolfowitz algorithm). Unlike Kiefer-Wolfowitz schemes, where parameter perturbations are performed along each co-ordinate direction separately (in order to estimate the corresponding partial derivatives), in SPSA, all component directions are perturbed simultaneously using perturbations that are vectors of independent random variables that are often assumed to be symmetric, zero-mean, ± 1-valued, and Bernoulli distributed.

In the following sections, we discuss in detail the original SPSA algorithm [26] as well as its variants that are based on one and two function measurements. In particular, we discuss an important variant of the SPSA algorithm that uses deterministic perturbations based on Hadamard matrices. We provide the convergence proofs of the SPSA algorithm and its variants that we discuss.

5.2 The Basic SPSA Algorithm

We present the SPSA algorithm here for the expected cost objective. Recall that the objective in this case is $J(\theta) = E_{\xi}[h(\theta, \xi)]$, where $h : \mathbb{R}^N \times \mathbb{R}^k \to \mathbb{R}$ is a given single-stage cost function. Here $h(\theta, \xi)$ denotes a noisy measurement of $J(\theta)$ and $\xi \in \mathbb{R}^k$ is a mean-zero, random variable that corresponds to the noise in the measurements. Also, as in previous chapters, we let $L(\theta) = \nabla J(\theta)$. Note that the parameter vector θ is N-dimensional, i.e., $\theta \stackrel{\Delta}{=} (\theta_1, \theta_2, \ldots, \theta_N)^T \in \mathbb{R}^N$.

S. Bhatnagar et al.: Stochastic Recursive Algorithms for Optimization, LNCIS 434, pp. 41–76.
springerlink.com

5.2.1 Gradient Estimate Using Simultaneous Perturbation

We first describe the gradient estimate $\nabla_\theta J(\theta)$ of $J(\theta)$ when using SPSA. The estimate is obtained from the following relation:

$$\nabla_\theta J(\theta(n)) = \lim_{\delta(n)\downarrow 0} E\left[\left(\frac{h(\theta(n)+\delta(n)\Delta(n),\xi^+) - h(\theta(n)+\delta(n)\Delta(n),\xi^-)}{2\delta(n)\Delta_i(n)}\right)\Big|\,\theta(n)\right]. \tag{5.1}$$

The above expectation is over the noise terms ξ^+ and ξ^- as well as the perturbation random vector $\Delta(n) \stackrel{\triangle}{=} (\Delta_1(n),\ldots,\Delta_N(n))^T$, where $\Delta_1(n),\ldots,\Delta_N(n)$ are independent, mean-zero random variables satisfying the conditions in Assumption 5.4 below. The idea here is to perturb all the coordinate components of the parameter vector *simultaneously* using $\Delta(n)$. The two perturbed parameters correspond to $\theta(n)+\delta(n)\Delta(n)$ and $\theta(n)-\delta(n)\Delta(n)$, respectively. Several remarks are in order.

Remark 5.1. $\Delta_i(n), i = 1,2,\ldots,N, n \geq 0$ satisfy an inverse moment bound, that is, $E[|\Delta_i(n)^{-1}|] < \infty$. Thus, these random variables assign zero probability mass to the origin. We will see later in Theorem 5.1 that such a choice of random variables for perturbing the parameter vector ensures that in the recursion (5.1), the estimate along *undesirable* gradient directions averages to zero.

Remark 5.2. In contrast to the Kiefer-Wolfowitz class of algorithms, one can see that, the SPSA updates have a common numerator for all the θ-components but a different denominator. The inverse moment condition and the step-size requirements ensure convergence to a local minimum. Hence, unlike the Kiefer-Wolfowitz class of algorithms which require $2N$ or $N+1$ samples of the objective function, SPSA algorithms need only two samples irrespective of the dimension of the parameter θ.

Remark 5.3. Most often, one assumes that the perturbation random variables are distributed according to the symmetric Bernoulli distribution with $\Delta_i(n) = \pm 1$ w.p. $1/2$, $i = 1,\ldots,N$, $n \geq 0$. In fact, it is found in [25] that under certain conditions, the optimal distribution on components of the simultaneous perturbation vector is a symmetric Bernoulli distribution. This result is obtained under two separate objectives (see [25]): (a) minimize the mean square error of the estimate, and (b) maximize the likelihood that the estimate remains in a symmetric bounded region around the true parameter.

5.2.2 The Algorithm

The update rule in the basic SPSA algorithm is as follows:

$$\begin{aligned}\theta_i(n+1) =& \theta_i(n) \\ & - a(n)\left(\frac{h(\theta(n)+\delta(n)\Delta(n),\xi^+(n)) - h(\theta(n)-\delta(n)\Delta(n),\xi^-(n))}{2\delta(n)\Delta_i(n)}\right),\end{aligned} \tag{5.2}$$

for $i = 1,\ldots,N$ and $n \geq 0$.

The overall flow of the basic SPSA algorithm is described in Fig. 5.1. In essence, it is a closed-loop procedure where the samples of the single stage cost function $h(\cdot,\cdot)$ are obtained for two perturbed parameter values $(\theta(n)+\delta(n)\Delta(n))$ and $(\theta(n)-\delta(n)\Delta(n))$, respectively. These samples are then used to update θ in the negative gradient descent direction using the estimate (5.1).

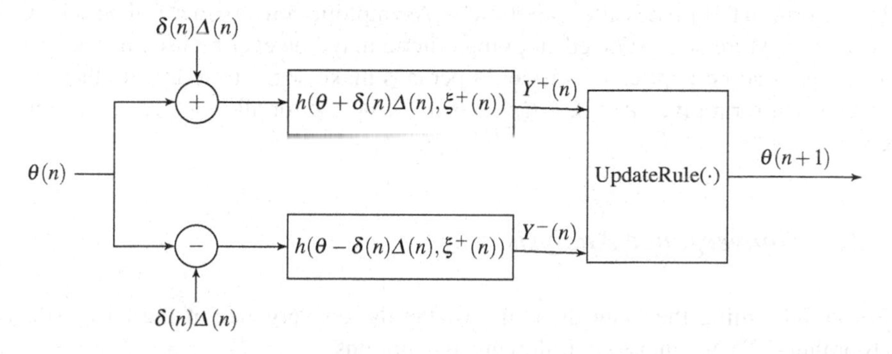

Fig. 5.1 Overall flow of the algorithm 5.1.

For the sake of completeness and because of its prominence in gradient estimation schemes, we describe below the SPSA algorithm in an algorithmic form.

Algorithm 5.1 The basic SPSA Algorithm for the Expected Cost Objective

Input:

- Q, a large positive integer;
- $\theta_0 \in C \subset \mathbb{R}^N$, initial parameter vector;
- Bernoulli(p), random independent Bernoulli ± 1 sampler with probability p for '+1' and $1-p$ for '−1';
- $h(\theta,\xi)$, noisy measurement of cost objective J;
- $a(n)$ and $\delta(n)$, step-size sequences chosen complying to assumption in (5.3);

Output: $\theta^* \triangleq \theta(Q)$.

$n \leftarrow 0$.
loop
  for $i = 1$ to N **do**
    $\Delta_i(n) \leftarrow$ Bernoulli(1/2).
  end for
  $Y(n)^+ \leftarrow h(\theta + \delta(n)\Delta(n), \xi^+(n))$.
  $Y(n)^- \leftarrow h(\theta - \delta(n)\Delta(n), \xi^-(n))$.
  for $i = 1$ to N **do**

$$\theta_i(n+1) \leftarrow \theta_i(n) - a(n)\frac{Y^+(n) - Y^-(n)}{2\delta(n)\Delta_i(n)}.$$

  end for
  $n \leftarrow n+1$
  if $n = Q$ **then**
    Terminate with $\theta(Q)$.
  end if
end loop

The algorithm terminates after Q iterations. Asymptotic convergence is then achieved as $Q \to \infty$. More sophisticated stopping criteria may however be used as well. For instance, in some applications it could perhaps make sense to terminate the algorithm when for a given $\varepsilon > 0$, $\|\theta(n) - \theta(n-m)\| < \varepsilon$ for all $m \in \{1, \ldots, R\}$, for a given $R > 1$.

5.2.3 Convergence Analysis

Before presenting the main theorem proving the convergence of the basic SPSA algorithm (5.2), we make the following assumptions:

Assumption 5.1. The map $J : \mathbb{R}^N \to \mathbb{R}$ is Lipschitz continuous and is differentiable with bounded second order derivatives. Further, the map $L : \mathbb{R}^N \to \mathbb{R}^N$ defined as $L(\theta) = -\nabla J(\theta), \forall \theta \in \mathbb{R}^N$ and the map $h : \mathbb{R}^N \times \mathbb{R}^k \to \mathbb{R}$ are both Lipschitz continuous.

The above is a technical requirement needed to push through a Taylor series expansion and is used in the analysis.

Assumption 5.2. The step-sizes $a(n), \delta(n) > 0, \forall n$ and

$$a(n), \delta(n) \to 0 \text{ as } n \to 0, \quad \sum_n a(n) = \infty, \quad \sum_n \left(\frac{a(n)}{\delta(n)}\right)^2 < \infty. \tag{5.3}$$

Thus, $a(n)$ and $\delta(n)$ are both diminishing sequences of positive numbers with $\delta(n)$ going to zero slower than $a(n)$. The second condition above is analogous to a similar condition in (3.4). The third condition is a stronger requirement. In [11], a relaxation is made and it is assumed that $a(n), \delta(n) \to 0$ as $n \to \infty$ and that

$$\sum_n a(n) = \infty, \sum_n a(n)^p < \infty$$

for some $p \in (1,2]$. Typically $a(n), \delta(n), n \geq 0$ can be chosen according to $a(n) = a/(A+n+1)^\alpha$ and $\delta(n) = c/(n+1)^\gamma$, for $a, A, c > 0$. The values of α and γ suggested in [13] and [15] are 1 and 1/6, respectively. In [28], it is observed that the choices $\alpha = 0.602$ and $\gamma = 0.101$ perform well in practical settings.

Assumption 5.3. $\xi^+(n)$, $\xi^-(n)$, $n \geq 0$ are $\mathbb{R}^k$-valued, independent random vectors having a common distribution and with finite second moments.

Note that the algorithm (5.2) can be rewritten as follows:

$$\theta_i(n+1) = \theta_i(n) - a(n)\left(\frac{J(\theta(n)+\delta(n)\Delta(n)) - J(\theta(n)-\delta(n)\Delta(n))}{2\delta(n)\Delta_i(n)} + \frac{\hat{\xi}^+(n) - \hat{\xi}^-(n)}{2\delta(n)\Delta_i(n)}\right), \tag{5.4}$$

where

$$\begin{aligned}\hat{\xi}^+(n) - \hat{\xi}^-(n) =& h(\theta(n)+\delta(n)\Delta(n), \xi^+(n)) - h(\theta(n)-\delta(n)\Delta(n), \xi^-(n)) \\ & - (J(\theta(n)+\delta(n)\Delta(n)) - J(\theta(n)-\delta(n)\Delta(n))).\end{aligned}$$

It is easy to see that $\dfrac{\hat{\xi}^+(n) - \hat{\xi}^-(n)}{2\delta(n)\Delta_i(n)}, n \geq 0$ forms a martingale difference sequence under an appropriate filtration.

Assumption 5.4. The random variables $\Delta_i(n)$, $n \geq 0$, $i = 1, \ldots, N$, are mutually independent, mean-zero, have a common distribution and satisfy $E[(\Delta_i(n))^{-2}] \leq \bar{K}$, $\forall n \geq 0$, for some $\bar{K} < \infty$.

In order for the inverse moment of $\Delta_i(n)$ to be uniformly bounded (see Assumption 5.4), it follows that the random variables $\Delta_i(n)$ must have zero probability mass at the origin. Many times, one simply lets $\Delta_i(n), n \geq 0$ to be independent, symmetric Bernoulli-distributed random variables with $\Delta_i(n) = \pm 1$ w.p. 1/2, $\forall i = 1, \ldots, N$.

Assumption 5.5. The iterates (5.2) remain uniformly bounded almost surely, i.e.,

$$\sup_n \|\theta(n)\| < \infty, \text{ a.s.} \tag{5.5}$$

Consider the ODE:

$$\dot{\theta}(t) = -\nabla J(\theta(t)). \tag{5.6}$$

Assumption 5.6. The set H containing the globally asymptotically stable equilibria of the ODE (5.6) (i.e., the local minima of J) is a compact subset of $\mathbb{R}^N$.

Theorem 5.1. *Under Assumptions 5.1-5.6, the parameter updates (5.2) satisfy $\theta(n) \to H$ with probability one.*

Proof. Let $\nabla_i J(\theta)$ represent the ith partial derivative of $J(\theta)$. The SPSA update rule (5.2) can be rewritten as follows:

$$\theta_i(n+1) = \theta_i(n) - a(n)\left(\nabla_i J(\theta(n)) + \eta_i(n) + \beta_i(n)\right), \tag{5.7}$$

where

$$\begin{aligned}\eta_i(n) =& \frac{h(\theta(n)+\delta(n)\Delta(n), \xi^+(n)) - h(\theta(n)-\delta(n)\Delta(n), \xi^-(n))}{2\delta(n)\Delta_i(n)} \\ &- \frac{J(\theta(n)+\delta(n)\Delta(n)) - J(\theta(n)-\delta(n)\Delta(n))}{2\delta(n)\Delta_i(n)},\end{aligned}$$

$$\beta_i(n) = \frac{J(\theta(n)+\delta(n)\Delta(n)) - J(\theta(n)-\delta(n)\Delta(n))}{2\delta(n)\Delta_i(n)} - \nabla_i J(\theta(n)),$$

for $i = 1, 2, \ldots, N$. Now,

$$|h(\theta,\xi)| - |h(0,0)| \leq |h(\theta,\xi) - h(0,0)| \leq \hat{L}\|(\theta,\xi) - (0,0)\|,$$

where $\hat{L} > 0$ is the Lipschitz constant of h. Since $\|\cdot\|$ is the Euclidean norm, it is easy to see that $\|(\theta,\xi) - (0,0)\| \leq \|\theta\| + \|\xi\|$. Thus, we have that

$$|h(\theta,\xi)| \leq \hat{K}(1 + \|\theta\| + \|\xi\|),$$

for some $\hat{K} > 0$. Now $\eta_i(n), n \geq 0$ forms a martingale difference sequence with respect to the sequence of sigma fields $\mathscr{F}(n) = \sigma(\theta(m), \Delta(m), \xi^+(m), \xi^-(m), m \leq n), n \geq 0$. Let

$$\bar{N}_i(n) = \sum_{m=0}^{n-1} a(m)\eta_i(m),\ n \geq 1,\ i = 1, \ldots, N.$$

It is easy to see as a consequence of Assumptions 5.1–5.5 and from the martingale convergence theorem (Theorem B.2) that $\bar{N}_i(n), n \geq 1$, is an almost surely convergent martingale sequence.

Now, Taylor's series expansions of $J(\theta(n) + \delta(n)\Delta(n))$ and $J(\theta(n) - \delta(n)\Delta(n))$, respectively, around the point $\theta(n)$ give,

$$J(\theta(n) + \delta(n)\Delta(n)) = J(\theta(n)) + \delta(n)\Delta(n)^T \nabla J(\theta(n)) + O(\delta(n)^2),$$
$$J(\theta(n) - \delta(n)\Delta(n)) = J(\theta(n)) - \delta(n)\Delta(n)^T \nabla J(\theta(n)) + O(\delta(n)^2).$$

Upon substitution of the above in the expression for $\beta_i(n)$, we get,

$$\beta_i(n) = \sum_{j=1, j \neq i}^{N} \frac{\Delta_j(n)}{\Delta_i(n)} \nabla_j J(\theta(n)) + O(\delta(n)).$$

Since $\Delta_j(n)$ are i.i.d., bounded and mean-zero random variables, the first term in the above is a square integrable mean-zero random noise for a given $\theta(n)$. The claim now follows from the Hirsch Lemma (see Lemma C.5). □

As suggested by equation (5.7), while the search direction is randomly chosen and need not follow a descent path, the algorithm is seen to make the right moves in the asymptotic average. In the next section, we discuss some of the variants of the basic SPSA algorithm.

5.3 Variants of the Basic SPSA Algorithm

The SPSA algorithm has evoked significant interest due to its good performance, ease of implementation and wide applicability. Moreover, it is observed to be scalable in that the computational effort does not increase significantly with the parameter dimension unlike the Kiefer-Wolfowitz algorithms (see for instance, the experiments in [7]). In the next few sections, we shall review some of the important variants of the SPSA algorithm.

5.3.1 One-Measurement SPSA Algorithm

Interestingly enough, it is possible to perform gradient estimation via just one measurement. In [27], a one-measurement version of SPSA has been presented. The simulation here is run with the parameter $\theta(n)+\delta(n)\Delta(n)$ where the update rule is of the form:

$$\theta_i(n+1) = \theta_i(n) - a(n)\left(\frac{h(\theta(n)+\delta(n)\Delta(n),\xi^+(n))}{\delta(n)\Delta_i(n)}\right), \tag{5.8}$$

for $i = 1,\ldots,N$ and $\Delta(n)$ as before.

Now, we present a proof of convergence of this scheme.

Theorem 5.2. *Under Assumptions 5.1-5.6, the parameter updates (5.8) satisfy $\theta(n) \to H$ with probability one.*

Proof. The proof follows in a similar manner as that of Theorem 5.1 except for the change that because of the presence of only one simulation, there is an additional bias term in the gradient estimate. Recall that a Taylor series expansion of $J(\theta(n)+\delta(n)\Delta(n))$ around $\theta(n)$ gives

$$J(\theta(n)+\delta(n)\Delta(n)) = J(\theta(n)) + \delta(n)\Delta(n)^T\nabla J(\theta(n)) + O(\delta(n)^2).$$

One can rewrite (5.8) in a manner similar to (5.7), where

$$\eta_i(n) = \frac{h(\theta(n)+\delta(n)\Delta(n),\xi^+(n))}{\delta(n)\Delta_i(n)} - \frac{J(\theta(n)+\delta(n)\Delta(n))}{\delta(n)\Delta_i(n)},$$

$$\beta_i(n) = \frac{J(\theta(n)+\delta(n)\Delta(n))}{\delta(n)\Delta_i(n)} - \nabla_i J(\theta(n)).$$

Thus,

$$\beta_i(n) = \frac{J(\theta(n))}{\delta(n)\Delta_i(n)} + \sum_{j=1,j\neq i}^{N} \frac{\Delta_j(n)}{\Delta_i(n)}\nabla_j J(\theta(n)) + O(\delta(n)). \tag{5.9}$$

The second term in the above, as previously discussed, is a square integrable mean-zero random noise (given $\theta(n)$). The first term above (for any $n \geq 0$) is also mean-zero for a given $\theta(n)$. Further, the product of $a(n)$ with the first term in the above can be seen to be square summable. Hirsch Lemma (see Lemma C.5) can now be applied to obtain the claim. □

As observed in [27] and other references, for instance, [8], the performance of the one-measurement SPSA algorithm is not as good as its two-measurement counterpart because of the presence of the additional bias term (above) that has a factor $\delta(n)$ in its denominator and which tends to zero asymptotically. However, it is noted in [27], that one-simulation SPSA may have better adaptability as compared to its two-simulation counterpart in non-stationary settings.

5.3.2 *One-Sided SPSA Algorithm*

A one-sided difference version of SPSA with two measurements has been considered in [11]. Here the two simulations are run with the parameters $\theta(n)+\delta(n)\Delta(n)$ and $\theta(n)$, respectively, and the update rule has the form:

$$\theta_i(n+1) = \theta_i(n) - a(n)\left(\frac{h(\theta(n)+\delta(n)\Delta(n),\xi^+(n)) - h(\theta(n),\xi(n))}{\delta(n)\Delta_i(n)}\right), \tag{5.10}$$

for $i = 1,\ldots,N$ with $\Delta(n)$ as before.

Also, $\xi^+(n),\xi(n)$ satisfy Assumption 5.3 with $\xi(n)$ in place of $\xi^-(n)$. One of the measurements of $h(\cdot,\cdot)$, here, is unperturbed which may be useful in certain applications [11]. A similar convergence result as that in 5.1 can be shown for this case as well. If higher order derivatives of J exist, then one can see that in the case of the original SPSA algorithm, all even order terms such as the second order terms involving the Hessian get directly cancelled. This is however not the case with the one-sided difference SPSA where such terms contribute to the overall bias. The two-sided form (5.2) is the most studied and used in applications.

5.3.3 *Fixed Perturbation Parameter*

In many applications [5, 6, 7] and also in discussions [20, pp. 15], of the SPSA algorithm, a constant value for the perturbation parameters $\delta(n) \equiv \delta > 0$, is often considered for convenience. The SPSA update rule would in this case take the form:

$$\theta_i(n+1) = \theta_i(n) - a(n)\left(\frac{h(\theta(n)+\delta\Delta(n),\xi^+(n)) - h(\theta(n)-\delta\Delta(n),\xi^-(n))}{2\delta\Delta_i(n)}\right), \tag{5.11}$$

for $i = 1,\ldots,N$ and $n \geq 0$.

As described in the theorem below, a suitable $\delta > 0$ can be chosen based on a desired $\varepsilon > 0$, to prove convergence of the update rule to an ε-neighborhood of the set H (the local minima of J). For $\varepsilon > 0$, let

$$H^\varepsilon = \{\theta \mid \|\theta - \theta^*\| < \varepsilon \text{ for some } \theta^* \in H\}.$$

Consider now the following requirement of the step-sizes $a(n), n \geq 0$, in place of Assumption 5.2:

Assumption 5.7. The step-sizes $a(n) > 0, \forall n$ and

$$\sum_n a(n) = \infty, \quad \sum_n a(n)^2 < \infty. \tag{5.12}$$

Theorem 5.3. *Under Assumptions 5.1, 5.3-5.7, given $\varepsilon > 0$, there exists $\bar{\delta} > 0$ such that for every $\delta \in (0, \bar{\delta}]$, the update rule (5.11) converges a.s. to H^{ε}.*

Proof. Proceeding along similar lines as in the proof of Theorem 5.1, the update rule (5.11) can be re-written as

$$\theta_i(n+1) = \theta_i(n) - a(n)\left(\nabla_i J(\theta(n)) + \eta_i(n) + \beta_i(n)\right), \tag{5.13}$$

where

$$\eta_i(n) = \frac{h(\theta(n)+\delta\Delta(n), \xi(n)^+) - h(\theta(n)-\delta\Delta(n), \xi(n)^-)}{2\delta\Delta_i(n)} - \frac{J(\theta(n)+\delta\Delta(n)) - J(\theta(n)-\delta\Delta(n))}{2\delta\Delta_i(n)},$$

$$\beta_i(n) = \frac{J(\theta(n)+\delta\Delta(n)) - J(\theta(n)-\delta\Delta(n))}{2\delta\Delta_i(n)} - \nabla_i J(\theta(n)),$$

for $i = 1, 2, \ldots, N$. As before, it is easy to see that $\eta_i(n), n \geq 0$, is a square integrable martingale difference sequence. Thus, $\left\{ \sum_{m=0}^{n-1} a(m)\eta_i(m), n \geq 1 \right\}$ can be seen from the martingale convergence theorem (Theorem B.2) to be an almost surely convergent martingale. Now, simplifying the expression of $\beta_i(n)$ using appropriate Taylor series expansions of $J(\theta(n) + \delta\Delta(n))$ and $J(\theta(n) - \delta\Delta(n))$, respectively, around $\theta(n)$, we get,

$$\beta_i(n) = \sum_{j=1, j\neq i}^{N} \frac{\Delta_j(n)}{\Delta_i(n)} \nabla_j J(\theta(n)) + O(\delta). \tag{5.14}$$

It is easy to see that $E[\beta_i(n) \mid \theta(n)] = O(\delta)$. The claim now follows by applying the Hirsch lemma (Lemma C.5). □

5.4 General Remarks on SPSA Algorithms

It is interesting to note that in each of the update rules (5.2), (5.8) and (5.10), the numerator is the same across all i, $i = 1,\ldots,N$, while the denominator has a quantity $\Delta_i(n)$ that depends on i. This is unlike the Kiefer-Wolfowitz algorithm (or FDSA) where the denominator is the same while the numerator is different for different i, see (4.2) and (4.12). Thus, in going from FDSA to SPSA, the complexity in estimating the gradient shifts (in a way) from the numerator of the estimator to its denominator. It should be noted that simulating N independent symmetric Bernoulli-distributed random variables is in general far less computationally expensive than obtaining $2N$ or $(N+1)$ objective function measurements or simulations, particularly when N is large. It has been seen both from theory and experiments [26], [7], [24] that two-sided, two-simulation SPSA (5.2) is computationally far more superior to FDSA. In [26], asymptotic normality results for SPSA and FDSA are used to establish the relative efficiency of SPSA. The asymptotic analysis for the Robbins-Monro algorithm can be adapted to prove almost sure convergence of the iterates in the SPSA algorithm [26]. Assuming that there is a unique globally asymptotically stable equilibrium θ^* for the associated ODE (i.e., a global minimum for the basic algorithm), the asymptotic normality result in [26] essentially says that

$$n^{r/2}(\theta(n) - \theta^*) \xrightarrow{D} N(\mu, \Sigma)$$

as $n \to \infty$, where $\xrightarrow{D}$ denotes convergence in distribution, $N(\mu, \Sigma)$ is a multi-variate Gaussian with mean μ and covariance matrix Σ that depends on the Hessian at θ^*. In general, $\mu \neq 0$. The quantity r depends upon the choice of the gain sequences $\{a(n)\}$ and $\{\delta(n)\}$.

Many interesting analyses of the SPSA algorithm have been reported in the literature. In [26], the above asymptotic normality result is used to argue the relative asymptotic efficiency of SPSA over FDSA. In particular, it is argued that SPSA results in an N-fold computational savings over FDSA. In [11], a projected version of SPSA where the projection region is gradually increased has been presented. This is a novel approach to take care of the issue of iterate stability in general. In [2] and [17], application of SPSA for optimization in the case of non-differentiable functions is considered. A detailed analysis of SPSA under general conditions can also be found in [16]. An analysis of SPSA and FDSA when common random numbers are used in the simulations is given in [18]. Different ways of gradient estimation in SPSA using past measurements have been reported in [1] and [21]. In [22], iterate averaging for stability of the SPSA recursions and improved algorithmic behaviour is explored. A case of weighted averaging of the Kiefer-Wolfowitz and SPSA iterates is considered in [14]. In [12] and [23], SPSA is proposed for use as a global search algorithm. In [13], SPSA is compared with a two-sided smoothed functional algorithm (see Chapter 6) and it is observed over the experiments considered there that SPSA is the better of the two algorithms. In [28], the general technique for implementing SPSA and the choice of gain sequences is discussed. In [10], non-Bernoulli distributions have been explored for the perturbation sequences.

In the next section, we discuss an important class of SPSA algorithms where the perturbation sequence is deterministic and regular (i.e., periodic) rather than a vector of independent Bernoulli-distributed random variables.

5.5 SPSA Algorithms with Deterministic Perturbations

The SPSA algorithms discussed in the previous sections used zero-mean and mutually independent random perturbations to obtain an estimate of the gradient of the objective function. We now consider the case when the perturbation sequences are constructed differently by a deterministic mechanism. These perturbations are obtained by cyclically passing through a certain construction based on Hadamard matrices. The principal idea behind the Hadamard matrix construction is to periodically cancel the bias terms aggregated over iterations where the length of the period over which such cancellation occurs is small. As a consequence, one expects an improved algorithmic performance. In [8], it is observed that in certain scenarios, the deterministic perturbations are theoretically sound and result in faster convergence empirically. For further discussions, we will use the setting of fixed perturbation parameter, that is, $\delta(n) \equiv \delta > 0$. Nevertheless, all the following discussions, can be suitably applied to the general setting with non-fixed perturbation sequences $\delta(n), n \geq 0$ satisfying the requirements in Assumption 5.2.

5.5.1 Properties of Deterministic Perturbation Sequences

We first explain the idea why such a construction can work in practice. Recall that a Taylor series expansion of $J(\theta(n)+\delta\Delta(n))$ around $\theta(n)$ is the following:

$$J(\theta(n)+\delta\Delta(n)) = J(\theta(n)) + \delta\Delta(n)^T \nabla J(\theta(n)) + o(\delta). \tag{5.15}$$

Similarly, an expansion of $J(\theta(n)-\delta\Delta(n))$ around $\theta(n)$ gives

$$J(\theta(n)-\delta\Delta(n)) = J(\theta(n)) - \delta\Delta(n)^T \nabla J(\theta(n)) + o(\delta). \tag{5.16}$$

Hence from (5.15) and (5.16), for $i = 1,\ldots,N$, one obtains in the case of a two-measurement algorithm with parameters $\theta(n)+\delta\Delta(n)$ and $\theta(n)-\delta\Delta(n)$,

$$\begin{aligned}\frac{J(\theta(n)+\delta\Delta(n))-J(\theta(n)-\delta\Delta(n))}{2\delta\Delta_i(n)} &= \frac{\Delta(n)^T\nabla J(\theta(n))}{\Delta_i(n)}+o(\delta)\\ &= \nabla_i J(\theta(n))+\sum_{j=1,j\neq i}^{N}\frac{\Delta_j(n)}{\Delta_i(n)}\nabla_j J(\theta(n))\\ &\quad +o(\delta). \end{aligned} \tag{5.17}$$

Note that the error terms (at the end) are still $o(\delta)$ above because the subsequent Hessian terms in the above expansions would directly cancel as well. Also, in the case of a two-measurement, but one-sided gradient estimate involving parameters $\theta(n)+\delta\Delta(n)$ and $\theta(n)$, one obtains

$$\frac{J(\theta(n)+\delta\Delta(n))-J(\theta(n))}{\delta\Delta_i(n)} = \nabla_i J(\theta(n))+\sum_{j=1,j\neq i}^{N}\frac{\Delta_j(n)}{\Delta_i(n)}\nabla_j J(\theta(n))+O(\delta). \tag{5.18}$$

As discussed before, unlike (5.17), the Hessian term would not cancel if it is considered in the expansion. Hence, the last term above is now $O(\delta)$.

Note that $\sum_{j=1,j\neq i}^{N}\frac{\Delta_j(n)}{\Delta_i(n)}\nabla_j J(\theta(n))$ constitutes the bias. When $\Delta_i(n)$, $i=1,\ldots,N$, $n\geq 0$ satisfy Assumption 5.4, for instance, if they are Bernoulli distributed independent random variables, $\Delta_i(n)=\pm 1$ w.p.1/2, $\forall i,n$, then it follows that

$$E\left[\sum_{j=1,j\neq i}^{N}\frac{\Delta_j(n)}{\Delta_i(n)}\nabla_j J(\theta(n))\,\middle|\,\theta(n)\right]=0. \tag{5.19}$$

The conditional expectation as such can be seen to be obtained in the asymptotic limit of the algorithm using a martingale argument. However, as we shall subsequently see, when the perturbations $\Delta_i(n)$ are not random but are obtained through a deterministic construction instead, it suffices that some finite sums of the bias terms tend to zero asymptotically.

In the case of a one-measurement algorithm with parameter $\theta(n)+\delta\Delta(n)$, on the other hand, a similar calculation shows

$$\frac{J(\theta(n)+\delta\Delta(n))}{\delta\Delta_i(n)} = \frac{J(\theta(n))}{\delta\Delta_i(n)}+\nabla_i J(\theta(n))+\sum_{j=1,j\neq i}^{N}\frac{\Delta_j(n)}{\Delta_i(n)}\nabla_j J(\theta(n))+O(\delta). \tag{5.20}$$

The first and the third terms on the RHS in (5.20) constitute the bias terms. In the case of random perturbations as described above, the following holds as well in addition to (5.19):

$$E\left[\frac{J(\theta(n))}{\delta\Delta_i(n)}\,\middle|\,\theta(n)\right]=0. \tag{5.21}$$

As mentioned before, the quantity $\delta > 0$ is usually chosen to be either a 'small' constant or else is *slowly diminishing to zero*. In either case, the variances of the estimates depend on δ. Nevertheless, in the case of one-measurement SPSA with deterministic perturbations, one wants that the bias contributed by the first term on the RHS of (5.20) tends to zero in addition to that contributed by the third term on the same RHS.

In general, a deterministic construction for the perturbation sequences should satisfy the following property in the case of two-measurement SPSA algorithms with both the two-sided balanced estimates (with parameters $\theta(n)+\delta\Delta(n)$ and $\theta(n)-\delta\Delta(n)$, $n \geq 0$) as well as the one-sided estimates (with parameters $\theta(n)+\delta\Delta(n)$ and $\theta(n)$, $n \geq 0$), respectively.

(P.1) There exists a $P \in \mathbb{N}$ such that for every $i, j \in \{1,\ldots,N\}$, $i \neq j$ and for any $s \in \mathbb{N}$,

$$\sum_{n=s}^{s+P} \frac{\Delta_i(n)}{\Delta_j(n)} = 0. \tag{5.22}$$

Further, in the case of one-measurement SPSA (with parameters $\theta(n) + \delta\Delta(n)$, $n \geq 0$), one requires the following property in addition to (P.1):

(P.2) There exists a $P \in \mathbb{N}$ such that for every $k \in \{1,\ldots,N\}$ and any $s \in \mathbb{N}$,

$$\sum_{n=s}^{s+P} \frac{1}{\Delta_k(n)} = 0. \tag{5.23}$$

Property (P.2) is not required to be satisfied by the aforementioned two-measurement SPSA algorithms while both (P.1) and (P.2) are required for one-measurement SPSA.

5.5.2 *Hadamard Matrix Based Construction*

Let H_{2^k}, $k \geq 1$ be matrices of order $2^k \times 2^k$ that are recursively obtained as:

$$H_2 = \begin{pmatrix} 1 & 1 \\ 1 & -1 \end{pmatrix} \text{ and } H_{2^k} = \begin{pmatrix} H_{2^{k-1}} & H_{2^{k-1}} \\ H_{2^{k-1}} & -H_{2^{k-1}} \end{pmatrix}, \; k > 1.$$

Such matrices are called normalized Hadamard matrices. These are characterized by all elements in their first row and column being 1.

5.5.2.1 Construction for Two-Measurement Algorithms

We now describe the construction of the perturbation sequences in the case when the gradient estimates have the form (5.17) or (5.18). Let $P = 2^{\lceil \log_2 N \rceil}$. (Note that $P \geq N$.) Consider now the matrix H_P (with P chosen as above). Let $h(1),\ldots,h(N)$, be any N columns of H_P. In case $P = N$, then $h(1),\ldots,h(N)$, will correspond to all N columns of H_P. Form a matrix H'_P of order $P \times N$ that has $h(1),\ldots,h(N)$ as its columns. Let $e(p), p = 1,\ldots,P$, be the P rows of H'_P. Now set $\Delta(n)^T = e(n \bmod P+1)$, $\forall n \geq 0$. The perturbations are thus generated by cycling through the rows of H'_P with $\Delta(0)^T = e(1), \Delta(1)^T = e(2), \ldots, \Delta(P-1)^T = e(P)$, $\Delta(P)^T = e(1)$, etc. The following result is obvious from the above construction.

Lemma 5.4. *The Hadamard matrix based perturbations $\Delta(n)$, $n \geq 0$ for two-measurement SPSA algorithms satisfy property (P.1).*

Here we give an example for the case when the parameter dimension N is 4. As per Lemma 5.4, we construct the perturbation sequence $\Delta(1),\ldots,\Delta(4)$, from H_4 as follows:

$$\Delta(1) = [1,1,1,1]^T,$$
$$\Delta(2) = [1,-1,1,-1]^T,$$
$$\Delta(3) = [1,1,-1,-1]^T,$$
$$\Delta(4) = [1,-1,-1,1]^T.$$

In this particular case where N was a power of 2, we ended up taking the row vectors of H_4 as the perturbations. If N is not a power of 2, the procedure would be similar to the above, except that we only pick N columns from the matrix H_P, where $P = 2^{\lceil \log_2 N \rceil}$. It can be easily checked that the perturbations generated above satisfy the property (P.1).

5.5.2.2 Construction for One-Measurement Algorithms

In the case when the gradient estimates are as in (5.20) and depend on a single measurement with parameter $\theta(n) + \delta\Delta(n)$, the value of P is set to $P = 2^{\lceil \log_2 (N+1) \rceil}$. Thus, $P \geq N+1$ in this case. Now let $h(1),\ldots,h(N)$ be any N columns of H_P other than the first column. Form the matrix H'_P of order $P \times N$ with $h(1),\ldots,h(N)$ as its N columns. As before, if $e(p), p = 1,\ldots,P$ are the P rows of H'_P, then the perturbation vectors $\Delta(n)$ are obtained again by cycling through the rows of H'_P. The following result is now easy to verify from the construction.

Lemma 5.5. *The Hadamard matrix based perturbations $\Delta(n)$, $n \geq 0$ for one-measurement SPSA algorithms satisfy both properties (P.1) and (P.2).*

We again consider an example where $N = 4$. Now, to construct perturbations in this case for one-simulation algorithms, we first form the normalized Hadamard matrix H_8 as follows:

$$H_8 = \left[\begin{array}{cccc|cccc} 1 & 1 & 1 & 1 & 1 & 1 & 1 & 1 \\ 1 & -1 & 1 & -1 & 1 & -1 & 1 & -1 \\ 1 & 1 & -1 & -1 & 1 & 1 & -1 & -1 \\ 1 & -1 & -1 & 1 & 1 & -1 & -1 & 1 \\ \hline 1 & 1 & 1 & 1 & -1 & -1 & -1 & -1 \\ 1 & -1 & 1 & -1 & -1 & 1 & -1 & 1 \\ 1 & 1 & -1 & -1 & -1 & -1 & 1 & 1 \\ 1 & -1 & -1 & 1 & -1 & 1 & 1 & -1 \end{array}\right].$$

Now, the perturbations $\Delta(i), i = 1,\ldots,8$ can be obtained by taking columns $2-5$ (or any 4 columns except the first) of H_8. For instance, taking the rows of columns $2-5$ from H_8 above, we obtain:

$$\begin{aligned}
\Delta(1) &= [1,1,1,1]^T, \\
\Delta(2) &= [-1,1,-1,1]^T, \\
\Delta(3) &= [1,-1,-1,1]^T, \\
\Delta(4) &= [-1,-1,1,1]^T, \\
\Delta(5) &= [1,1,1,-1]^T, \\
\Delta(6) &= [-1,1,-1,-1]^T, \\
\Delta(7) &= [1,-1,-1,-1]^T, \\
\Delta(8) &= [-1,-1,1,-1]^T.
\end{aligned}$$

Any other choice of four columns other than the first can be seen to work as well. Properties P.1–P.2 are seen to be satisfied here.

5.5.3 *Two-Sided SPSA with Hadamard Matrix Perturbations*

Let $\theta(n) = (\theta_1(n),\ldots,\theta_N(n))^T$, $n \geq 0$ be a sequence of parameters that are tuned according to the algorithm below (cf. (5.24)). Also, let $\Delta(n), n \geq 0$ be a sequence of perturbations obtained from the Hadamard matrix construction described in Section 5.5.2.1. Then, the update rule of the two-sided SPSA algorithm is given by

$$\theta_i(n+1) = \theta_i(n) - a(n)\left(\frac{h(\theta(n)+\delta\Delta(n),\xi^+(n)) - h(\theta(n)-\delta\Delta(n),\xi^-(n))}{2\delta\Delta_i(n)}\right), \tag{5.24}$$

for $i = 1,\ldots,N$ and $n \geq 0$.

Remark 5.4. In (5.24), δ is a fixed positive real number. The convergence analysis of the earlier SPSA schemes established that they converge to the set H of asymptotically stable equilibrium points of the corresponding ODE. However, with a fixed δ, it is later established that (5.24) converges in the limit to a set that can be made arbitrarily close to H by the choice of δ. Further, a decreasing δ-sequence can also be incorporated in (5.24) as well as the one-sided and one-measurement variants discussed in the later sections.

Remark 5.5. The overall flow and the algorithm structure of the two-sided SPSA (5.24) is similar to Fig. 5.1 and Algorithm 5.1 respectively, except that $\{\Delta(n)\}$ are obtained here using Hadamard perturbations.

5.5.3.1 Convergence Analysis

Recall that H is the set of globally asymptotically stable equilibria of the ODE (cf. Assumption 5.6):

$$\dot{\theta}(t) = L(\theta(t)) = -\nabla J(\theta(t)). \tag{5.25}$$

Given $\eta > 0$, let $H^\eta \triangleq \{\theta \in C \mid \|\theta - \theta_0\| < \eta,\ \theta_0 \in H\}$ be the set of points that are within a distance η from the set H. We first provide the main convergence result of the two-sided SPSA scheme (5.24):

Theorem 5.6. *Given $\eta > 0$, there exists $\delta_0 > 0$ such that for all $\delta \in (0, \delta_0]$, $\theta(n), n \geq 0$ obtained according to (5.24) satisfy $\theta(n) \to H^\eta$ almost surely.*

In the rest of the section, we provide a sequence of lemmas which would we used to prove the above theorem. The outline of the steps in the process of proving Theorem 5.6 is as follows:

(i) Using the equivalent update rule (5.26), the associated martingale difference sequence is extracted and shown to diminish to zero asymptotically.
(ii) Lemmas 5.7 and 5.8 together establish that certain bias terms in the algorithm obtained upon writing $\theta_i(n+P)$ in terms of $\theta_i(n)$, go to zero asymptotically.
(iii) Finally, using suitable Taylor expansions and neglecting the terms corresponding to the bias and the martingale difference, the proof of Theorem 5.6 establishes that the algorithm (5.26) tracks the ODE (5.25).
(iv) The last step of the proof is proven by invoking the Hirsch lemma.

The formal proof of Theorem 5.6 is provided at the end of this section.

The update recursion (5.24) could be revised into the following: $\forall n \geq 0$, $\forall i = 1,\ldots,N$,

$$\theta_i(n+1) = \theta_i(n) - a(n)\left(\frac{J(\theta(n)+\delta\Delta(n)) - J(\theta(n)-\delta\Delta(n))}{2\delta\Delta_i(n)} + M^i(n+1)\right), \tag{5.26}$$

where $M^i(n+1)$, $n \geq 0$ is a martingale difference sequence for each $i = 1,\ldots,N$, with respect to the sigma fields $\mathcal{F}(n) = \sigma(\theta(m), M^1(m),\ldots,M^N(m), m \leq n)$, $n \geq 0$.

We shall analyze (5.26) below. Let Assumptions 5.1, 5.5 and 5.6 continue to hold. We also make the following assumptions in addition:

Assumption 5.8. The step-sizes $a(n), n \geq 0$ satisfy the requirements

$$\sum_n a(n) = \infty, \quad \sum_n a(n)^2 < \infty. \tag{5.27}$$

Further, $\dfrac{a(j)}{a(n)} \to 1$ as $n \to \infty$, for all $j \in \{n, n+1,\ldots,n+M\}$ for any given $M > 0$.

Assumption 5.9. The sequence $(M(n), \mathcal{F}(n))$, $n \geq 0$ forms a martingale difference sequence. Further, $M(n)$, $n \geq 0$ are square integrable random variables satisfying

$$E[\|M(n+1)\|^2 \mid \mathcal{F}(n)] \leq K(1 + \|\theta(n)\|^2) \text{ a.s., } n \geq 0,$$

for a given constant $K > 0$.

Note that the initial requirements in Assumption 5.8 are the same as in Assumption 5.7. The last condition in Assumption 5.8 is seen to be satisfied by most diminishing step-size sequences. Assumption 5.9 is the same as Assumption 3.3.

Remark 5.6. As noted before, each function measurement is, in general, independently noise corrupted. Thus, the two measurements corresponding to parameters $\theta(n)+\delta\Delta(n)$ and $\theta(n)-\delta\Delta(n)$ may correspond to $X^1(n) \equiv J(\theta(n)+\delta\Delta(n)) + \xi^1(n+1)$ and $X^2(n) \equiv J(\theta(n)-\delta\Delta(n)) + \xi^2(n+1)$, respectively, where $\xi^1(n+1)$, $\xi^2(n+1)$, $n \geq 0$ themselves are independent martingale difference sequences. In such a case,

$$M^i(n+1) = \frac{\xi^1(n+1) - \xi^2(n+1)}{2\delta\Delta_i(n)}, \; n \geq 0, \; i = 1,\ldots,N,$$

are also martingale difference sequences since

$$E\left[M^i(n+1) \mid \mathscr{F}(n)\right] = E\left[\frac{\xi^1(n+1)-\xi^2(n+1)}{2\delta\Delta_i(n)} \mid \mathscr{F}(n)\right]$$
$$= \frac{1}{2\delta\Delta_i(n)}\left(E\left[(\xi^1(n+1)-\xi^2(n+1)) \mid \mathscr{F}(n)\right]\right)$$
$$= 0. \tag{5.28}$$

Further, if we assume that

$$E[|\xi^1(n+1)|^2 \mid \mathscr{F}(n)] \leq K(1+\|\theta(n)+\delta\Delta(n)\|^2),$$
$$E[|\xi^2(n+1)|^2 \mid \mathscr{F}(n)] \leq K(1+\|\theta(n)-\delta\Delta(n)\|^2),$$

then

$$E\left[\left|\frac{\xi^1(n+1)-\xi^2(n+1)}{2\delta\Delta_i(n)}\right|^2 \middle| \mathscr{F}(n)\right] \leq C_0(1+\|\theta(n)\|^2),$$

for some $C_0 > 0$ and since $\delta > 0$ is a constant. Moreover, $\|\Delta(n)\| = C_1$, for some $C_1 > 0$, $\forall n$, because $\Delta(n), n \geq 0$, are vectors with only $+1$s and -1s. Thus, Assumption 5.9 holds on $M^i(n+1)$, $n \geq 0$, if a similar requirement holds for $\xi^1(n+1), \xi^2(n+1), n \geq 0$, respectively.

A result similar to Theorem 3.3 would hold if one can show that the bias terms in the expansion in (5.17) vanish asymptotically in the limit as $\delta \to 0$.

Lemma 5.7. *Given any fixed integer $P > 0$, $\|\theta(m+k)-\theta(m)\| \to 0$ w.p. 1, as $m \to \infty$, for all $k \in \{1,\ldots,P\}$.*

Proof. Fix a $k \in \{1,\ldots,P\}$. Note that the algorithm (5.26) can be rewritten as

$$\theta_i(n+k) = \theta_i(n) - \sum_{j=n}^{n+k-1} a(j)\left(\frac{J(\theta(j)+\delta\Delta(j)) - J(\theta(j)-\delta\Delta(j))}{2\delta\Delta_i(j)}\right)$$
$$- \sum_{j=n}^{n+k-1} a(j)M^i(j+1). \tag{5.29}$$

Thus,

$$|\theta_i(n+k)-\theta_i(n)| \leq \sum_{j=n}^{n+k-1} a(j)\left|\frac{J(\theta(j)+\delta\Delta(j)) - J(\theta(j)-\delta\Delta(j))}{2\delta\Delta_i(j)}\right|$$
$$+ \left|\sum_{j=n}^{n+k-1} a(j)M^i(j+1)\right|. \tag{5.30}$$

It is easy to see that (for each $i = 1,\ldots,N$),

$$N^i(n) = \sum_{j=0}^{n-1} a(j)M^i(j+1),\ n \geq 1,$$

forms a martingale sequence. Further, it follows from Assumption 5.9 that

$$\sum_{m=0}^{n} E\left[(N^i(m+1) - N^i(m))^2 \mid \mathscr{F}(m)\right] = \sum_{m=0}^{n} E\left[a(n)^2(M^i(n+1))^2 \mid \mathscr{F}(m)\right]$$

$$\leq \sum_{m=0}^{n} a(n)^2 K(1 + \|\theta(n)\|^2).$$

From Assumptions 5.8 and 5.5, it follows that the quadratic variation process of $N^i(n), n \geq 0$ converges almost surely. Hence, by the martingale convergence theorem (Theorem B.2), it follows that $N^i(n), n \geq 0$ converges almost surely. Hence, $\left|\sum_{j=n}^{n+k-1} a(j)M^i(j+1)\right| \to 0$ almost surely as $n \to \infty$. Now observe that

$$\left|\frac{J(\theta(j)+\delta\Delta(j)) - J(\theta(j)-\delta\Delta(j))}{2\delta\Delta_i(j)}\right| \leq \left(\frac{|J(\theta(j)+\delta\Delta(j)) - J(\theta(j)-\delta\Delta(j))|}{2\delta|\Delta_i(j)|}\right)$$

$$\leq \left(\frac{|J(\theta(j)+\delta\Delta(j))| + |J(\theta(j)-\delta\Delta(j))|}{2\delta}\right),$$

since $|\Delta_i(j)| = 1, \forall j \geq 0, i = 1,\ldots,N$. Now note that

$$|J(\theta(j)+\delta\Delta(j))| - |J(0)| \leq |J(\theta(j)+\delta\Delta(j)) - J(0)|$$

$$\leq \hat{B}\|\theta(j)+\delta\Delta(j)\|$$

where $\hat{B} > 0$ is the Lipschitz constant of the function $J(\cdot)$. Hence,

$$|J(\theta(j)+\delta\Delta(j))| \leq \tilde{B}(1 + \|\theta(j)+\delta\Delta(j)\|),$$

for $\tilde{B} = \max(|J(0)|, \hat{B})$. Similarly,

$$|J(\theta(j)-\delta\Delta(j))| \leq \tilde{B}(1 + \|\theta(j)-\delta\Delta(j)\|).$$

From Assumption 5.5, it follows that

$$\sup_j \left|\frac{J(\theta(j)+\delta\Delta(j)) - J(\theta(j)-\delta\Delta(j))}{2\delta\Delta_i(j)}\right| \leq \tilde{K} < \infty,$$

for some $\tilde{K} > 0$. Thus, from (5.30),

$$|\theta_i(n+k) - \theta_i(n)| \le \tilde{K} \sum_{j=n}^{n+k-1} a(j) + \left| \sum_{j=n}^{n+k-1} a(j) M^i(j+1) \right| \to 0 \text{ a.s. with } n \to \infty.$$

The claim follows. □

For compact notation, let $\nabla_k(\cdot) = \frac{\partial(\cdot)}{\partial \theta_k}$. For instance $\nabla_k J(\theta(m)) = \frac{\partial J(\theta(m))}{\partial \theta_k}$.

Lemma 5.8. *The following holds for any $m \ge 0$, $k,l \in \{1,\ldots,N\}$, $k \ne l$:*

$$\left\| \sum_{n=m}^{m+P-1} \frac{a(n)}{a(m)} \frac{\Delta_k(n)}{\Delta_l(n)} \nabla_k J(\theta(n)) \right\| \to 0,$$

almost surely, as $m \to \infty$.

Proof. From Lemma 5.7, $\|\theta(m+s) - \theta(m)\| \to 0$ as $m \to \infty$, for all $s = 1,\ldots,P$. Also, from Assumption 5.1, we have $\|\nabla_k J(\theta(m+s)) - \nabla_k J(\theta(m))\| \to 0$ as $m \to \infty$, for all $s = 1,\ldots,P$. Now from Lemma 5.4, $\sum_{n=m}^{m+P-1} \frac{\Delta_k(n)}{\Delta_l(n)} = 0 \; \forall \, m \ge 0$. Note that by construction, P is an even positive integer. Hence, one can split any set of the type $A(m) \stackrel{\Delta}{=} \{m, m+1, \ldots, m+P-1\}$ into two disjoint subsets $A_{k,l}(m)^+$ and $A_{k,l}(m)^-$ each having the same number of elements, with $A_{k,l}(m)^+ \cup A_{k,l}(m)^- = A(m)$ and such that $\frac{\Delta_k(n)}{\Delta_l(n)}$ takes value $+1$ on $A_{k,l}(m)^+$ and -1 on $A_{k,l}(m)^-$, respectively. Thus,

$$\left\| \sum_{n=m}^{m+P-1} \frac{a(n)}{a(m)} \frac{\Delta_k(n)}{\Delta_l(n)} \nabla_k J(\theta(n)) \right\|$$

$$= \left\| \sum_{n \in A_{k,l}(m)^+} \frac{a(n)}{a(m)} \nabla_k J(\theta(n)) - \sum_{n \in A_{k,l}(m)^-} \frac{a(n)}{a(m)} \nabla_k J(\theta(n)) \right\|.$$

The claim now follows as a consequence of the above and Assumption 5.8 (applied with $M = P-1$). □

Proof of Theorem 5.6. Note that the recursion (5.26) can be iteratively written as

$$\theta_i(n+P) = \theta_i(n) - \sum_{l=n}^{n+P-1} a(l) \left(\frac{J(\theta(l) + \delta\Delta(l)) - J(\theta(l) - \delta\Delta(l))}{2\delta\Delta_i(l)} + M^i(l+1) \right) \tag{5.31}$$

From (5.17), it follows that

$$\theta_i(n+P) = \theta_i(n) - \sum_{l=n}^{n+P-1} a(l) \nabla_i J(\theta(l)) - \sum_{l=n}^{n+P-1} a(l) \sum_{j=1, j\ne i}^{N} \frac{\Delta_j(l)}{\Delta_i(l)} \nabla_j J(\theta(l))$$
$$- \sum_{l=n}^{n+P-1} a(l) o(\delta) - \sum_{l=n}^{n+P-1} a(l) M^i(l+1). \tag{5.32}$$

Now the third term on the RHS of (5.32) can be rewritten as

$$a(n) \sum_{l=n}^{n+P-1} \frac{a(l)}{a(n)} \sum_{j=1, j\neq i}^{N} \frac{\Delta_j(l)}{\Delta_i(l)} \nabla_j J(\theta(l)) = a(n)\xi_i^1(n),$$

where $\xi_i^1(n) = o(1)$ from Lemma 5.8. Thus, the algorithm (5.26) can be seen to be asymptotically analogous to the following algorithm:

$$\theta_i(n+1) = \theta_i(n) - a(n)\left(\nabla_i J(\theta(n)) + o(\delta) + M^i(n+1)\right). \tag{5.33}$$

Now from convergence of the martingale sequence $N^i(n)$, it follows that $\sum_{l=n}^{\infty} a(l)M^i(l+1) \to 0$ as $n \to \infty$, almost surely. The rest now follows from the Hirsch lemma (Lemma C.5). □

5.5.4 One-Sided SPSA with Hadamard Matrix Perturbations

As in the case of the two-sided SPSA algorithm in the previous section, assume that the sequence of perturbations $\Delta(n), n \geq 0$ is obtained from the Hadamard matrix construction described in Section 5.5.2.1. Then, the update rule of the one-sided SPSA algorithm is given by

$$\theta_i(n+1) = \theta_i(n) - a(n)\left(\frac{h(\theta(n)+\delta\Delta(n), \xi^+(n)) - h(\theta(n))}{\delta\Delta_i(n)}\right), \tag{5.34}$$

for $i = 1, \ldots, N$ and $n \geq 0$.

The above recursion can be seen to be equivalent to:

$$\theta_i(n+1) = \theta_i(n) - a(n)\left(\frac{J(\theta(n)+\delta\Delta(n)) - J(\theta(n))}{\delta\Delta_i(n)} + \bar{M}^i(n+1)\right), i = 1, \ldots, N. \tag{5.35}$$

In the above, $\bar{M}^i(n+1)$, $n \geq 0$ is a martingale difference sequence for each $i = 1, \ldots, N$, with respect to the sigma fields $\mathcal{F}(n) = \sigma(\theta(m), \bar{M}^1(m), \ldots, \bar{M}^N(m), m \leq n), n \geq 0$. The conclusions of Remark 5.6 can be seen to hold here as well with $\bar{M}^i(n)$ in place of $M^i(n)$, $n \geq 0$, $i = 1, \ldots, N$. The proof of Lemma 5.7 goes through with minor changes. Further, Lemma 5.8 continues to hold.

Theorem 5.9. *Given $\eta > 0$, there exists $\delta_0 > 0$ such that for all $\delta \in (0, \delta_0]$, $\theta(n), n \geq 0$ obtained according to (5.34) satisfy $\theta(n) \to H^\eta$ almost surely.*

Proof. The proof follows in a similar manner as Theorem 5.6, except that the Taylor's series expansion (5.18) is now used instead of (5.17), as a result of which the term $a(n)o(\delta)$ in (5.33) is replaced with $a(n)O(\delta)$. The rest follows as in Theorem 5.6. □

5.5.5 One-Measurement SPSA with Hadamard Matrix Perturbations

The perturbations $\Delta(n), n \geq 0$ are obtained here from the Hadamard matrix construction described in Section 5.5.2.2. Recall that this construction results in a perturbation sequence with a period $P = 2^{\lceil \log_2(N+1) \rceil}$ that is, in general, larger than the corresponding period for the perturbation sequence for two-sided SPSA algorithm. Further, this construction satisfies both properties (P.1) and (P.2). In the case of two-measurement SPSA algorithms satisfying (P.1) alone was sufficient to ensure convergence. The update rule of one-measurement SPSA algorithm is given by

$$\theta_i(n+1) = \theta_i(n) - a(n)\left(\frac{h(\theta(n)+\delta\Delta(n),\xi^+(n))}{\delta\Delta_i(n)}\right), \tag{5.36}$$

for $i = 1,\ldots,N$ and $\delta > 0$ as before.

5.5.5.1 Convergence Analysis

The algorithm (5.36) can be seen as equivalent to:

$$\theta_i(n+1) = \theta_i(n) - a(n)\left(\frac{J(\theta(n)+\delta\Delta(n))}{\delta\Delta_i(n)} + \hat{M}^i(n+1)\right), \tag{5.37}$$

where $\hat{M}^i(n+1)$, $n \geq 0$ is a martingale difference sequence for each $i = 1,\ldots,N$, with respect to the sigma fields $\mathscr{F}(n) = \sigma(\theta(m), \hat{M}^1(m), \ldots, \hat{M}^N(m), m \leq n), n \geq 0$. The conclusions of Remark 5.6 can be seen to hold here as well with $\hat{M}^i(n)$ in place of $M^i(n)$, $n \geq 0$, $i = 1,\ldots,N$. The proof of Lemma 5.7 can again be seen to hold with minor changes. As discussed before in Section 5.5.1, the one-measurement SPSA algorithms involve additional bias terms in comparison to their two-measurement counterparts and the following lemma proves that the bias terms that result from a Taylor series expansion of the second term on the RHS of (5.37) go down to zero asymptotically in the norm.

Lemma 5.10. *The following holds for any $m \geq 0$, $i,k,l \in \{1,\ldots,N\}$, $k \neq l$:*

$$\left\| \sum_{n=m}^{m+P-1} \frac{a(n)}{a(m)} \frac{1}{\Delta_i(n)} J(\theta(n)) \right\|, \left\| \sum_{n=m}^{m+P-1} \frac{a(n)}{a(m)} \frac{\Delta_k(n)}{\Delta_l(n)} \nabla_k J(\theta(n)) \right\| \to 0,$$

as $m \to \infty$, almost surely.

Proof. From Lemma 5.5, the sequence $\Delta(n), n \geq 0$ obtained as per the construction described in Section 5.5.2.2 satisfies both (P.1) and (P.2). It can be shown in a similar manner as Lemma 5.8 that $\left\| \sum_{n=m}^{m+P-1} \frac{a(n)}{a(m)} \frac{\Delta_k(n)}{\Delta_l(n)} \nabla_k J(\theta(n)) \right\| \to 0$ almost surely as $m \to \infty$. Now since $J : \mathbb{R}^N \to \mathbb{R}$ is continuously differentiable, it is in particular continuous. It thus follows from Lemma 5.7 that

$$\|J(\theta(m+k)) - J(\theta(m))\| \to 0 \text{ as } m \to \infty,$$

for all $k \in \{1, \ldots, P\}$. It can now be shown in a similar manner as Lemma 5.8 (using (P.2)) that

$$\left\| \sum_{n=m}^{m+P-1} \frac{a(n)}{a(m)} \frac{1}{\Delta_i(n)} J(\theta(n)) \right\| \to 0,$$

almost surely as $m \to \infty$. The claim follows. □

We now have the main convergence result for the one-measurement SPSA with Hadamard perturbations.

Theorem 5.11. *Given $\eta > 0$, there exists $\delta_0 > 0$ such that for all $\delta \in (0, \delta_0]$, $\theta(n), n \geq 0$ obtained according to (5.37) satisfy $\theta(n) \to H^\eta$ almost surely.*

Proof. Note that the recursion (5.37) can be iteratively written as

$$\theta_i(n+P) = \theta_i(n) - \sum_{l=n}^{n+P-1} a(l) \left(\frac{J(\theta(l) + \delta\Delta(l))}{\delta\Delta_i(l)} \right) - \sum_{l=n}^{n+P-1} a(l) M^i(l+1). \quad (5.38)$$

From (5.20), it follows that

$$\begin{aligned} \theta_i(n+P) = \theta_i(n) &- \sum_{l=n}^{n+P-1} a(l) \nabla_i J(\theta(l)) - \sum_{l=n}^{n+P-1} a(l) \frac{J(\theta(l))}{\delta\Delta_i(l)} \\ &- \sum_{l=n}^{n+P-1} a(l) \sum_{j=1, j\neq i}^{N} \frac{\Delta_j(l)}{\Delta_i(l)} \nabla_j J(\theta(l)) - \sum_{l=n}^{n+P-1} a(l) O(\delta) \\ &- \sum_{l=n}^{n+P-1} a(l) M^i(l+1). \end{aligned} \quad (5.39)$$

Now note that

$$\sum_{l=n}^{n+P-1} a(l) \frac{J(\theta(l))}{\delta\Delta_i(l)} = a(n) \sum_{l=n}^{n+P-1} \frac{a(l)}{a(n)} \frac{J(\theta(l))}{\delta\Delta_i(l)} = a(n)\xi_i^2(n),$$

where $\xi_i^2(n) = o(1)$ by Lemma 5.10. Similarly,

$$\sum_{l=n}^{n+P-1} a(l) \sum_{j=1, j\neq i}^{N} \frac{\Delta_j(l)}{\Delta_i(l)} \nabla_j h(\theta(l)) = a(n)\xi_i^3(n),$$

with $\xi_i^3(n) = o(1)$ from Lemma 5.10. The rest follows as in Theorem 5.6. □

5.6 SPSA Algorithms for Long-Run Average Cost Objective

We now present a multi-timescale version of two-measurement SPSA (with random perturbations) for the case when the underlying process is Markovian and depends on a parameter. The states of this process can either be directly observed or obtained through simulation. We will assume for simplicity that the states are simulated even though the same framework also works for the case of real observations. The single-stage cost function in this case depends on the (simulated) system state and the goal is to find a parameter (on which the state depends) that optimizes a long-run average cost objective. Even though we present here only the two-simulation SPSA with random perturbations, the analogs of the other SPSA algorithms for the expected cost criterion presented previously can similarly be described. We now present the basic framework in more detail below.

5.6.1 The Framework

Let $\{X(n), n \geq 1\}$ be an $\mathbb{R}^d$-valued parameterized Markov process with a tunable parameter θ that takes values in $\mathbb{R}^N$. Let for any given $\theta \in \mathbb{R}^N$, $\{X(n)\}$ be ergodic Markov. Let $p(\theta, x, dy)$ and $\nu_\theta(dx)$, respectively, denote the transition kernel and stationary distribution of $\{X(n)\}$ when θ is the operative parameter. When the process is in state x, let $h(x)$ be the single-stage cost incurred. The aim is to find a $\theta^* \in \mathbb{R}^N$ that minimizes (over all θ) the long-run average cost

$$J(\theta) = \lim_{l\to\infty} \frac{1}{l} \sum_{j=0}^{l-1} h(X_j). \tag{5.40}$$

5.6.2 The Two-Simulation SPSA Algorithm

Let $\{X^+(n)\}, \{X^-(n)\}$ be two simulated Markov processes that are respectively governed by the parameter sequences $(\theta(n) + \delta\Delta(n))$ and $(\theta(n) - \delta\Delta(n))$, respectively, where $\Delta(n) \triangleq (\Delta_1(n), \ldots, \Delta_N(n))^T$ with $\Delta_i(n), n \geq 0, i = 1, \ldots, N$ satisfying

Assumption 5.4 and $\delta > 0$ is a given small positive scalar. The algorithm is as follows: For $i = 1, \ldots, N$,

$$\theta_i(n+1) = \theta_i(n) - a(n)\left(\frac{Z^+(n) - Z^-(n)}{2\delta\Delta_i(n)}\right), \quad (5.41)$$

$$Z^+(n+1) = Z^+(n) + b(n)\left(h(X^+(n)) - Z^+(n)\right), \quad (5.42)$$

$$Z^-(n+1) = Z^-(n) + b(n)\left(h(X^-(n)) - Z^-(n)\right). \quad (5.43)$$

The quantities $Z^+(n)$ and $Z^-(n)$ in (5.42)–(5.43) are used to recursively estimate the long-run average costs corresponding to the simulations $\{X^+(n)\}$ and $\{X^-(n)\}$, respectively. Because of the difference in timescales with the recursions (5.42)–(5.43) proceeding on the faster timescale as compared to the recursion (5.41), the former recursions appear equilibrated when viewed from the timescale of the latter.

Remark 5.7. In practice, it is usually observed that an additional averaging over L instants (for some $L > 1$) of the recursions (5.42)–(5.43) improves performance. In other words, for practical implementations, it is suggested to run the above recursions for L instants in an inner loop, in between two successive updates of (5.41). The value of L is however arbitrary. It is generally observed, see for instance, [7, 3, 4], that a value of L in between 50 and 500 works well. While for our analysis, we focus on the case of $L = 1$, the analysis for general L is available in [7].

5.6.3 Assumptions

We make the following assumptions for average cost SPSA algorithms:

Assumption 5.10. The single-stage cost function $h : \mathbb{R}^N \times \mathbb{R}^k \to \mathbb{R}$ is Lipschitz continuous.

Assumption 5.11. The long-run average cost $J(\theta)$ is continuously differentiable in θ with bounded second derivatives.

Assumptions 5.10 and 5.11 are standard requirements. In particular, Assumption 5.11 is a technical requirement that ensures that the Hessian of the objective exists and is bounded, and is used to push through suitable Taylor series arguments in the proof.

Next, let $\{\theta(n)\}$ be a sequence of random parameters obtained using (say) an iterative scheme on which the process $\{X(n)\}$ depends. Let $\mathscr{H}(n) = \sigma(\theta(m), X(m),$

$m \leq n$), $n \geq 1$ denote a sequence of associated σ-fields. We call $\{\theta(n)\}$ *non-anticipative* if for all Borel sets $A \subset \mathbb{R}^d$,

$$P(X(n+1) \in A \mid \mathscr{H}(n)) = p(\theta(n), X(n), A).$$

Under a non-anticipative $\{\theta(n)\}$, the process $\{(X(n), \theta(n))\}$ is Markov. It can be easily seen that sequences $\{\theta(n)\}$ obtained using the algorithms below are non-anticipative. We shall assume the existence of a stochastic Lyapunov function (below).

Assumption 5.12. There exist $\varepsilon_0 > 0$, $K \subset \mathbb{R}^d$ compact and $V \in C(\mathbb{R}^d)$ such that $\lim_{\|x\| \to \infty} V(x) = \infty$ and under any non-anticipative $\{\theta(n)\}$,

1. $\sup_n E[V(X(n))^2] < \infty$ and
2. $E[V(X(n+1)) \mid \mathscr{H}(n)] \leq V(X(n)) - \varepsilon_0$, whenever $X(n) \notin K$, $n \geq 0$.

Assumption 5.12 is required to ensure that the system remains stable under a tunable parameter. It is not required if the cost function $h(\cdot)$ is bounded in addition. Here and elsewhere, we let $\|\cdot\|$ denotes the Euclidean norm.

The algorithm in Section 5.6.2 relies on two different step-size schedules, $a(n)$, $b(n)$, $n \geq 0$ that satisfy the following requirements:

Assumption 5.13. The step-sizes $a(n), b(n), n \geq 0$ satisfy the following requirements:

$$\sum_n a(n) = \sum_n b(n) = \infty, \tag{5.44}$$

$$\sum_n (a(n)^2 + b(n)^2) < \infty, \tag{5.45}$$

$$\lim_{n \to \infty} \frac{a(n)}{b(n)} = 0. \tag{5.46}$$

Assumption 5.14. The iterates $\{\theta(n)\}$ stay uniformly bounded, i.e., $\sup_n \|\theta(n)\| < \infty$, with probability one.

Assumption 5.14 essentially ensures that the θ-update remains stable. An alternative here is to assume that θ can only take values in some compact subset C of $\mathbb{R}^N$, whereby after each update, θ is projected to the set C, thereby enforcing stability. Such a projection-based scheme is considered in Section 5.6.5.

5.6.4 Convergence Analysis

Consider the ODE

$$\dot{\theta}(t) = -\nabla J(\theta(t)), \tag{5.47}$$

which is the same as (5.6), except that $J(\cdot)$ is now defined according to (5.40). Let $F = \{\theta \mid \nabla J(\theta) = 0\}$ be the set of fixed points of (5.47). Further, let $H \subset F$ be the set of globally asymptotically stable attractors of (5.47). Also, given $\varepsilon > 0$, let $H^\varepsilon = \{\theta \mid \|\theta - \theta_0\| < \varepsilon, \theta_0 \in H\}$ denotes the ε-neighborhood of the set H. We give first the main convergence result for the algorithm (5.41)-(5.43).

Theorem 5.12. *Under Assumptions 5.10–5.14, given $\varepsilon > 0$, there exists a $\delta_0 > 0$ such that the sequence of parameter iterates $\theta(n), n \geq 0$ satisfy $\theta(n) \to H^\varepsilon$ with probability one as $n \to \infty$.*

The proof of Theorem 5.12 involves steps similar to those used for proving Theorem 5.6, except that in this case of the long-run average cost setting, it is also necessary to establish that the iterates $Z^+(\cdot)$ and $Z^-(\cdot)$ asymptotically converge to the average cost estimates $J(\theta(n) + \delta\Delta(n)$ and $J(\theta(n) - \delta\Delta(n)$, respectively. We will address the latter in Lemma 5.16. Further, Lemma 5.19 will establish using suitable Taylor expansions that the conditional average of the SPSA estimate, i.e.,

$$E\left[\frac{J(\theta(n) + \delta\Delta(n)) - J(\theta(n) - \delta\Delta(n))}{2\delta\Delta_i(n)} \mid \mathscr{F}(n)\right]$$

is asymptotically close to the gradient of the objective function $J(\theta(n))$. The final step is again to invoke Hirsch lemma to complete the proof. The formal proof of Theorem 5.12 is provided at the end of this section.

Let $\mathscr{G}(n) = \sigma(\theta(p), X^+(p), X^-(p), \Delta(p), p \leq n)$, $n \geq 1$, denote σ-fields generated by the quantities above. Define sequences $N^+(p), N^-(p), p \geq 0$ as follows:

$$N^+(p) = \sum_{m=1}^{p} b(m)\left(h(X^+(m)) - E\left[h(X^+(m)) \mid \mathscr{G}(m-1)\right]\right),$$

$$N^-(p) = \sum_{m=1}^{p} b(m)\left(h(X^-(m)) - E\left[h(X^-(m)) \mid \mathscr{G}(m-1)\right]\right),$$

respectively.

Lemma 5.13. *The sequences $(N^+(p), \mathscr{G}(p))$, $(N^-(p), \mathscr{G}(p))$, $p \geq 0$ are almost surely convergent martingale sequences.*

Proof. We show the proof for the case of $N_p^+, p \geq 0$ as the same for $N_p^-, p \geq 0$ is completely analogous. It is easy to see that almost surely, $E[N^+(p+1) \mid \mathscr{G}(p)] = N^+(p)$, for all $p \geq 0$. Now note that

$$E[(N^+(p))^2] \leq C_p \sum_{m=1}^{p} b^2(m)(E[h^2(X^+(m)) + E^2[h(X^+(m)) \mid \mathcal{G}(m-1)]]),$$

for some constant $C_p > 0$ (that however depends on p). For the second term on RHS above, note that by the conditional Jensen's inequality, we have that almost surely,

$$E^2[h(X^+(m)) \mid \mathcal{G}(m-1)] \leq E[h^2(X^+(m)) \mid \mathcal{G}(m-1)].$$

Hence,

$$E[(N^+(p))^2] \leq 2C_p \sum_{m=1}^{p} b^2(m) E[h^2(X^+(m))].$$

Now, since $h(\cdot)$ is a Lipschitz continuous function, we have

$$|h(X^+(m))| - |h(0)| \leq |h(X^+(m)) - h(0)| \leq K\|X^+(m)\|,$$

where $K > 0$ is the Lipschitz constant. Thus,

$$|h(X^+(m))| \leq C_1(1 + \|X^+(m)\|),$$

for $C_1 = \max(K, |h(0)|) < \infty$. Hence, one gets

$$E[h^2(X^+(m))] \leq 2C_1^2(1 + E[\|X^+(m)\|^2]).$$

As a consequence of Assumption 5.12, $\sup_m E[\|X^+(m)\|^2] < \infty$. Thus, $E[(N^+(p))^2] < \infty$, for all $p \geq 1$. Now note that

$$\begin{aligned}\sum_p E[(N^+(p+1) - N^+(p))^2 \mid \mathcal{G}(p)] &\leq \sum_p b^2(p+1)(E[h^2(X^+(p+1)) \mid \mathcal{G}(p)] \\ &\quad + E[(E[h(X^+(p+1)) \mid \mathcal{G}(p)])^2 \mid \mathcal{G}(p)]) \\ &\leq \sum_p 2b^2(p+1)E[h^2(X^+(p+1)) \mid \mathcal{G}(p)],\end{aligned}$$

almost surely. The last inequality above again follows from the conditional Jensen's inequality. It can now be easily seen as before, using Assumption 5.12, that

$$\sup_p E[h^2(X^+_{p+1}) \mid \mathcal{G}(p)] < \infty \text{ w.p.1.}$$

Hence,

$$\sum_p E[(N^+(p+1) - N^+(p))^2 \mid \mathcal{G}(p)] < \infty$$

almost surely. Thus, by the martingale convergence theorem (Theorem B.2), $N^+(p), p \geq 0$ is an almost surely convergent martingale sequence. □

Lemma 5.14. *The updates $Z^+(p), Z^-(p), p \geq 0$ are uniformly bounded with probability one.*

Proof. We show the proof for the updates $Z^+(p), p \geq 0$ as the same for the other sequence is completely analogous. Note that (5.42) can be rewritten as

$$Z^+(p+1) = Z^+(p) + b(p)(E[h(X_p^+) \mid \mathscr{G}(p-1)] - Z^+(p))$$
$$+ b(p)(h(X_p^+) - E[h(X_p^+) \mid \mathscr{G}(p-1)]). \tag{5.48}$$

From Lemma 5.13, $N^+(p) \to N^+(\infty) < \infty$ almost surely. Hence,

$$\sum_p b(p)(h(X_p^+) - E[h(X_p^+) \mid \mathscr{G}(p-1)]) < \infty, \text{ a.s.}$$

Thus, it is enough to show the uniform boundedness of the following alternate recursion:

$$Z^+(p+1) = Z^+(p) + b(p)(E[h(X_p^+) \mid \mathscr{G}(p-1)] - Z^+(p)).$$

Note that

$$\begin{aligned} |E[h(X_p^+) \mid \mathscr{G}(p-1)]| &\leq E[|h(X_p^+)| \mid \mathscr{G}(p-1)] \\ &\leq C_1(1 + E[\|X_p^+\| \mid \mathscr{G}(p-1)]) \\ &< \infty, \end{aligned}$$

almost surely. The first inequality above follows from the conditional Jensen's inequality, while the second inequality follows as a consequence of the function h being Lipschitz continuous, see the proof in Lemma 5.13. Further, the last inequality follows from Assumption 5.12. The claim now easily follows from the Borkar and Meyn theorem (Theorem D.1). □

Now define two sequences of time points $\{s(n)\}$ and $\{t(n)\}$, respectively, as follows: $s(0) = t(0) = 0$, $s(n) = \sum_{j=0}^{n-1} a(j)$ and $t(n) = \sum_{j=0}^{n-1} b(j)$, $n \geq 1$. Then, the timescale corresponding to $\{s(n)\}$ (resp. $\{t(n)\}$) is the slower (resp. faster) of the two timescales.

Consider the following system of ordinary differential equations (ODEs):

$$\dot{\theta}(t) = 0, \tag{5.49}$$

$$\dot{Z}^+(t) = J(\theta(t) + \delta\Delta(t)) - Z^+(t), \tag{5.50}$$

$$\dot{Z}^-(t) = J(\theta(t) - \delta\Delta(t)) - Z^-(t). \tag{5.51}$$

From Lemma 5.14, $\sup_n |Z^+(n)|$, $\sup_n |Z^-(n)| < \infty$ almost surely. Consider the functions $\hat{Z}^+(t), \hat{Z}^-(t)$ defined according to $\hat{Z}^+(t(n)) = Z^+(n)$ and $\hat{Z}^-(t(n)) = Z^-(n)$ with the maps $t \to \hat{Z}^+(t)$ and $t \to \hat{Z}^-(t)$ corresponding to continuous linear interpolations on the intervals $[t(n), t(n+1)]$.

Given $\bar{T} > 0$, define $\{\bar{T}(n)\}$ as follows: $\bar{T}(0) = 0$ and for $n \geq 1$, $\bar{T}(n) = \min\{t(m) \mid t(m) \geq \bar{T}(n-1) + \bar{T}\}$. Let $\bar{I}(n) = [\bar{T}(n), \bar{T}(n+1)]$. It is clearly the case that there exists some integer $q(n) > 0$ such that $\bar{T}(n) = t(q(n))$.

Define also the functions $\theta^n(t), Z^{+,n}(t), Z^{-,n}(t), t \in I(n), n \geq 0$, that are obtained as trajectories of the following ODEs:

$$\dot{\theta}^n(t) = 0, \tag{5.52}$$

$$\dot{Z}^{+,n}(t) = J(\theta(t) + \delta\Delta(t)) - Z^{+,n}(t), \tag{5.53}$$

$$\dot{Z}^{-,n}(t) = J(\theta(t) - \delta\Delta(t)) - Z^{-,n}(t), \tag{5.54}$$

with $\theta^n(\bar{T}(n)) = \theta(q(n)), Z^{+,n}(\bar{T}(n)) = \hat{Z}^+(t(q(n))) = Z^+(q(n))$ and $Z^{-,n}(\bar{T}(n)) = \hat{Z}^-(t(q(n))) = Z^-(q(n))$, respectively. Further, $\Delta(t) \stackrel{\Delta}{=} (\Delta_1(t), \ldots, \Delta_N(t))^T$ is defined according to $\Delta(t) = \Delta(n)$, for $t \in [s(n), s(n+1))$.

Let $\hat{\theta}(t), \hat{Z}^+(t), \hat{Z}^-(t), t \geq 0$ be defined according to $\hat{\theta}(t(n)) = \theta(n)$, $\hat{Z}^+(t(n)) = Z^+(n)$ and $\hat{Z}^-(t(n)) = Z^-(n), n \geq 0$ with continuous linear interpolation in between points, i.e., for all $t \in (t(n), t(n+1)), n \geq 0$.

Lemma 5.15. *Given $\bar{T}, \varepsilon > 0$, $(\hat{\theta}(t(n)+\cdot), \hat{Z}^+(t(n)+\cdot), \hat{Z}^-(t(n)+\cdot))$, is a bounded $(\bar{T}, \varepsilon)$-perturbation of (5.49)-(5.51) for n sufficiently large.*

Proof. Note that the recursion (5.41) can be rewritten as follows: For $i = 1, \ldots, N$,

$$\theta_i(n+1) = \theta_i(n) - b(n)\bar{\xi}_i(n), \tag{5.55}$$

where $\bar{\xi}_i(n) = \dfrac{a(n)}{b(n)} \left(\dfrac{Z^+(n) - Z^-(n)}{2\delta\Delta_i(n)} \right) = o(1)$ because $a(n) = o(b(n))$ from Assumption 5.13.

Now note that the recursion (5.42) can be rewritten as

$$Z^+(n+1) = Z^+(n) + b(n)(J(\theta(n) + \delta\Delta(n)) + \xi_1^+(n) + \xi_2^+(n) - Z^+(n)), \tag{5.56}$$

where $\xi_1^+(n) = E[h(X^+(n)) \mid \mathcal{G}_{n-1}] - J(\theta(n) + \delta\Delta(n))$ and $\xi_2^+(n), n \geq 1$ is the martingale difference $\xi_2^+(n) = h(X^+(n)) - E[h(X^+(n)) \mid \mathcal{G}_{n-1}]$, respectively. Recall that $\bar{T}(n) = t(q(n))$. Also, let $\bar{T}(n+1) = t(q(n+1))$. Then, from Lemma 5.13,

$$\sum_{j=q(n)}^{q(n+1)} b(j)\xi_2^+(j) \to 0 \text{ as } n \to \infty.$$

Now $\xi_1^+(n) \to 0$ as $n \to \infty$ almost surely because $\{X^+(n)\}$ is ergodic Markov for a fixed parameter. Hence, the Markov noise vanishes on the 'natural' timescale where $t(n) = n$ that is faster than the timescale of the algorithm as in the latter, $t(n) - t(n-1) \to 0$ as $n \to \infty$. Thus, the algorithm will see the averaged effect of the iterate on the natural timescale, see Section 6.2 of [9] for a detailed treatment of averaging on the natural timescale. It is thus easy to see that with probability one,

$$\lim_{n \to \infty} \sup_{t \in \bar{I}(n)} \|Z^{+,n}(t) - \hat{Z}^+(t)\| = 0.$$

A similar argument holds for the recursion $Z^-(n), n \geq 0$. The claim follows. □

Lemma 5.16. *As $n \to \infty$, we have with probability one,*

$$\|Z^+(n) - J(\theta(n) + \delta\Delta(n))\|,\ \|Z^-(n) - J(\theta(n) - \delta\Delta(n))\| \to 0.$$

Proof. Follows from Lemma 5.15 and an application of the Hirsch lemma (Lemma C.5) for every $\varepsilon > 0$. □

We now concentrate on the slower timescale recursion. Let $\mathscr{F}(n) = \sigma(X^+(n), X^-(n), \theta(m), m \le n; \Delta(m), m < n), n \ge 1$, be a sequence of sigma fields. One can rewrite (5.41) as

$$\begin{aligned}\theta_i(n+1) =& \theta_i(n) - a(n)\Bigg(E\left[\frac{J(\theta(n)+\delta\Delta(n)) - J(\theta(n)-\delta\Delta(n))}{2\delta\Delta_i(n)} \mid \mathscr{F}(n)\right] \\ &+ \zeta_i^1(n) + \zeta_i^2(n)\Bigg),\end{aligned} \tag{5.57}$$

where

$$\begin{aligned}\zeta_i^1(n) =& \frac{J(\theta(n)+\delta\Delta(n)) - J(\theta(n)-\delta\Delta(n))}{2\delta\Delta_i(n)} \\ &- E\left[\frac{J(\theta(n)+\delta\Delta(n)) - J(\theta(n)-\delta\Delta(n))}{2\delta\Delta_i(n)} \mid \mathscr{F}(n)\right],\end{aligned}$$

$$\zeta_i^2(n) = \frac{Z^+(n) - Z^-(n)}{2\delta\Delta_i(n)} - \frac{J(\theta(n)+\delta\Delta(n)) - J(\theta(n)-\delta\Delta(n))}{2\delta\Delta_i(n)},$$

respectively.

Let $\chi_i(n), n \ge 0$ be defined according to $\chi_i(n) = \sum_{m=0}^{n} a(m)\zeta_i^1(m)$.

Lemma 5.17. *The sequence $(\chi_i(n), \mathscr{F}(n)), n \ge 0$ forms a convergent martingale sequence.*

Proof. It is easy to see that $(\chi_i(n), \mathscr{F}(n)), n \ge 0$ forms a martingale sequence. By Assumption 5.14, $M(w) \overset{\Delta}{=} \sup_n \|\theta(n)\| < \infty$ w.p.1. Here w denotes the particular sample point in the probability space corresponding to the given $\theta(n)$-trajectory. Note that $\theta(n), n \ge 0$ take values in the sample-path-dependent compact set $D(w) = \{\theta \mid \|\theta\| \le M(w)\}$. Now as a consequence of Assumption 5.11, since $\theta(n) \in D(w), \forall n$, $\sup_n |\zeta_i^1(n)| < \infty$. Further, since $P(w \mid M(w) < \infty) = 1$, we have that $\sup_n |\zeta_i^1(n)| < \infty$ with probability one. It is now easy to see from an application of the martingale convergence theorem (Theorem B.2) that $\{\chi_i(n)\}$ converges almost surely. □

Lemma 5.18. *As $n \to \infty$, $\zeta_i^2(n) \to 0$ with probability one.*

Proof. The proof follows easily from Lemma 5.16. □

Lemma 5.19. *With probability one,*

$$\left| E\left[\frac{J(\theta(n)+\delta\Delta(n)) - J(\theta(n)-\delta\Delta(n))}{2\delta\Delta_i(n)} \mid \mathscr{F}(n)\right] - \nabla_i J(\theta(n))\right| \to 0,$$

as $\delta \to 0$.

Proof. It follows from suitable Taylor series expansions of $J(\theta(n)+\delta\Delta(n))$ and $J(\theta(n)-\delta\Delta(n))$ around the point $\theta(n)$ that

$$\frac{J(\theta(n)+\delta\Delta(n)) - J(\theta(n)-\delta\Delta(n))}{2\delta\Delta_i(n)} = \nabla_i J(\theta(n))$$

$$+ \sum_{j=1, j\neq i}^{N} \frac{\Delta_j(n)}{\Delta_i(n)} \nabla_j J(\theta(n)) + o(\delta).$$

It follows from the properties of $\Delta_j(n), j = 1, \ldots, N$ that

$$E\left[\frac{J(\theta(n)+\delta\Delta(n)) - J(\theta(n)-\delta\Delta(n))}{2\delta\Delta_i(n)} \mid \mathscr{F}(n)\right] = \nabla_i J(\theta(n)) + o(\delta).$$

The claim follows. □

In a similar manner as (5.57), one can now rewrite (5.41) as

$$\theta_i(n+1) = \theta_i(n) - a(n)\left(\nabla_i J(\theta(n)) + \zeta_i^3(n) + \zeta_i^1(n) + \zeta_i^2(n)\right), \tag{5.58}$$

where, as a consequence of Lemma 5.19,

$$\zeta_i^3(n) \triangleq E\left[\frac{J(\theta(n)+\delta\Delta(n)) - J(\theta(n)-\delta\Delta(n))}{2\delta\Delta_i(n)} \mid \mathscr{F}(n)\right] - \nabla_i J(\theta(n)) \to 0 \text{ as } n \to \infty.$$

Proof of Theorem 5.12. Recall that the recursions (5.41) can be rewritten as (5.58). Now define $\bar{\theta}(t), t \geq 0$ according to $\bar{\theta}(t) = \theta(n)$ for $t \in [s(n), s(n+1))$. As a consequence of Lemmas 5.17–5.19, $\bar{\theta}(t)$ can be viewed as a (T,γ)–perturbation of the ODE (5.47). The claim now follows by the Hirsch lemma (Lemma C.5). □

5.6.5 *Projected SPSA Algorithm*

We now consider the case when after each update, the parameter θ is projected onto a compact and convex subset C of $\mathbb{R}^N$. This ensures that the parameter updates remain stable as they do not escape the set C and thus Assumption 5.14 is automatically satisfied. Let $\Gamma : \mathbb{R}^N \to C$ denotes an operator that projects any $x = (x_1, \ldots, x_N)^T \in \mathbb{R}^N$ to its nearest point in C. In particular, if $x \in C$, then $\Gamma(x) \in C$ as well. For given $x = (x_1, \ldots, x_N)^T \in \mathbb{R}^N$, one may identify $\Gamma(x)$ via the tuple

$\Gamma(x) = (\Gamma_1(x_1),\ldots,\Gamma_N(x_N))^T$ for suitable $\mathbb{R}$-valued operators $\Gamma_1,\ldots,\Gamma_N$. We consider here the projected variant of the two-simulation (two-sided) SPSA algorithm for the long-run average cost objective that was presented in Section 5.6. A detailed treatment of projected stochastic approximation can be found in [19] and has been summarized in Appendix E.

Let $\{X^+(n)\},\{X^-(n)\}$ be two simulated Markov processes that are respectively governed by the parameter sequences $(\theta(n)+\delta\Delta(n))$ and $(\theta(n)-\delta\Delta(n))$, respectively, where $\Delta(n) \stackrel{\Delta}{=} (\Delta_1(n),\ldots,\Delta_N(n))^T$ with $\Delta_i(n), n \geq 0, i = 1,\ldots,N$ satisfying Assumption 5.4 and $\delta > 0$ is a given small positive scalar. The algorithm is as follows:

For $i = 1,\ldots,N$,

$$\theta_i(n+1) = \Gamma_i\left(\theta_i(n) - a(n)\left(\frac{Z^+(n)-Z^-(n)}{2\delta\Delta_i(n)}\right)\right), \tag{5.59}$$

$$Z^+(n+1) = Z^+(n) + b(n)\left(h(X^+(n)) - Z^+(n)\right), \tag{5.60}$$

$$Z^-(n+1) = Z^-(n) + b(n)\left(h(X^-(n)) - Z^-(n)\right). \tag{5.61}$$

Note that recursions (5.60)-(5.61) are the same as (5.42)-(5.43). Hence, the analysis of these recursions proceeds along the same lines as the latter (described in Section 5.6).

Let $\mathscr{C}(C)$ denotes the space of all continuous functions from C to $\mathbb{R}^N$. The operator $\bar{\Gamma} : \mathscr{C}(C) \to \mathscr{C}(\mathbb{R}^N)$ is defined according to

$$\bar{\Gamma}(v(x)) = \lim_{\eta \to 0}\left(\frac{\Gamma(x+\eta v(x)) - x}{\eta}\right), \tag{5.62}$$

for any continuous $v : C \to \mathbb{R}^N$. The limit in (5.62) exists and is unique since C is a convex set. In case the limit does not exist, one may consider the set of all limit points of (5.63). From its definition, $\bar{\Gamma}(v(x)) = v(x)$ if $x \in C^o$ (the interior of C). By an abuse of notation, let H denote the set of all asymptotically stable attractors of the ODE (5.63) and H^ε be the ε-neighborhood of H (given $\varepsilon > 0$).

$$\dot{\theta}(t) = \bar{\Gamma}(-\nabla J(\theta(t))). \tag{5.63}$$

Theorem 5.20. *Under Assumptions 5.10–5.13, given $\varepsilon > 0$, there exists a $\delta_0 > 0$ such that the sequence of parameter iterates $\theta(n), n \geq 0$ satisfy $\theta(n) \to H^\varepsilon$ with probability one as $n \to \infty$.*

Proof. The result follows from the Kushner and Clark theorem (see Theorem E.1). The assumptions there are seen to hold here, see Remark E.1. □

5.7 Concluding Remarks

In this chapter, we described the idea of simultaneous perturbation for estimating the gradient of an objective function, for both the expected as well as the long run average cost settings. Using two alternative constructions - random and Hadamard matrix-based - several SPSA algorithms including the one-measurement variants were presented. Detailed convergence proofs were given for the various algorithms discussed. The SPSA algorithms along with the smoothed functional algorithms presented in the next chapter are widely applied gradient estimation techniques in a variety of applications, some of which are discussed in the later chapters of this book. This is probably because these algorithms are simple and can be implemented in an on-line manner; further, they require very less computational resources and are provably convergent.

References

1. Abdulla, M.S., Bhatnagar, S.: SPSA with measurement reuse. In: Proceedings of Winter Simulation Conference, Monterey, CA, pp. 320–328 (2006)
2. Bartkute, V., Sakalauskas, L.: Application of stochastic approximation in technical design. European Journal of Operational Research 181(3), 1174–1188 (2007)
3. Bhatnagar, S.: Multiscale stochastic approximation algorithms with applications to ABR service in ATM networks. Ph.D. thesis, Department of Electrical Engineering. Indian Institute of Science, Bangalore, India (1997)
4. Bhatnagar, S.: Adaptive Newton-based smoothed functional algorithms for simulation optimization. ACM Transactions on Modeling and Computer Simulation 18(1), 2:1–2:35 (2007)
5. Bhatnagar, S., Borkar, V.S.: Multiscale stochastic approximation for parametric optimization of hidden Markov models. Prob. Engg. and Info. Sci. 11, 509–522 (1997)
6. Bhatnagar, S., Borkar, V.S.: A two time scale stochastic approximation scheme for simulation based parametric optimization. Prob. Engg. and Info. Sci. 12, 519–531 (1998)
7. Bhatnagar, S., Fu, M.C., Marcus, S.I., Bhatnagar, S.: Two timescale algorithms for simulation optimization of hidden Markov models. IIE Transactions 33(3), 245–258 (2001)
8. Bhatnagar, S., Fu, M.C., Marcus, S.I., Wang, I.J.: Two-timescale simultaneous perturbation stochastic approximation using deterministic perturbation sequences. ACM Transactions on Modelling and Computer Simulation 13(2), 180–209 (2003)
9. Borkar, V.S.: Stochastic Approximation: A Dynamical Systems Viewpoint. Cambridge University Press and Hindustan Book Agency (Jointly Published), Cambridge and New Delhi (2008)
10. Cao, X.: Preliminary results on non-Bernoulli distribution of perturbation for simultaneous perturbation stochastic approximation. In: Proceedings of the American Control Conference, San Francisco, CA, pp. 2669–2670 (2011)
11. Chen, H.F., Duncan, T.E., Pasik-Duncan, B.: A Kiefer-Wolfowitz algorithm with randomized differences. IEEE Trans. Auto. Cont. 44(3), 442–453 (1999)
12. Chin, D.C.: A more efficient global optimization algorithm based on Styblinski and Tang. Neural Networks 7, 573–574 (1994)

13. Chin, D.C.: Comparative study of stochastic algorithms for system optimization based on gradient approximation. IEEE Trans. Sys. Man. Cyber. Part B 27(2), 244–249 (1997)
14. Dippon, J., Renz, J.: Weighted means in stochastic approximation of minima. SIAM J. Contr. and Optim. 35, 1811–1827 (1997)
15. Fabian, V.: Stochastic approximation. In: Rustagi, J.J. (ed.) Optimizing Methods in Statistics, pp. 439–470. Academic Press, New York (1971)
16. Gerencsér, L.: Convergence rate of moments in stochastic approximation with simultaneous perturbation gradient approximation and resetting. IEEE Trans. Auto. Cont. 44(5), 894–906 (1999)
17. He, Y., Fu, M.C., Marcus, S.I.: Convergence of simultaneous perturbation stochastic approximation for nondifferentiable optimization. IEEE Transactions on Automatic Control 48, 1459–1463 (2003)
18. Kleinman, N.L., Spall, J.C., Naiman, D.Q.: Simulation-based optimization with stochastic approximation using common random numbers. Management Science 45, 1570–1578 (1999)
19. Kushner, H.J., Clark, D.S.: Stochastic Approximation Methods for Constrained and Unconstrained Systems. Springer, New York (1978)
20. Kushner, H.J., Yin, G.G.: Stochastic Approximation Algorithms and Applications. Springer, New York (1997)
21. Maeda, Y.: Time difference simultaneous perturbation method. Electronics Letters 32, 1016–1018 (1996)
22. Maryak, J.L.: Some guidelines for using iterate averaging in stochastic approximation. In: Proceedings of the IEEE Conference on Decision and Control, San Diego, CA, pp. 2287–2290 (1997)
23. Maryak, J.L., Chin, D.C.: Global random optimization by simultaneous perturbation stochastic approximation. IEEE Transactions on Automatic Control 53, 780–783 (2008)
24. Poyiadjis, G., Singh, S.S., Doucet, A.: Gradient-free maximum likelihood parameter estimation with particle filters. In: Proceedings of the American Control Conference, Minneapolis, MN, pp. 3052–3067 (2006)
25. Sadegh, P., Spall, J.C.: Optimal random perturbations for stochastic approximation using a simultaneous perturbation gradient approximation. In: Proceedings of the American Control Conference, Albuquerque, NM, pp. 3582–3586 (1997)
26. Spall, J.C.: Multivariate stochastic approximation using a simultaneous perturbation gradient approximation. IEEE Trans. Auto. Cont. 37(3), 332–341 (1992)
27. Spall, J.C.: A one-measurement form of simultaneous perturbation stochastic approximation. Automatica 33, 109–112 (1997)
28. Spall, J.C.: Implementation of the simultaneous perturbation algorithm for stochastic optimization. IEEE Transactions on Aerospace and Electronic Systems 34, 817–823 (1998)
29. Spall, J.C.: Introduction to Stochastic Search and Optimization. John Wiley and Sons, New York (2003)

Chapter 6
Smoothed Functional Gradient Schemes

6.1 Introduction

We studied the gradient SPSA algorithm in Chapter 5. A remarkable feature of that algorithm is that it estimates the gradient of the objective by simultaneously perturbing all parameter components and requires only one or two measurements of the objective function for this purpose. Smoothed functional (SF) algorithms also belong to the class of simultaneous perturbation methods, because they update the gradient/Hessian of the objective using function measurements involving parameter updates that are perturbed simultaneously in all component directions. The SF gradient estimates were originally developed by Katkovnik and Kulchitsky [7, 8]. The original idea was to approximate the gradient of expected performance by its convolution with a multivariate Gaussian distribution. This results in the objective function getting smoothed because of the convolution. The objective function smoothing that results from the convolution with a smoothing density function can in fact help the algorithm to converge to a global minimum or to a point close to it. This fact has been observed in [9]. We illustrate this in Fig. 6.1. As shown in the Figure, the smoothing might offset the global minimum slightly. But that problem is more than compensated by the fact that other local minima may have disappeared because of the smoothing.

While the original SF algorithm in [7] uses only one simulation, in [9] and [5], a related two-simulation SF algorithm based on a finite difference gradient estimate is presented. The latter algorithm has been shown in [9] to have lower variability as compared to the one-simulation SF algorithm. It has been observed in [8] that the Cauchy and the Uniform density functions can also be used for the perturbation random variables in addition to Gaussian.

The objective function used in the aforementioned references that discuss the SF algorithm is largely an expectation over noisy cost samples. In [3, 2], the SF algorithm with Gaussian perturbations has been explored when the objective is a

S. Bhatnagar et al.: Stochastic Recursive Algorithms for Optimization, LNCIS 434, pp. 77–102.
springerlink.com

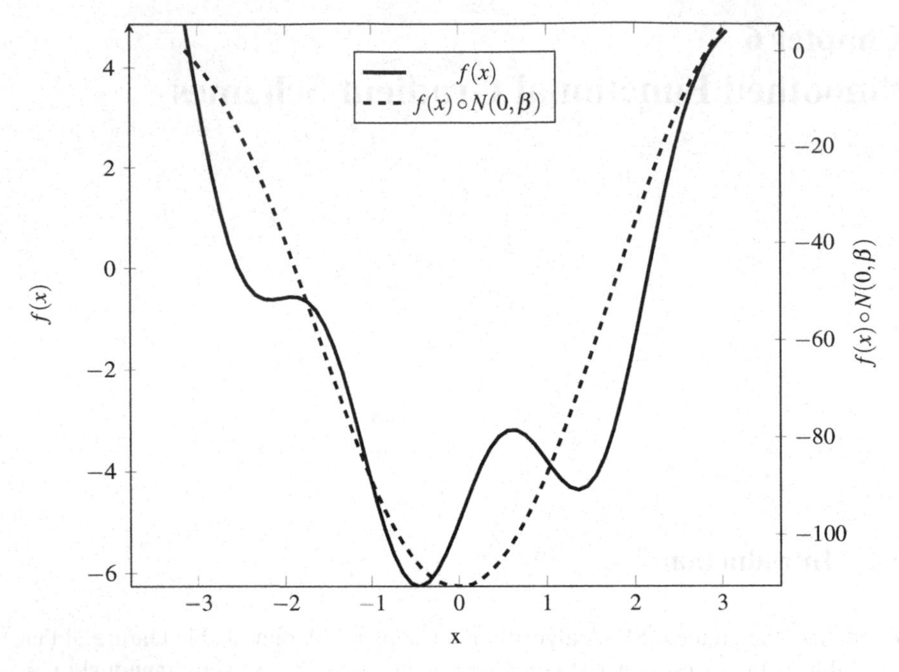

Fig. 6.1 A sample function $f(\cdot)$ with multiple local minima and a global minimum and its convolution with a Gaussian density of standard deviation $\beta = 0.08$. Here, we use '$\circ$' to denote the convolution operation.

long-run average cost function and the basic underlying process is an ergodic Markov process for any given parameter value. In [2], two different Hessian estimates for the objective function using the smoothed functional technique have been obtained as well. It is interesting to note that these (Hessian estimates) also require only one and two system simulations, which are the same as those used to estimate the gradients. We discuss the Hessian estimators and the resulting Newton SF algorithms in Chapter 8. The focus of the current chapter is on the gradient SF algorithms.

The remaining part of this chapter is organized as follows: In Section 6.2, we present for the expected cost objective, the SF gradient algorithm with perturbations distributed according to the Gaussian distribution. In Section 6.3, we present general conditions for any candidate p.d.f. to be used for smoothing and hence gradient estimation. Cauchy density function satisfies the necessary properties for smoothing and also offers better exploration of the parameter space owing to its more heavy tailed nature in comparison to Gaussian density. In Section 6.4, we present SF algorithms using Cauchy density for smoothing. In Section 6.5, we discuss multi-timescale versions of Gaussian SF algorithms for the long-run average cost objective including a variant with projection. Finally, we present the concluding remarks in Section 6.6.

6.2 Gaussian Based SF Algorithm

We consider here the expected cost objective where the objective function is $J(\theta) = E_{\xi}[h(\theta,\xi)]$, where $h(\theta,\xi)$ denotes a noisy measurement of $J(\theta)$ with ξ as the noise that is assumed to be zero mean and independent of θ. Also, as in previous chapters, we let $L(\theta) = \nabla J(\theta)$. Note that the parameter vector θ is N-dimensional, i.e., $\theta \stackrel{\triangle}{=} (\theta_1, \theta_2, \ldots, \theta_N)^T \in \mathbb{R}^N$.

Gaussian-based SF techniques, originally developed by [7], have been proposed for solving stochastic optimization problems. Section 6.2.1 discusses the basic idea of smoothing the gradient of a function using a Gaussian density. The Gaussian SF algorithm using this basic idea is then discussed in Section 6.2.2 followed by its detailed convergence analysis in Section 6.2.3.

6.2.1 Gradient Estimation via Smoothing

We first illustrate the idea of obtaining an estimate of the gradient of the objective function $J : \mathbb{R}^N \to \mathbb{R}$ using a Gaussian density function for smoothing. Later, we extend this idea to the case when Cauchy density is used for the purpose of smoothing. The gradient estimate that we describe now requires only one measurement with a certain perturbed parameter update.

The SF estimate of the gradient of an N-dimensional objective is obtained by the following steps:

- Define the SF estimate as the convolution of a multivariate Gaussian density with the gradient of the objective function. This step is illustrated in Fig. 6.1, where a sample function $f(\cdot)$ is convolved with a Gaussian random variable with mean 0 and variance $\beta = 0.08$.
- Argue that the same is equivalent to expectation of product of a scaling term and the objective with perturbed parameter where the perturbation is with a multivariate standard Gaussian random vector. The scaling term is a function of the multivariate standard Gaussian random vector itself.
- In the limit as the spread parameter (denoted β below) goes to zero, the SF estimate becomes equal to the true gradient of the objective.

In what follows, we make this intuition precise. For some scalar constant $\beta > 0$, let

$$D_{\beta,1}J(\theta) = \int G_{\beta}(\theta - \eta)\nabla_{\eta}J(\eta)d\eta \qquad (6.1)$$

be the convolution of the gradient of the objective function $J(\cdot)$ with the N-dimensional multivariate Gaussian density function $G_\beta(\cdot)$ (i.e., the p.d.f. of N independent $N(0,\beta^2)$-distributed Gaussian random variables) defined by

$$G_\beta(\theta-\eta) = \frac{1}{(2\pi)^{N/2}\beta^N}\exp\left(-\frac{1}{2}\sum_{i=1}^{N}\frac{(\theta_i-\eta_i)^2}{\beta^2}\right),$$

where $\theta,\eta \in \mathbb{R}^N$ with $\eta \stackrel{\triangle}{=} (\eta_1,\ldots,\eta_N)^T$. The quantity $D_{\beta,1}J(\theta)$ can be viewed as a smoothed gradient of the objective, which in fact converges to the true gradient ($\nabla J(\theta)$) in the limit as $\beta \to 0$.

Integrating by parts in (6.1), it is easy to see that

$$\begin{aligned} D_{\beta,1}J(\theta) &= -\int \nabla_\eta G_\beta(\theta-\eta)J(\eta)d\eta \\ &= \int \nabla_\eta G_\beta(\eta)J(\theta-\eta)d\eta. \end{aligned} \tag{6.2}$$

It is easy to verify that $\nabla_\eta G_\beta(\eta) = \frac{-\eta}{\beta^2}G_\beta(\eta)$. Substituting the last and $\eta' = \frac{\eta}{\beta}$ in (6.2), one obtains

$$D_{\beta,1}J(\theta) = \frac{1}{\beta}\int -\eta'\frac{1}{(2\pi)^{N/2}}\exp\left(-\frac{1}{2}\sum_{i=1}^{N}(\eta_i')^2\right)J(\theta-\beta\eta')d\eta'. \tag{6.3}$$

In the above we use the fact that $\eta = \beta\eta' = (\beta\eta_1',\ldots,\beta\eta_N')^T$ (written component-wise), and hence $d\eta = \beta^N d\eta_1'\cdots d\eta_N' = \beta^N d\eta'$. Upon substituting $\bar{\eta} = -\eta'$, one obtains

$$D_{\beta,1}J(\theta) = E\left[\frac{1}{\beta}\bar{\eta}J(0+\beta\bar{\eta})\right], \tag{6.4}$$

where the expectation above is taken w.r.t. the N-dimensional multivariate Gaussian p.d.f.

$$G(\bar{\eta}) = \frac{1}{(2\pi)^{N/2}}\exp\left(-\frac{1}{2}\sum_{i=1}^{N}(\bar{\eta}_i)^2\right),$$

with $\beta = 1$ (i.e., the joint p.d.f. of N independent $N(0,1)$-distributed random variables).

The form of the gradient estimator suggested by (6.4) (for a large positive integer M and a small scalar $\beta > 0$) is

$$\nabla J(\theta) \approx \frac{1}{\beta}\frac{1}{M}\sum_{n=1}^{M}\eta(n)J(\theta+\beta\eta(n)). \tag{6.5}$$

Here $\eta(n) \stackrel{\triangle}{=} (\eta_1(n), \ldots, \eta_N(n))^T$, with $\eta_i(n), i = 1, \ldots, N, n \geq 0$, being independent $N(0,1)$-distributed random variables.

6.2.2 *The Basic Gaussian SF Algorithm*

In this section, we discuss an incremental gradient descent algorithm that finds the optimal parameter θ that minimizes the expected cost objective. The basic SF algorithm's update rule is given by

$$\theta_i(n+1) = \theta_i(n) - a(n)\left(\frac{\eta_i(n)}{\beta} h(\theta(n) + \beta\eta(n), \xi(n))\right), \tag{6.6}$$

for $i = 1, \ldots, N$ and $\beta > 0$.

In the above, $\eta(n) = (\eta_1(n), \ldots, \eta_N(n))^T, n \geq 0$, denotes a sequence of independent $N(0,1)$-distributed random variables $\eta_1(n), \ldots, \eta_N(n), n \geq 0$.

The term in brackets in the above update rule is motivated by the smoothed gradient estimator $D_{\beta,1}$ in (6.4). While (6.4) has an expectation, in (6.6) a sample evaluation of the estimate is used following the idea from (6.5). This coupled with the fact that a stochastic approximation algorithm *sees* the asymptotic average ensures that we are indeed performing a negative descent in the long run w.r.t. the objective function $J(\cdot)$. In fact, we prove later in Section 6.2.3 that the above recursion (6.6) eventually tracks the ODE,

$$\dot{\theta}(t) = -\nabla J(\theta(t)). \tag{6.7}$$

The algorithm flow is diagrammatically described in Fig. 6.2. As evident in (6.4), each step of the algorithm (6.6) involves a perturbed simulation using the parameter $\theta + \beta\eta$, and the output of the simulation is used to tune the parameter θ in the negative gradient descent direction. An algorithmic view of the basic SF scheme is provided in Algorithm 6.1.

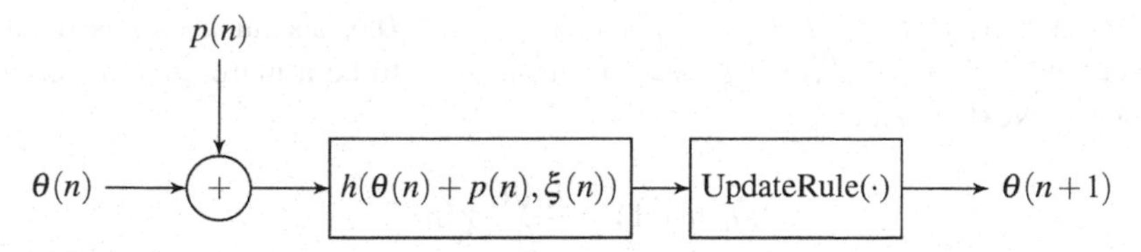

Fig. 6.2 Overall flow of the basic SF algorithm.

Algorithm 6.1 The Complete Algorithm Structure

Input:

- R, a large positive integer representing the number of iterations;
- $\theta(0)$, initial parameter vector;
- $\beta > 0$ is a fixed smoothing control parameter;
- $K \geq 1$ is fixed integer used to control the duration of the average cost accumulation (c.f. (12.10));
- $\{\eta(n), n \geq 1\}$, N-dimensional i.i.d. Gaussian random variables.
- UpdateRule(), the stochastic update rule (6.6).
- Simulate(θ) $\rightarrow X$, the simulator of the system. X represents the state of the underlying Markov process at the end of the simulation.

Output: θ^*, the parameter vector after R iterations.

$\theta \leftarrow \theta(0)$, $n \leftarrow 1$
loop
  $\hat{X} \leftarrow$ Simulate($\theta + \beta\eta(n)$).
  $\theta \leftarrow$UpdateRule($\hat{X}, \theta$).
  if $n = R$ **then**
    Terminate with θ.
  end if
  $n \leftarrow n+1$.
end loop

6.2.3 *Convergence Analysis of Gaussian SF Algorithm*

The basic SF algorithm (6.6) can be rewritten as follows: For $i = 1, \ldots, N$, $n \geq 0$,

$$\theta_i(n+1) = \theta_i(n) - a(n)\frac{\eta_i(n)}{\beta}\left(J(\theta(n) + \beta\eta(n)) + \chi(n)\right), \tag{6.8}$$

where

$$\chi(n) = h(\theta(n) + \beta\eta(n), \xi(n)) - J(\theta(n) + \beta\eta(n)).$$

Let $\mathscr{F}(n) = \sigma(\theta(m), \chi(m), m \leq n; \eta(m), m < n)$, $n > 0$ be a sequence of associated sigma fields. Now $(\chi(n), \mathscr{F}(n))$, $n \geq 0$ can be seen to be a martingale difference sequence. Next consider

$$\hat{M}_i(n+1) = \frac{\eta_i(n)}{\beta}\chi(n).$$

It is easy to see that $E\left[\hat{M}_i(n+1) \mid \mathscr{F}(n)\right] = 0$, $\forall n \geq 0$. Thus, (6.8) can be rewritten as

$$\theta_i(n+1) = \theta_i(n) - a(n)\left(\frac{\eta_i(n)}{\beta}J(\theta(n) + \beta\eta(n)) + \hat{M}_i(n+1)\right), \tag{6.9}$$

where $(\hat{M}_i(n), \mathscr{F}(n)), n \geq 0$ are suitable martingale difference sequences, $i = 1, \ldots, N$.

We now analyze the algorithm (6.9) under the following assumptions:

Assumption 6.1. The function $J : \mathbb{R}^N \to \mathbb{R}$ is Lipschitz continuous and continuously differentiable with bounded second derivatives.

Assumption 6.2. The sequence $(\hat{M}_i(n), \mathscr{F}(n))$, $n \geq 0$ forms a martingale difference sequence. Further, $(\hat{M}_i(n), n \geq 0)$ are square integrable random variables satisfying

$$E[\|\hat{M}_i(n+1)\|^2 \mid \mathscr{F}(n)] \leq K(1 + \|\theta(n)\|^2) \text{ a.s., } n \geq 0,$$

for a given constant $K > 0$.

Assumption 6.3. The step-sizes $a(n), n \geq 0$ satisfy the requirements

$$\sum_n a(n) = \infty, \quad \sum_n a(n)^2 < \infty. \tag{6.10}$$

Assumption 6.4. The iterates (6.9) remain almost surely bounded, i.e.,

$$\sup_n \|\theta(n)\| < \infty, \text{ a.s.} \tag{6.11}$$

Assumption 6.5. The ODE (6.7) has H as a compact set of globally asymptotically stable equilibria.

Given $\varepsilon > 0$, let H^ε denote the set of points that are in an open ε-neighborhood of H, i.e.,

$$H^\varepsilon = \{\theta \mid \|\theta - \theta_0\| < \varepsilon, \ \theta_0 \in H\}.$$

The main convergence result of the SF scheme (6.9) is as follows:

Theorem 6.1. *Under Assumptions 6.1 to 6.5, given $\varepsilon > 0$, there exists $\beta_0 > 0$, such that for all $\beta \in (0, \beta_0]$, the iterates $\theta(n)$ obtained from (6.9) satisfy $\theta(n) \to H^\varepsilon$ almost surely as $n \to \infty$.*

In order to prove Theorem 6.1, we provide a sequence of Lemmas and Propositions in the following order:

(i) Proposition 6.2 and Lemma 6.3 together analyze the martingale difference sequence associated with the algorithm (6.9). Lemma 6.4 shows that the resulting martingale is almost surely convergent.

(ii) Proposition 6.5 establishes that the conditional average of the SF estimate, i.e., $E\left[\frac{\eta(n)}{\beta}J(\theta(n)+\beta\eta(n)) \mid \mathscr{F}(n)\right]$ is asymptotically close to the gradient of the objective function $J(\theta(\cdot))$.
(iii) Proposition 6.6 proves that the interpolated trajectory $\bar{\theta}(t)$ of the algorithm (6.9) tracks the ODE (6.7).
(iv) The last step of the proof is proven by invoking the Hirsch lemma.

The formal proof of Theorem 6.1 is given at the end of this section.

Note that (6.9) can be rewritten as

$$\theta_i(n+1) = \theta_i(n) - a(n)\left(E\left[\frac{\eta_i(n)}{\beta}J(\theta(n)+\beta\eta(n)) \mid \mathscr{F}(n)\right] + \bar{M}_i(n+1) + \hat{M}_i(n+1)\right), \tag{6.12}$$

where $\bar{M}_i(n+1) = \frac{\eta_i(n)}{\beta}J(\theta(n)+\beta\eta(n)) - E\left[\frac{\eta_i(n)}{\beta}J(\theta(n)+\beta\eta(n)) \mid \mathscr{F}(n)\right]$.

Proposition 6.2. *We have that* $(\bar{M}_i(n), \mathscr{F}(n)), n \geq 0$ *is a martingale difference sequence with*

$$E\left[|\bar{M}_i(n+1)|^2 \mid \mathscr{F}(n)\right] \leq \hat{K}\left(1 + \|\theta(n)\|^2\right), \ \forall n \geq 0,$$

for some $\hat{K} > 0$.

Proof. It is easy to see that $(\bar{M}_i(n), \mathscr{F}(n)), n \geq 0$ is a martingale difference sequence. Now note that

$$\left.\begin{aligned} E\left[|\bar{M}_i(n+1)|^2 \mid \mathscr{F}(n)\right] &\leq 2\Bigg(E\left[\tfrac{\eta_i^2(n)}{\beta^2}J^2(\theta(n)+\beta\eta(n)) \mid \mathscr{F}(n)\right] \\ &\quad + E\left[E\left[\tfrac{\eta_i(n)}{\beta}J(\theta(n)+\beta\eta(n)) \mid \mathscr{F}(n)\right]^2 \mid \mathscr{F}(n)\right]\Bigg) \\ &\leq 4E\left[\tfrac{\eta_i^2(n)}{\beta^2}J^2(\theta(n)+\beta\eta(n)) \mid \mathscr{F}(n)\right]. \end{aligned}\right\} \tag{6.13}$$

The first inequality above follows because for any $x, y \in \mathbb{R}$, $(x-y)^2 \leq 2(x^2+y^2)$, while the second inequality follows from the conditional Jensen's inequality.

From Assumption 6.1, $J(\cdot)$ is Lipschitz continuous. Hence,

$$|J(\theta)| - |J(0)| \leq |J(\theta) - J(0)| \leq \hat{L}\|\theta\|,$$

where $\hat{L} > 0$ is the Lipschitz constant of the function $J(\cdot)$. Hence,

$$|J(\theta)| \leq C_0(1 + \|\theta\|),$$

where $C_0 = \max(|J(0)|, \hat{L}) > 0$. Hence,

$$J^2(\theta) \leq C_0^2(1 + \|\theta\|)^2 \leq 2C_0^2(1 + \|\theta\|^2),$$

where we use the inequality that for any $x,y \in \mathbb{R}$, $(x+y)^2 \leq 2(x^2+y^2)$. It now follows from (6.13) that

$$E\left[|\bar{M}_i(n+1)|^2 \mid \mathscr{F}(n)\right] \leq \frac{8C_0^2}{\beta^2} E\left[\eta_i^2(n)(1+\|\theta(n)+\beta\eta(n)\|^2) \mid \mathscr{F}(n)\right]. \quad (6.14)$$

Now, note that

$$\begin{aligned}\|\theta(n)+\beta\eta(n)\|^2 &= (\theta(n)+\beta\eta(n))^T(\theta(n)+\beta\eta(n)) \\ &= \|\theta(n)\|^2 + \beta^2\|\eta(n)\|^2 + 2\beta\theta(n)^T\eta(n).\end{aligned}$$

Hence,

$$\begin{aligned}&E\left[\eta_i^2(n)(1+\|\theta(n)+\beta\eta(n)\|^2) \mid \mathscr{F}(n)\right] \\ &= E\left[\eta_i^2(n)(1+\|\theta(n)\|^2+\beta^2\|\eta(n)\|^2+2\beta\theta(n)^T\eta(n) \mid \mathscr{F}(n)\right] \\ &= 1+\|\theta(n)\|^2+\beta^2 E\left[\textstyle\sum_{j\neq i}\eta_j^2(n)\eta_i^2(n)+\eta_i^4(n)\right] \\ &\qquad\qquad +2\beta\theta(n)^T E\left[\textstyle\sum_{j\neq i}\eta_i^2(n)\eta_j(n)+\eta_i^3(n)\right].\end{aligned} \quad (6.15)$$

In the above, we make use of the fact that $\theta(n)$ is measurable $\mathscr{F}(n)$ while $\eta(n)$ is independent of $\mathscr{F}(n)$. Moreover, $\eta_j(n), j = 1,\ldots,N, n \geq 0$ are independent random variables. We now make use of the fact that $E[\eta_j(n)] = E[\eta_j^3(n)] = 0$, $E[\eta_j^2(n)] = 1$ and $E[\eta_j^4(n)] = 3$, $j = 1,\ldots,N, n \geq 0$. Thus, (6.15) can be rewritten as

$$\begin{aligned}E\left[\eta_i^2(n)(1+\|\theta(n)+\beta\eta(n)\|^2) \mid \mathscr{F}(n)\right] &= 1+\|\theta(n)\|^2+\beta^2(N+2) \\ &\leq (1+\beta^2(N+2))(1+\|\theta(n)\|^2).\end{aligned}$$

It now follows from (6.14) that

$$\begin{aligned}E\left[|\bar{M}_i(n+1)|^2 \mid \mathscr{F}(n)\right] &\leq \tfrac{8C_0^2}{\beta^2}(1+\beta^2(N+2))(1+\|\theta(n)\|^2) \\ &= \hat{K}(1+\|\theta(n)\|^2),\end{aligned}$$

where $\hat{K} = \dfrac{8C_0^2}{\beta^2}(1+\beta^2(N+2))$. The claim follows. □

Now (6.9) can be rewritten as

$$\theta_i(n+1) = \theta_i(n) - a(n)E\left[\frac{\eta_i(n)}{\beta}J(\theta(n)+\beta\eta(n)) \mid \mathscr{F}(n)\right] + a(n)M_i(n+1), \quad (6.16)$$

where $M_i(n+1) = \bar{M}_i(n+1) + \hat{M}_i(n+1)$.

Lemma 6.3. *For each $i = 1,\ldots,N$, $(M_i(n),\mathscr{F}(n)), n \geq 0$ form martingale difference sequences with*

$$E[|M_i(n+1)|^2 \mid \mathscr{F}(n)] \leq \tilde{K}(1+\|\theta(n)\|^2),\ \forall n,$$

for some $\tilde{K} > 0$.

Proof. The proof follows from Proposition 6.2 and Assumption 6.2. □

Now, let $M(n) = (M_1(n),\ldots,M_N(n))^T, n \geq 0$. Then $(M(n),\mathscr{F}(n)), n \geq 0$ is a vector martingale sequence. Also,

$$\begin{aligned}
E\left[\|M(n+1)\|^2 \mid \mathscr{F}(n)\right] &= E\left[(M_1(n+1))^2 + \ldots + (M_N(n+1))^2 \mid \mathscr{F}(n)\right],\\
&\leq N\tilde{K}(1+\|\theta(n)\|^2),\\
&= K_1(1+\|\theta(n)\|^2),
\end{aligned}$$

where $K_1 = N\tilde{K}$. Let $Z(n), n \geq 0$ be defined according to

$$Z(n) = \sum_{m=0}^{n-1} a(m)M(m+1).$$

Lemma 6.4. *The sequence $(Z(n),\mathscr{F}(n))$, $n \geq 0$ is a zero-mean, square integrable, almost surely convergent martingale.*

Proof. The proof follows from an application of the martingale convergence theorem (Theorem B.2) using the result in Lemma 6.3 and from Assumptions 6.3 and 6.4, see for instance, the proof of Lemma 3.1. □

We now have the following result:

Proposition 6.5. *Almost surely,*

$$\left\| E\left[\frac{\eta(n)}{\beta} J(\theta(n)+\beta\eta(n)) \mid \mathscr{F}(n)\right] - \nabla J(\theta(n)) \right\| \to 0 \text{ as } \beta \to 0.$$

Proof. Recall that $\eta(n) = (\eta_1(n),\ldots,\eta_N(n))^T$ is a vector of independent $N(0,1)$ distributed random variates. Using a Taylor series expansion of $J(\theta(n)+\beta\eta(n))$ around $\theta(n)$, one obtains

$$J(\theta(n)+\beta\eta(n)) = J(\theta(n)) + \beta\eta(n)^T\nabla J(\theta(n)) + \frac{\beta^2}{2}\eta(n)^T\nabla^2 J(\theta(n))\eta(n) + o(\beta^2).$$

Thus,

$$E\left[\frac{\eta(n)}{\beta} J(\theta(n)+\beta\eta(n)) \mid \mathscr{F}(n)\right] = \frac{1}{\beta}E[\eta(n)J(\theta(n)) \mid \mathscr{F}(n)]$$

$$+E[\eta(n)\eta(n)^T\nabla J(\theta(n)) \mid \mathscr{F}(n)]+\frac{\beta}{2}E[\eta(n)\eta(n)^T\nabla^2 J(\theta(n))\eta(n) \mid \mathscr{F}(n)]+o(\beta). \tag{6.17}$$

Now,

$$E[\eta(n)J(\theta(n)) \mid \mathscr{F}(n)] = E[\eta(n) \mid \mathscr{F}(n)]J(\theta(n)) = 0,$$

since $\eta(n)$ is independent of $\mathscr{F}(n)$. Also,

$$E[\eta(n)\eta(n)^T\nabla J(\theta(n)) \mid \mathscr{F}(n)] = E[\eta(n)\eta(n)^T \mid \mathscr{F}(n)]\nabla J(\theta(n))$$

$$= \nabla J(\theta(n)),$$

again since $\eta(n)\eta(n)^T$ is independent of $\mathscr{F}(n)$ and $E[\eta(n)\eta(n)^T] = I$ (the identity matrix). Consider, now the third term on the RHS of (6.17). Note that

$$\eta(n)^T\nabla^2 J(\theta(n))\eta(n) = \sum_{j=1}^{N}\sum_{i=1}^{N}\eta_j(n)\eta_i(n)\nabla^2_{ji}J(\theta(n)).$$

Thus,

$$E\left[\eta(n)\eta(n)^T\nabla^2 J(\theta(n))\eta(n) \mid \mathscr{F}(n)\right] = E\left[\eta(n)\sum_{j=1}^{N}\sum_{i=1}^{N}\eta_j(n)\eta_i(n)\right]\nabla^2_{ji}J(\theta(n))$$

$$= E\left[\eta_1(n)\sum_{j=1}^{N}\sum_{i=1}^{N}\eta_j(n)\eta_i(n),\ldots,\eta_N(n)\sum_{j=1}^{N}\sum_{i=1}^{N}\eta_j(n)\eta_i(n)\right]\nabla^2_{ji}J(\theta(n))$$

$$= 0,$$

since, $E[\eta_k(n)] = E[\eta_k^3(n)] = 0$ and $E[\eta_k^2(n)] = 1$ and $\eta_i(n)$ is independent of $\eta_j(n)$ for $i \neq j$. The claim follows. □

In lieu of Proposition 6.5, the update rule (6.8) can be rewritten as

$$\theta(n+1) = \theta(n) - a(n)\left(\nabla J(\theta(n)) + (Z(n+1) - Z(n))\right). \tag{6.18}$$

Now as with Chapter 3, consider a sequence of time points $t(n)$, $n \geq 0$ in the following manner: $t(0) = 0$ and for $n \geq 1$, $t(n) = \sum_{m=0}^{n-1} a(m)$. Define now a continuously interpolated trajectory $\bar{\theta}(t), t \geq 0$ (obtained from the algorithm's updates) as follows: Let $\bar{\theta}(t(n)) = \theta(n), n \geq 0$, with linear interpolation on the interval $[t(n), t(n+1)]$. By Assumption 6.4, it follows that $\sup_{t\geq 0}\|\bar{\theta}(t)\| = \sup_n \|\theta(n)\| < \infty$ a.s. Let $T > 0$ be a given real number. Define another sequence $\{T(n), n \geq 0\}$ as follows: $T(0) = t(0) = 0$ and for $n \geq 1$,

$$T(n) = \min\{t(m) \mid t(m) \geq T(n-1) + T\}.$$

Let $I(n)$ denote the interval $[T(n), T(n+1))$. From its definition, there exists an increasing subsequence $\{m(n)\}$ of $\{n\}$ such that $T(n) = t(m(n))$, $n \geq 0$. Also, let

$\theta^n(t), t \geq t(n)$ denote the trajectory of the ODE (6.7) starting at time $t(n)$ and under the initialization $\theta^n(t(n)) = \bar{\theta}(t(n)) = \theta(n)$.

Proposition 6.6. *We have*

$$\lim_{n\to\infty} \sup_{t\in I(n)} \|\bar{\theta}(t) - \theta^n(t)\| = 0, \ a.s.$$

Proof. Follows as in [4, Chapter 2, Lemma 1]. □

Proof of Theorem 6.1. It follows from Proposition 6.6 that $\bar{\theta}(t)$ serves as a (T,Δ)-perturbation for the ODE (6.7) (see Appendix C for definition of (T,Δ)-perturbation). The claim follows by applying the Hirsch lemma (Lemma C.5), for every $\varepsilon > 0$. □

6.2.4 Two-Measurement Gaussian SF Algorithm

The algorithms discussed so far require one measurement of the objective function to estimate the gradient. We now discuss a two-sided finite-difference SF estimate [9, 5, 2] - a variant that has the advantage of a lower estimation bias in comparison to the one-sided form described in Section 6.2.1.

6.2.4.1 Gradient Estimate

Recall the one-sided gradient estimate obtained in (6.4). The two-sided form of the gradient estimate will be described by $D_{\beta,2}J(\theta)$ where

$$D_{\beta,2}J(\theta) \stackrel{\triangle}{=} E\left[\frac{\bar{\eta}}{2\beta}(J(\theta+\beta\bar{\eta}) - J(\theta-\beta\bar{\eta}))\right]. \tag{6.19}$$

As with $D_{\beta,1}J(\theta)$, $\bar{\eta}$ above is an N-dimensional vector of independent $N(0,1)$ random variates and the expectation in (6.19) is taken w.r.t. the distribution of $\bar{\eta}$. It will be seen using suitable Taylor series expansions that $D_{\beta,2}J(\theta)$ is a valid SF gradient estimate that has a lower bias as compared to its one-measurement counterpart (6.4).

The form of the two-measurement SF gradient estimator is thus

$$\nabla J(\theta) \approx \frac{1}{2\beta}\frac{1}{M}\sum_{n=1}^{M} \eta(n)(J(\theta+\beta\eta(n)) - J(\theta-\beta\eta(n))), \tag{6.20}$$

where $M > 0$ is a large positive integer and $\beta > 0$ is a small scalar. Also, in (6.20), $\eta(n) \stackrel{\triangle}{=} (\eta_1(n),\ldots,\eta_N(n))^T$ is a vector of independent $N(0,1)$-distributed random variables as before.

6.2.4.2 The Algorithm

Based on (6.20), the two-measurement version of the gradient Gaussian SF algorithm is as follows: For $i = 1,\ldots,N$, $n \geq 0$,

$$\theta_i(n+1) = \theta_i(n) - a(n)\left(\frac{\eta_i(n)}{2\beta}\left(h(\theta(n)+\beta\eta(n),\xi^+(n)) - h(\theta(n)-\beta\eta(n),\xi^-(n))\right)\right), \tag{6.21}$$

for $i = 1,\ldots,N$ and $\beta > 0$.

The above recursion can be seen as equivalent to

$$\theta_i(n+1) = \theta_i(n) - a(n)\frac{\eta_i(n)}{2\beta}\left(J(\theta(n)+\beta\eta(n)) - J(\theta(n)-\beta\eta(n)) + \chi^1(n) - \chi^2(n)\right), \tag{6.22}$$

where $(\chi^1(n),\mathscr{F}(n)), n \geq 0$ and $(\chi^2(n),\mathscr{F}(n)), n \geq 0$ are two martingale difference sequences that are independent of one another. In particular, for the measurement corresponding to $(\theta(n)+\beta\eta(n))$, the measurement noise is $\chi^1(n)$, while for the measurement with $(\theta(n)-\beta\eta(n))$, the same is $\chi^2(n)$.

6.2.4.3 Convergence Analysis

Let $\mathscr{F}(n) = \sigma(\theta(m),\chi^1(m),\chi^2(m), m \leq n; \eta(m), m < n), n \geq 1$ be a sequence of sigma fields. By an abuse of notation, let $\hat{M}_i(n) = \dfrac{\eta_i(n)}{2\beta}(\chi^1(n) - \chi^2(n))$. It is easy to see that $(\hat{M}_i(n),\mathscr{F}(n)), n \geq 0$ is also a martingale difference sequence. We now consider the following analogous algorithm:

$$\theta_i(n+1) = \theta_i(n) - a(n)\left(\frac{\eta_i(n)}{2\beta}(J(\theta(n)+\beta\eta(n)) - J(\theta(n)-\beta\eta(n))) + \hat{M}_i(n+1)\right). \tag{6.23}$$

We let Assumptions 6.1–6.5 continue to hold with the following changes: In Assumption 6.2, $\hat{M}_i(n)$ and $\mathscr{F}(n), n \geq 0$ are as defined above (i.e., for two-measurement algorithms). Further, Assumption 6.4 holds with (6.23) in place of (6.9). Following the same sequence of steps as for the one-measurement algorithm (cf. Section 6.2.4), one can rewrite (6.23) as

$$\left.\begin{aligned}\theta_i(n+1) = \theta_i(n) - a(n)\Big(E\left[\tfrac{\eta_i(n)}{2\beta}(J(\theta(n)+\beta\eta(n)) - J(\theta(n)-\beta\eta(n))) \mid \mathscr{F}(n)\right]\\ +M_i(n+1)\Big),\end{aligned}\right\} \tag{6.24}$$

where $(M_i(n), \mathcal{F}(n)), n \geq 0$ is a martingale difference sequence satisfying the conclusions of Lemma 6.3. We now have the following result for the two-measurement SF algorithm.

Proposition 6.7. *Almost surely,*

$$\left\| E\left[\frac{\eta(n)}{2\beta}(J(\theta(n)+\beta\eta(n)) - J(\theta(n)-\beta\eta(n))) \mid \mathcal{F}(n)\right] - \nabla J(\theta(n))\right\| \to 0 \text{ as } \beta \to 0.$$

Proof. Recall that $\eta(n) = (\eta_1(n), \ldots, \eta_N(n))^T$ is a vector of independent $N(0,1)$-distributed random variates. Using a Taylor series expansion of $J(\theta(n)+\beta\eta(n))$ around $\theta(n)$, one obtains

$$J(\theta(n)+\beta\eta(n)) = J(\theta(n)) + \beta\eta(n)^T\nabla J(\theta(n)) + \frac{\beta^2}{2}\eta(n)^T\nabla^2 J(\theta(n))\eta(n) + o(\beta^2).$$

Similarly, a Taylor series expansion of $J(\theta(n)-\beta\eta(n))$ around $\theta(n)$ gives

$$J(\theta(n)-\beta\eta(n)) = J(\theta(n)) - \beta\eta(n)^T\nabla J(\theta(n)) + \frac{\beta^2}{2}\eta(n)^T\nabla^2 J(\theta(n))\eta(n) + o(\beta^2).$$

Thus,

$$\begin{aligned} E&\left[\tfrac{\eta(n)}{2\beta}(J(\theta(n)+\beta\eta(n)) - J(\theta(n)-\beta\eta(n))) \mid \mathcal{F}(n)\right] \\ &= E[\eta(n)\eta(n)^T\nabla J(\theta(n)) \mid \mathcal{F}(n)] + o(\beta). \end{aligned} \tag{6.25}$$

Now, it can be seen as in the proof of Proposition 6.5 that

$$\begin{aligned} E[\eta(n)\eta(n)^T\nabla J(\theta(n)) \mid \mathcal{F}(n)] &= E[\eta(n)\eta(n)^T \mid \mathcal{F}(n)]\nabla J(\theta(n)) \\ &= \nabla J(\theta(n)), \text{ a.s.} \end{aligned}$$

The claim follows. □

Remark 6.1. Regarding the bias term resulting in the basic update rule (6.22), we observe the following:

- Note that from the Taylor series expansions of $J(\theta(n)+\beta\eta(n))$ and $J(\theta(n)-\beta\eta(n))$ around $\theta(n)$, it can be seen that in the expression

$$\frac{\eta(n)}{2\beta}(J(\theta(n)+\beta\eta(n)) - J(\theta(n)-\beta\eta(n))),$$

  the bias terms $\frac{1}{2\beta}\eta(n)J(\theta(n))$ and $\beta\eta(n)\eta(n)^T\nabla^2 J(\theta(n))\eta(n)$ resulting from the expansions of $J(\theta(n)+\beta\eta(n))$ and $J(\theta(n)-\beta\eta(n))$, respectively, around $\theta(n)$, directly cancel (cf. Proposition 6.7), and so do not contribute to the bias.

On the other hand, the aforementioned bias terms average to zero in the case of one-measurement SF (cf. Proposition 6.5).

- For small $\beta > 0$, the term $\frac{1}{\beta}\eta(n)J(\theta(n))$ can result in a much higher bias in the one-measurement algorithm as compared to the two-measurement case because of the presence of β in the denominator of that term. It has also been observed in [9, 2] that the two-measurement algorithm performs better than its one-measurement counterpart. In particular, in the one-measurement SF algorithm, a low value of β results in a large bias and a high β results in inaccuracies in the estimate. The two-measurement counterpart, on the other hand, is more robust to different values of β largely because of the direct cancellation of the bias terms that results from the use of two measurements.

The remainder of the analysis follows along similar lines as in Section 6.2.3. The main convergence result again is the following:

Theorem 6.8. *Under Assumptions 6.1 to 6.5, given $\varepsilon > 0$, there exists $\beta_0 > 0$, such that for all $\beta \in (0, \beta_0]$, the iterates $\theta(n)$ obtained from (6.23) satisfy $\theta(n) \to H^{\varepsilon}$ almost surely as $n \to \infty$.*

6.3 General Conditions for a Candidate Smoothing Function

While Gaussian density function has been a popular choice as a smoothing function to estimate the gradient, it is possible to achieve the same effect using other candidate functions as well. In this section, we discuss general conditions that have to be met by any candidate function, to be used as a smoothing function. Let F_β be an operator defined such that

$$F_\beta J(\theta) = \int_\eta h_\beta(\eta)\nabla_\eta J(\theta - \eta)d\eta = \int_\eta h_\beta(\theta - \eta)\nabla_\eta J(\eta)d\eta,$$

is the SF estimate of $J(\theta)$ with $h_\beta(\cdot)$ as a smoothing function with smoothing parameter $\beta > 0$. [8, pp. 471] lists a set of four conditions for a smoothing function. They are as follows:

(a) $h_\beta(\eta) = \frac{1}{\beta^N}h(\eta/\beta)$, is a piece-wise differentiable function w.r.t. η;

(b) $\lim_{\beta\to 0} h_\beta(\eta) = \delta(\eta)$, where $\delta(\cdot)$ is the Dirac-Delta function;

(c) $\lim_{\beta\to 0} F_\beta J(\theta) = \nabla_\theta J(\theta)$ and

(d) $h_\beta(\cdot)$ is a probability density function (p.d.f.), i.e., $F_\beta J(\theta) = E_\eta \nabla_\eta J(\theta - \eta)$.

Remark 6.2. One can easily verify that Gaussian, Cauchy and uniform distributions satisfy the above set of conditions. In a recent work, [6], the q-Gaussian family of distributions have also been shown to satisfy the aforementioned conditions. (The q-Gaussians are a generalized class of density functionals than the Gaussian density. Here, q represents a real-valued parameter and in fact, a q-Gaussian density with $q = 1$ is the standard Gaussian density.)

Remark 6.3. It is worth noting here that smoothing with uniform density results in an iteration procedure very similar to that of the Kiefer-Wolfowitz algorithm discussed in Chapter 4. We leave the verification of this fact as an exercise to the readers. However, as discussed earlier, Kiefer-Wolfowitz algorithm does not scale well for higher dimensions. On the contrary, gradient estimates based on Gaussian and Cauchy smoothing procedures can be obtained with only one or two simulations regardless of the parameter dimension.

6.4 Cauchy Variant of the SF Algorithm

In this section, we use the Cauchy density function to obtain an estimate of the gradient and derive a stochastic iterative algorithm that performs negative gradient descent w.r.t. the objective function $J(\theta)$. The idea of obtaining the gradient estimate here is similar to the case when Gaussian density was used (see Section 6.2.1), and the specifics are handled in the next section.

6.4.1 Gradient Estimate

Let Λ be a hypercube centered at the origin such that $C \subseteq \Lambda$, i.e., the set of all admissible values of θ is contained in Λ. Now the N-dimensional Cauchy p.d.f., truncated to Λ, can be written as follows:

$$H_\beta(\theta - \eta) = \begin{cases} \dfrac{\Gamma\left(\frac{N+1}{2}\right)}{\pi^{\frac{N+1}{2}} \beta^n \Omega} \dfrac{1}{\left(1 + \dfrac{(\theta-\eta)^T(\theta-\eta)}{\beta^2}\right)^{\frac{N+1}{2}}} & \text{for } \eta \in \Lambda^0, \\ 0 & \text{otherwise,} \end{cases} \tag{6.26}$$

where $\theta, \eta \in \mathbb{R}^N$, $\beta > 0$, Λ^0 represents the interior of the set Λ, $\Gamma(\cdot)$ is the standard gamma function (that must not be confused with the projection operator Γ that we use in projected algorithms) and Ω is a scaling factor to ensure that $\int_\eta H_\beta(\theta - \eta)d\eta = 1$. Cauchy distribution without truncation has no moments defined. However, truncation to a bounded set, in this case Λ, ensures that all moments of the Cauchy distribution are well defined. Like in the case of Gaussian smoothing, we now define the Cauchy smoothing operator $F_{\beta,1}$ below:

$$F_{\beta,1}J(\theta) = \int_{\eta} H_\beta(\theta - \eta)\nabla_\eta J(\eta)d\eta. \tag{6.27}$$

Upon integration by parts of (6.27), we get,

$$\begin{aligned} F_{\beta,1}J(\theta) &= -\int_{\eta} \nabla_\eta H_\beta(\theta - \eta)J(\eta)d\eta \\ &= \int_{\eta} \nabla_\eta H_\beta(\eta)J(\theta - \eta)d\eta. \end{aligned}$$

Now, observing that

$$\nabla_\eta H_\beta(\eta) = \begin{cases} -\dfrac{\eta(N+1)}{(\beta^2+\eta^T\eta)} H_\beta(\eta) & \text{for } \eta \in \Lambda^0, \\ 0 & \text{otherwise,} \end{cases}$$

the expression for $F_{\beta,1}J(\theta)$ can be updated to

$$F_{\beta,1}J(\theta) = -\int_{\eta\in\Lambda} \frac{\eta(N+1)}{(\beta^2+\eta^T\eta)} H_\beta(\eta)J(\theta-\eta)d\eta.$$

Now, by simple substitution of $\eta' = -\dfrac{\eta}{\beta}$, we get,

$$F_{\beta,1}J(\theta) = \int_{\eta\in\Lambda} \frac{\eta'(N+1)}{\beta(1+\eta'^T\eta')} H_1(\eta)J(\theta+\beta\eta')d\eta',$$

which could be compactly written as

$$F_{\beta,1}J(\theta) = E_\eta\left[\frac{\eta(N+1)}{\beta(1+\eta^T\eta)} J(\theta+\beta\eta)\right], \tag{6.28}$$

where the expectation is over η which has standard truncated multivariate Cauchy distribution, $H_1(\cdot)$. Now, since truncated Cauchy distribution satisfies condition (c) for a smoothing function, it is possible to write a sample average estimate of the above as an approximate estimate of the gradient of $J(\theta)$, i.e.,

$$\nabla_\theta J(\theta) \approx \frac{1}{M}\sum_{n=1}^{M} \frac{\eta(n)(N+1)}{\beta(1+\eta(n)^T\eta(n))} J(\theta+\beta\eta(n)), \tag{6.29}$$

where $\eta(1), \eta(2), \ldots, \eta(M)$ are i.i.d. samples with standard truncated multivariate Cauchy distribution $H_1(\cdot)$. For the estimate to be accurate, M is chosen to be a large integer with β close to zero.

6.4.2 Cauchy SF Algorithm

Let $\eta(n) = (\eta_1(n),\ldots,\eta_N(n))^T$, $n \geq 0$ be a sequence of independent standard truncated multivariate Cauchy distributed random variables. Note that unlike multivariate Gaussian distributed random variables, here for a given n, $\eta_1(n),\ldots,\eta_N(n)$, are not independent of each other. Based on the SF estimate (6.29), the algorithm that we consider is the following: For $i = 1,\ldots,N$, $n \geq 0$,

$$\theta_i(n+1) = \theta_i(n) - a(n)\left(\frac{\eta_i(n)(N+1)}{\beta(1+\eta(n)^T\eta(n))}(J(\theta(n)+\beta\eta(n)))\right). \quad (6.30)$$

The overall algorithm is similar to Algorithm 6.1, except that Cauchy-based perturbations are used in this case. One can now follow a similar sequence of steps as in the previous section to conclude with the following theorem.

Theorem 6.9. *Under Assumptions 6.1 to 6.5, given $\varepsilon > 0$, there exists $\beta_0 > 0$, such that for all $\beta \in (0,\beta_0]$, the iterates $\theta(n)$ obtained from (6.30) satisfy $\theta(n) \to H^\varepsilon$ almost surely as $n \to \infty$.*

Remark 6.4. The single measurement version of the Cauchy SF estimate (6.28) can be extended for two measurements as follows:

$$F_{\beta,1}J(\theta) = E_\eta\left[\frac{\eta(N+1)}{2\beta(1+\eta^T\eta)}(J(\theta+\beta\eta) - J(\theta-\beta\eta))\right]. \quad (6.31)$$

Based on (6.31), the two-measurement version of the Cauchy gradient SF algorithm is as follows:

$$\theta_i(n+1) = \theta_i(n) - a(n)\left(\frac{\eta_i(n)(N+1)}{2\beta(1+\eta(n)^T\eta(n))}(J(\theta(n)+\beta\eta(n)) - J(\theta(n)-\beta\eta(n))) + \chi^1(n) - \chi^2(n)\right), \quad i = 1,\ldots,N, \quad n \geq 0, \quad (6.32)$$

where as before $\eta(n), n \geq 0$, are independent standard truncated multivariate Cauchy distributed random variables (even though for a given n, $\eta_1(n),\ldots,\eta_N(n)$ are not independent). The convergence analysis follows in a similar manner as that of the two-measurement Gaussian SF algorithm discussed previously.

6.5 SF Algorithms for the Long-Run Average Cost Objective

We now present gradient SF algorithms for the case when the underlying process is Markovian and depends on a parameter. The basic framework is the same as in

Chapter 5.6. The single-stage cost function depends on the simulated system state. Let $\{X(n), n \geq 1\}$ be an $\mathbb{R}^d$-valued parameterized Markov process with a tunable parameter θ that takes values in $\mathbb{R}^N$. Let for any given $\theta \in \mathbb{R}^N$, $\{X(n)\}$ be ergodic Markov. Let $p(\theta, x, dy)$ and $\nu_\theta(dx)$, respectively, denote the transition kernel and stationary distribution of $\{X(n)\}$ when θ is the operative parameter. When the process is in state x, let $h(x)$ be the single-stage cost incurred. The aim is to find a $\theta^* \in \mathbb{R}^N$ that minimizes (over all θ) the long-run average cost (5.40). As with the average cost setting for SPSA, we let Assumptions 5.10–5.14 hold in this setting as well.

The algorithms that we present below are the analogs of the algorithms presented in Section 6.2.2 and are taken from [3, 2]. In [2], they are referred to as G-SF1 and G-SF2, respectively. We use the same abbreviations here. Note that G-SF1 refers to the one-simulation variant of the SF algorithm, while G-SF2 refers to the two-simulation variant.

6.5.1 The G-SF1 Algorithm

The algorithm that we present below is the analog of the one-measurement Gaussian SF algorithm described in Section 6.2.2. Let $X(n), n \geq 0$ be a simulated Markov process governed by the parameter sequence $(\theta(n) + \beta\eta(n)), n \geq 0$, where $\eta(n) \stackrel{\triangle}{=} (\eta_1(n), \ldots, \eta_N(n))^T$ with $\eta_i(n), n \geq 0, i = 1, \ldots, N$ being independent $N(0,1)$-distributed random variables and $\beta > 0$ is a given small positive scalar. The algorithm is as follows: For $i = 1, \ldots, N, n \geq 0$,

$$\theta_i(n+1) = \theta_i(n) - a(n)Z_i(n), \tag{6.33}$$

$$Z_i(n+1) = Z_i(n) + b(n)\left(\frac{\eta_i(n)}{\beta}h(X(n)) - Z_i(n)\right). \tag{6.34}$$

Remark 6.5. As with the simulation-based SPSA algorithms for the long-run average cost objective, it is seen that in practice, an additional averaging over L instants (for some $L > 1$) of the recursion (6.34) improves performance. In other words, for practical implementations, it is suggested to run (6.34) for L instants in an inner loop, in between two successive updates of (6.33). The value of L is, however, arbitrary. It is generally observed, see for instance, [1, 2] that a value of L in between 50 and 500 works well. While for our analysis, we focus on the case of $L = 1$, the analysis for general L is available in [1, 2].

6.5.1.1 Convergence Analysis of the G-SF1 Algorithm

Let $\mathcal{G}(n) = \sigma(\theta(p), X_p, \eta(p), p \leq n), n \geq 1$, denote σ-fields generated by the quantities above. Define sequences $N_i(p), p \geq 0, i \in \{1, \ldots, N\}$ as follows:

$$N_i(p) = \sum_{m=1}^{p} b(m) \left(\frac{\eta_i(m)}{\beta} h(X_m) - E\left[\frac{\eta_i(m)}{\beta} h(X_m) \mid \mathcal{G}(m-1) \right] \right).$$

Lemma 6.10. *The sequences $(N_i(p), \mathcal{G}(p)), p \geq 0$, $i = 1, \ldots, N$ are almost surely convergent martingale sequences.*

Proof. It is easy to see that almost surely, $E[N_i(p+1) \mid \mathcal{G}(p)] = N_i(p)$, for all $p \geq 0$. Now, note that

$$E[N_i^2(p)] \leq \frac{C_p}{\beta^2} \sum_{m=1}^{p} b^2(m) (E[\eta_i^2(m) h^2(X_m) + E^2[\eta_i(m) h(X_m) \mid \mathcal{G}(m-1)]])$$

for some constant $C_p > 0$ (that however depends on p). For the second term on the RHS above, note that almost surely,

$$E^2[\eta_i(m) h(X_m) \mid \mathcal{G}(m-1)] \leq E[\eta_i^2(m) h^2(X_m) \mid \mathcal{G}(m-1)],$$

by the conditional Jensen's inequality. Hence,

$$\begin{aligned} E[N_i^2(p)] &\leq \tfrac{2C_p}{\beta^2} \textstyle\sum_{m=1}^{p} b^2(m) E[\eta_i^2(m) h^2(X_m)] \\ &\leq \tfrac{2C_p}{\beta^2} \textstyle\sum_{m=1}^{p} b^2(m) E[\eta_i^4(m)]^{1/2} E[h^4(X_m)]^{1/2} \end{aligned}$$

by the Cauchy-Schwartz inequality. Since, $h(\cdot)$ is a Lipschitz continuous function, we have

$$|h(X_m)| - |h(0)| \leq |h(X_m) - h(0)| \leq K \|X_m\|,$$

where $K > 0$ is the Lipschitz constant for the function h. Thus,

$$|h(X_m)| \leq C_1 (1 + \|X_m\|),$$

for $C_1 = \max(K, |h(0)|) < \infty$. Hence, one gets

$$E[h^4(X_m)] \leq C_2 (1 + E[\|X_m\|^4])$$

for (constant) $C_2 = 8C_1^4$.

As a consequence of Assumption 5.12, $\sup_m E[\|X_m\|^4] < \infty$. Thus, $E[N_i^2(p)] < \infty$, for all $p \geq 1$. Now, note that

$$\sum_p E[(N_i(p+1) - N_i(p))^2 \mid \mathcal{G}(p)] \leq \sum_p b^2(p+1) (E[(\frac{\eta_i^2(p+1)}{\beta^2} h(X_{p+1}))^2 \mid \mathcal{G}(p)]$$

$$+E[E^2[\frac{\eta_i^2(p+1)}{\beta^2}h(X_{p+1}) \mid \mathscr{G}(p)] \mid \mathscr{G}(p)])$$

$$\leq \sum_p 2b^2(p+1)E[(\frac{\eta_i^2(p+1)}{\beta^2}h(X_{p+1}))^2 \mid \mathscr{G}(p)],$$

almost surely. The last inequality above again follows from the conditional Jensen's inequality. It can now be easily seen as before, using Assumption 5.12, that

$$\sup_p \frac{1}{\beta^2}E[(\eta_i^2(p+1)h(X_{p+1}))^2 \mid \mathscr{G}(p)] < \infty \text{ w.p.1.}$$

Now, from Assumption 5.13,

$$\sum_p E[(N_i(p+1) - N_i(p))^2 \mid \mathscr{G}(p)] < \infty$$

almost surely. Thus, by the martingale convergence theorem (Theorem B.2), $N_i(p)$, $p \geq 0$ are almost surely convergent for each $i = 1,\ldots,N$. □

Lemma 6.11. *The sequences of updates* $\{Z_i(p)\}, i = 1,\ldots,N$ *in (6.34) are uniformly bounded with probability one.*

Proof. Note that (6.34) can be rewritten as

$$Z_i(p+1) = Z_i(p) + b(p)(\frac{1}{\beta}E[\eta_i(p)h(X_p) \mid \mathscr{G}(p-1)] - Z_i(p))$$

$$+b(p)\frac{1}{\beta}(\eta_i(p)h(X_p) - E[\eta_i(p)h(X_p) \mid \mathscr{G}(p-1)]). \quad (6.35)$$

From Lemma 6.10, it follows that

$$\sum_p b(p)\frac{1}{\beta}(\eta_i(p)h(X_p) - E[\eta_i(p)h(X_p) \mid \mathscr{G}(p-1)]) < \infty, \text{ a.s.}$$

Thus, it is enough to consider the boundedness of the following alternate recursion:

$$Z_i(p+1) = Z_i(p) + b(p)(\frac{1}{\beta}E[\eta_i(p)h(X_p) \mid \mathscr{G}(p-1)] - Z_i(p)).$$

It can be seen as in the proof of Lemma 6.10 that $\sup_p E[\eta_i(p)h(X_p) \mid \mathscr{G}(p-1)] < \infty$ with probability one. The claim now easily follows from the Borkar and Meyn theorem (Theorem D.1). □

Define $\{s(n)\}$ and $\{t(n)\}$ as follows: $s(0) = t(0) = 0$, $s(n) = \sum_{j=0}^{n-1} a(j)$ and $t(n) = \sum_{j=0}^{n-1} b(j)$, $n \geq 1$, respectively. Then, the timescale corresponding to $\{s(n)\}$ (resp. $\{t(n)\}$) is the slower (resp. faster) of the two timescales.

Consider the following system of ordinary differential equations (ODEs):

$$\dot{\theta}(t) = 0, \tag{6.36}$$

$$\dot{Z}(t) = D_{\beta,1}J(\theta(t)) - Z(t). \tag{6.37}$$

Let $Z(n) \triangleq (Z_1(n),\ldots,Z_N(n))^T$, $n \geq 0$. From Lemma 6.11, $\sup_n \|Z(n)\| < \infty$. Consider the functions $\hat{Z}(t)$ defined according to $\hat{Z}(t(n)) = Z(n)$ with the maps $t \to \hat{Z}(t)$ corresponding to continuous linear interpolations on intervals $[t(n), t(n+1)]$. Given $\bar{T} > 0$, define $\{\bar{T}(n)\}$ as follows: $\bar{T}(0) = 0$ and for $n \geq 1$, $\bar{T}(n) = \min\{t(m) \mid r(m) \geq \bar{T}(n-1) + \bar{T}\}$. Let $\bar{I}(n) = [\bar{T}(n), \bar{T}(n+1)]$. There exists some integer $q(n) > 0$ such that $\bar{T}(n) = t(q(n))$. Define also functions $Z^n(t)$, $t \in I(n)$, $n \geq 0$, that are obtained as trajectories of the following ODEs:

$$\dot{Z}^n(t) = D_{\beta,1}J(\theta) - Z^n(t), \tag{6.38}$$

with $Z^n(\bar{T}(n)) = \hat{Z}(t(q(n))) = Z(q(n))$.

Lemma 6.12. $\lim_{n\to\infty} \sup_{t\in\bar{I}(n)} \|Z^n(t) - \hat{Z}(t)\| = 0$ *w.p.1.*

Proof. Follows in the same manner as [4, Chapter 2, Lemma 1]. □

Lemma 6.13. *Given $\bar{T}, \gamma > 0$, $((\theta(t(n) + \cdot), Z(t(n) + \cdot))$, is a bounded $(\bar{T}, \gamma)$-perturbation of (6.36)-(6.37) for n sufficiently large.*

Proof. Note that the parameter update recursion (6.33) can be rewritten as

$$\theta(n+1) = \theta(n) + b(n)\hat{\xi}_1(n), \tag{6.39}$$

where $\hat{\xi}_1(n) = o(1)$, since $a(n) = o(b(n))$. The rest follows from Lemma 6.12. □

Lemma 6.14. $\|Z(n) - D_{\beta,1}J(\theta(n))\| \to 0$ *w.p. 1, as $n \to \infty$.*

Proof. Follows from an application of the Hirsch lemma (Lemma C.5) for every $\varepsilon > 0$. □

Finally, we consider the slower recursion (6.33). In lieu of Lemma 6.14, the slower recursion can be rewritten as

$$\theta(n+1) = \theta(n) - a(n)(D_{\beta,1}J(\theta(n) + \hat{\varepsilon}(n)), \tag{6.40}$$

where $\hat{\varepsilon}(n) = o(1)$. Now from Proposition 6.5,

$$\|D_{\beta,1}J(\theta(n)) - \nabla J(\theta(n))\| \to 0 \text{ as } \beta \to 0 \text{ a.s.}$$

Consider now the ODE (6.7), but with J being the long-run average cost. Let H be as in Assumption 6.5 with J as above. The following is the main result that follows again from an application of the Hirsch lemma (Lemma C.5).

Theorem 6.15. *Under Assumptions 5.10–5.13 and Assumption 6.5, given $\varepsilon > 0$, there exists $\beta_0 > 0$, such that for all $\beta \in (0, \beta_0]$, the iterates $\theta(n)$ obtained from (6.33) to (6.34) satisfy $\theta(n) \to H^{\varepsilon}$ almost surely as $n \to \infty$.*

6.5.2 The G-SF2 Algorithm

We now present the two-simulation gradient SF algorithm (G-SF2) that is the analog of the two-measurement algorithm described in Section 6.2.4. Let $X^1(n), n \geq 0$ and $X^2(n), n \geq 0$ be two simulated Markov processes that are respectively governed by the parameter sequences $(\theta(n) + \beta\eta(n))$ and $(\theta(n) - \beta\eta(n)), n \geq 0$, where $\eta(n) \stackrel{\triangle}{=} (\eta_1(n), \ldots, \eta_N(n))^T$ with $\eta_i(n), n \geq 0, i = 1, \ldots, N$ being independent $N(0,1)$-distributed random variables and $\beta > 0$ is a given small positive scalar. The two processes $X^1(n)$ and $X^2(n), n \geq 0$ evolve independently of one another. The algorithm is as follows: For $i = 1, \ldots, N, n \geq 0$,

$$\theta_i(n+1) = \theta_i(n) - a(n)Z_i(n), \tag{6.41}$$

$$Z_i(n+1) = Z_i(n) + b(n)\left(\frac{\eta_i(n)}{\beta}\left(h(X^1(n)) - h(X^2(n))\right) - Z_i(n)\right). \tag{6.42}$$

The convergence of the algorithm proceeds along the same lines as G-SF1 and one could prove an analogue of Theorem 6.15 by following the same steps as in Section 6.5.1.1.

Finally, as described in Remark 6.5, the empirical performance of the G-SF2 algorithm also improves with an additional averaging over L instants of the iterates (6.42) in between two successive updates of the parameter (6.41).

Remark 6.6. It is worth noting here that one can easily derive two algorithms analogous to G-SF1 and G-SF1 described above, based on the Cauchy SF estimates from (6.28) and (6.31), respectively. We leave it to the interested reader to derive these (multi-timescale Cauchy-perturbation SF) algorithms.

6.5.3 Projected SF Algorithms

As with Section 5.6.5 of Chapter 5, we consider here the case where the parameter θ can take values in a predefined compact set C that is a subset of $\mathbb{R}^N$. Let $\Gamma : \mathscr{R}^N \to C$ denote the 'projection operator' that projects any $x = (x_1,\ldots,x_N)^T \in \mathscr{R}^N$ to its nearest point in C, i.e., for any given $x = (x_1,\ldots,x_N)^T \in \mathscr{R}^N$, $\Gamma(x) \stackrel{\triangle}{=} (\Gamma_1(x_1),\ldots,\Gamma_N(x_N))^T \in \mathbb{R}^N$. The operator Γ ensures that Assumption 5.14 is automatically enforced.

6.5.3.1 The G-SF1 Algorithm with Projection

Let $X(n), n \geq 0$ be a simulated Markov process governed by the parameter sequence $(\theta(n) + \beta\eta(n)), n \geq 0$, where $\eta(n) \stackrel{\triangle}{=} (\eta_1(n),\ldots,\eta_N(n))^T$ with $\eta_i(n), n \geq 0, i = 1,\ldots,N$ being independent $N(0,1)$-distributed random variables and $\beta > 0$ is a given small positive scalar. The algorithm is as follows: For $i = 1,\ldots,N, n \geq 0$,

$$\theta_i(n+1) = \Gamma_i(\theta_i(n) - a(n)Z_i(n)), \tag{6.43}$$

$$Z_i(n+1) = Z_i(n) + b(n)\left(\frac{\eta_i(n)}{\beta}h(X(n)) - Z_i(n)\right). \tag{6.44}$$

Let $\bar{\Gamma} : \mathscr{C}(C) \to \mathscr{C}(\mathscr{R}^N)$ be defined as in (5.62). Also, let H denote the set of all asymptotically stable attractors of the ODE (5.63) and H^ε its ε-neighborhood. An application of the Kushner-Clark theorem (Theorem E.1) shows the following result:

Theorem 6.16. *Under Assumptions 5.10–5.13, given $\varepsilon > 0$, there exists a $\delta_0 > 0$ such that the sequence of parameter iterates $\theta(n), n \geq 0$ governed by (6.43)-(6.44) satisfy $\theta(n) \to H^\varepsilon$ with probability one as $n \to \infty$.*

6.5.3.2 The G-SF2 Algorithm with Projection

Let $X^1(n), n \geq 0$ and $X^2(n), n \geq 0$ be two simulated Markov processes that are, respectively, governed by the parameter sequences $(\theta(n) + \beta\eta(n))$ and $(\theta(n) - \beta\eta(n)), n \geq 0$, where $\eta(n) \stackrel{\triangle}{=} (\eta_1(n),\ldots,\eta_N(n))^T$ with $\eta_i(n), n \geq 0, i = 1,\ldots,N$ being independent $N(0,1)$-distributed random variables and $\beta > 0$ is a given small positive scalar. The two processes $X^1(n)$ and $X^2(n), n \geq 0$ evolve independently of one another. The algorithm is as follows: For $i = 1,\ldots,N, n \geq 0$,

$$\theta_i(n+1) = \Gamma_i(\theta_i(n) - a(n)Z_i(n)), \tag{6.45}$$

$$Z_i(n+1) = Z_i(n) + b(n)\left(\frac{\eta_i(n)}{\beta}\left(h(X^1(n)) - h(X^2(n))\right) - Z_i(n)\right). \tag{6.46}$$

The convergence result of (6.45) follows in a similar manner as in the case of the projected G-SF1 algorithm (6.43).

6.6 Concluding Remarks

In this chapter, we introduced the smoothed functional technique for estimating the gradient, and showed how it can be used to develop convergent stochastic recursive algorithms for both the expected cost as well as the long-run average cost objectives. As in the previous chapter, the algorithms presented in this chapter also incorporated the simultaneous perturbation approach to estimate the gradient. However, unlike SPSA algorithms, the algorithms here used certain smoothing functions - most commonly Gaussian and Cauchy densities - for simultaneous perturbation and in turn for gradient estimation. All the SF algorithms presented here are online implementable. Further, as with the SPSA algorithms, they require only one or two samples of the objective function for any N-dimensional parameter. We demonstrate the empirical usefulness of both SPSA as well as SF algorithms in various application contexts in Part V of this book.

References

1. Bhatnagar, S.: Multiscale stochastic approximation algorithms with applications to ABR service in ATM networks. Ph.D. thesis, Department of Electrical Engineering. Indian Institute of Science, Bangalore, India (1997)
2. Bhatnagar, S.: Adaptive Newton-based smoothed functional algorithms for simulation optimization. ACM Transactions on Modeling and Computer Simulation 18(1), 2:1–2:35 (2007)
3. Bhatnagar, S., Borkar, V.S.: Multiscale chaotic SPSA and smoothed functional algorithms for simulation optimization. Simulation 79(10), 568–580 (2003)
4. Borkar, V.S.: Stochastic Approximation: A Dynamical Systems Viewpoint. Cambridge University Press and Hindustan Book Agency (Jointly Published), Cambridge and New Delhi (2008)
5. Chin, D.C.: Comparative study of stochastic algorithms for system optimization based on gradient approximation. IEEE Trans. Sys. Man. Cyber. Part B 27(2), 244–249 (1997)

6. Ghoshdastidar, D., Dukkipati, A., Bhatnagar, S.: q-Gaussian based smoothed functional algorithms for stochastic optimization. In: Proceedings of the IEEE International Symposium on Information Theory. IEEE Press, Boston (to appear, 2012)
7. Katkovnik, V.Y., Kulchitsky, Y.: Convergence of a class of random search algorithms. Automation Remote Control 8, 1321–1326 (1972)
8. Rubinstein, R.Y.: Simulation and the Monte Carlo Method. Wiley, New York (1981)
9. Styblinski, M.A., Tang, T.S.: Experiments in nonconvex optimization: stochastic approximation with function smoothing and simulated annealing. Neural Networks 3, 467–483 (1990)

Part III
Hessian Estimation Schemes

Newton-based algorithms estimate both the gradient and the Hessian of the objective, and are in general seen to be more efficient than gradient-based algorithms as they exhibit fast convergence (in terms of the number of iterates). However, they require more computation than gradient-based schemes, because of the need to project the associated Hessian matrix to the space of positive definite and symmetric matrices at each update epoch and to invert the same. During the course of the last twelve years, there has been significant work done on developing efficient simultaneous perturbation-based Hessian estimators and Newton-based schemes. This part consists of two chapters – on Newton SPSA and Newton SF methods.

In a paper in 2000, Spall presented a Newton-based procedure involving an efficient four-simulation Hessian estimation procedure that relies on two independent perturbation sequences, each of which have similar properties to the one used for obtaining gradient SPSA estimates. This algorithm was presented for the case when the cost is an expectation over noisy cost samples. Subsequently, in a paper in 2005, Bhatnagar obtained three other SPSA-based Hessian estimates, that require three, two and one simulation(s), respectively. Together with the four simulation estimate presented by Spall, Bhatnagar proposed four Newton SPSA algorithms (using the aforementioned estimates in addition) for the long-run average cost objective. We discuss the Newton SPSA schemes in detail in Chapter 7.

Bhatnagar, in a paper in 2007, presented two Newton SF algorithms. He presented both algorithms for the case when the perturbation sequence used is i.i.d. and Gaussian distributed. By convolving the Hessian of the objective function with the N-dimensional Gaussian density and applying the integration-by-parts argument twice, one obtains the convolved Hessian as a convolution of the Hessian with a scaled N-dimensional Gaussian where the scaling matrix is obtained from the components of the N-vector Gaussian. The same simulation (with a Gaussian-perturbed parameter) is seen to estimate both the gradient as well as the Hessian of the objective. A two-simulation version of the same is seen to exhibit better performance in general. We discuss in detail the Newton SF schemes in Chapter 8.

Chapter 7
Newton-Based Simultaneous Perturbation Stochastic Approximation

7.1 Introduction

In Chapters 4–6, we discussed gradient-based approaches. Whereas, the finite-difference Kiefer-Wolfowitz algorithm is seen to require $2N$ (resp. $N+1$) simulations for an N-dimensional parameter when two-sided balanced (resp. one-sided) estimates are used, the gradient SPSA algorithms are based on exactly two or one simulation samples at each update epoch regardless of the parameter dimension N. Hessian estimates, on the other hand, are harder to obtain than the gradient and typically require more samples of the objective function. In [4], the Hessian is estimated using finite differences that are in turn based on finite difference estimates of the gradient. This requires $O(N^2)$ samples of the objective function at each update epoch. In [9], for the case where the objective function gradients are known, the Hessian is estimated using finite gradient differences.

The first Newton SPSA algorithm that estimates the Hessian in addition to the gradient using simultaneous perturbation estimates has been proposed in [11]. The estimates in this algorithm are obtained using four objective function samples at each update epoch in cases where the gradient estimates are not known and three samples in cases where the latter are known. This is achieved using two independent perturbation sequences with random variables in these assumed bounded, zero-mean, symmetric, having a common distribution and mutually independent of one another. This method is an extension of the random perturbation gradient SPSA algorithm of [10] that uses only one such perturbation sequence. In [13], a similar algorithm that, however, uses the geometric mean of the eigen-values of the Hessian (suitably projected so that the eigen-values remain positive) in place of the Hessian itself, has been proposed.

In [2], four multi-timescale Newton SPSA algorithms that require four, three, two and one simulation(s), respectively, have been proposed for the long-run average cost objective. All of these algorithms incorporate random perturbation Hessian

S. Bhatnagar et al.: Stochastic Recursive Algorithms for Optimization, LNCIS 434, pp. 105–131.
springerlink.com

estimates. The four-simulation algorithm incorporates a similar Hessian estimate as the one proposed in [11]. While unbiased and convergent Hessian estimates based on a deterministic construction for the perturbation variables are not yet available, a Hadamard matrix-based deterministic Hessian estimate has been proposed and incorporated in [8] in the context of adaptive optimization of parameters in the random early detection (RED) flow control scheme for the Internet, see Chapter 14.2. Even though the Hessian estimates there are not unbiased, the resulting scheme still tracks a local minimum of the objective and is seen to perform well. The current chapter is largely based on [2].

We first summarize the long-run average cost framework in Section 7.2, along with the necessary assumptions. In Section 7.3, the four multi-timescale Newton SPSA algorithms are described. Also in Section 7.4, we describe a means to improve the performance of these four algorithms using the Woodbury's identity. Convergence analyses of all these algorithms are then provided in Section 7.5.

7.2 The Framework

We describe in this section the problem framework and assumptions. Let $\{X(n), n \geq 1\}$ be an $\mathbb{R}^d$-valued parameterized Markov process with a tunable parameter θ that takes values in a given compact and convex set $C \subset \mathbb{R}^N$. We assume that for any given $\theta \in C$, the process $\{X(n)\}$ is ergodic Markov. Let $p(\theta, x, dy)$ and ν_θ, respectively, denote the transition kernel and stationary distribution of $\{X(n)\}$, when θ is the operative parameter. When the process is in state x, let $h(x)$ be the single-stage cost incurred. The aim is to find a $\theta^* \in C$ that minimizes (over all $\theta \in C$) the long-run average cost

$$J(\theta) = \lim_{l \to \infty} \frac{1}{l} \sum_{j=0}^{l-1} h(X(j)). \tag{7.1}$$

7.3 Newton SPSA Algorithms

We now describe the Newton SPSA algorithms from [2]. The analysis of these algorithms will be subsequently shown. The four algorithms that we present below are based on different estimates of the Hessian and require four, three, two and one simulation(s), respectively. Two of these algorithms, based on two and one simulation(s), respectively, also incorporate different simultaneous perturbation gradient estimates (than the gradient SPSA algorithms presented in Chapter 5). The Hessian estimates in all the four algorithms depend on two independent sequences of perturbation random variables. Let $C \subset \mathbb{R}^N$ be a compact and convex set in which the parameters θ take values. To ensure that the θ-updates take values in this set, we use the projection operator $\Gamma : \mathbb{R}^N \to C$. For $x = (x_1, \ldots, x_N)^T \in \mathbb{R}^N$, let $\Gamma(x) = (\Gamma_1(x_1), \ldots, \Gamma_N(x_N))^T$ be the closest point in C to x in terms of the Euclidean norm distance. We also define a matrix-valued projection operator $P : \mathbb{R}^{N\times N} \to \mathbb{R}^{N\times N}$ in a way that for any $N \times N$-matrix A, $P(A)$ is positive definite and symmetric. In particular, if A is positive definite and symmetric, then $P(A) = A$.

7.3.1 Four-Simulation Newton SPSA (N-SPSA4)

We first present the Hessian estimate for the N-SPSA4 algorithm. This will be followed by a description of the algorithm itself. The gradient estimate here is the same as that of regular (two-simulation) SPSA.

7.3.1.1 The Hessian Estimate

Let $\Delta_1, \ldots, \Delta_N, \hat{\Delta}_1, \ldots, \hat{\Delta}_N$ be independent, bounded, zero-mean, symmetric random variables having a common distribution and mutually independent of one another. Let $\Delta = (\Delta_1, \ldots, \Delta_N)^T$ and $\Delta^{-1} = (1/\Delta_1, \ldots, 1/\Delta_N)^T$. Also, let $\hat{\Delta} = (\hat{\Delta}_1, \ldots, \hat{\Delta}_N)^T$ and $\hat{\Delta}^{-1} = (1/\hat{\Delta}_1, \ldots, 1/\hat{\Delta}_N)^T$, respectively.

Then the four-simulation estimate of the Hessian of $J(\theta)$ with respect to θ is based on the following relationship.

$$\nabla^2_\theta J(\theta) = \lim_{\delta_1,\delta_2 \to 0} E\Delta^{-1} \left(\frac{\begin{array}{r} J(\theta + \delta_1\Delta + \delta_2\hat{\Delta}) - J(\theta + \delta_1\Delta) \\ -(J(\theta - \delta_1\Delta + \delta_2\hat{\Delta}) - J(\theta - \delta_1\Delta)) \end{array}}{2\delta_1\delta_2} \right) (\hat{\Delta}^{-1})^T. \tag{7.2}$$

The expectation above is taken w.r.t. the common distribution of Δ and $\hat{\Delta}$. The relationship in (7.2) will be used in the next section to construct a stochastic approximation algorithm. In [11], the form of the Hessian estimate is slightly different from (7.2). It has the form

$$\nabla^2_\theta J(\theta) = \lim_{\delta_1,\delta_2 \to 0} E\left[\frac{1}{2}\Delta^{-1}\left(\frac{\begin{array}{c}J(\theta+\delta_1\Delta+\delta_2\hat{\Delta})-J(\theta+\delta_1\Delta)\\ -(J(\theta-\delta_1\Delta+\delta_2\hat{\Delta})-J(\theta-\delta_1\Delta))\end{array}}{2\delta_1\delta_2}\right)\left(\hat{\Delta}^{-1}\right)^T\right]$$

$$+\lim_{\delta_1,\delta_2 \to 0} E\left[\frac{1}{2}\hat{\Delta}^{-1}\left(\frac{\begin{array}{c}J(\theta+\delta_1\Delta+\delta_2\hat{\Delta})-J(\theta+\delta_1\Delta)\\ -(J(\theta-\delta_1\Delta+\delta_2\hat{\Delta})-J(\theta-\delta_1\Delta))\end{array}}{2\delta_1\delta_2}\right)\left(\Delta^{-1}\right)^T\right].$$

In Section 7.5.2, it will be shown that the relationship (7.2) is a valid Hessian estimator. In all the algorithms below, $\Delta_1(n),\ldots,\Delta_N(n)$, $\hat{\Delta}_1(n),\ldots,\hat{\Delta}_N(n), n \geq 0$ will be treated to independent random variables requiring standard assumptions on such perturbations (see Assumption 7.6). Also, $\delta_1, \delta_2 > 0$ will be considered to be given small constants. (A discussion on how these constants can be chosen in practice can be found in [11].)

7.3.1.2 The N-SPSA4 Algorithm

Consider four simulated Markov processes $\{X^-(n)\}$, $\{X^+(n)\}$, $\{X^{-+}(n)\}$ and $\{X^{++}(n)\}$, that are governed by the parameter sequences $\{\theta(n)-\delta_1\Delta(n)\}$, $\{\theta(n)+\delta_1\Delta(n)\}$, $\{\theta(n)-\delta_1\Delta(n)+\delta_2\hat{\Delta}(n)\}$, and $\{\theta(n)+\delta_1\Delta(n)+\delta_2\hat{\Delta}(n)\}$, respectively. The aforementioned Markov processes are assumed to be independent of one another. Let $Z^w(n)$, $w \in \{-,+,-+,++\}$, be quantities defined via recursions (7.3)-(7.6) below that are used for averaging the cost function in the four simulations. We initialize $Z^w(0) = 0$, $\forall w \in \{-,+,-+,++\}$.

The algorithm is given as follows: For $n \geq 0$, $j,i = 1,\ldots,N$, $j \leq i$, $k = 1,\ldots,N$,

$$Z^-(n+1) = Z^-(n) + b(n)(h(X^-(n)) - Z^-(n)), \tag{7.3}$$

$$Z^+(n+1) = Z^+(n) + b(n)(h(X^+(n)) - Z^+(n)), \tag{7.4}$$

$$Z^{-+}(n+1) = Z^{-+}(n) + b(n)(h(X^{-+}(n)) - Z^{-+}(n)), \tag{7.5}$$

$$Z^{++}(n+1) = Z^{++}(n) + b(n)(h(X^{++}(n)) - Z^{++}(n)), \tag{7.6}$$

$$H_{j,i}(n+1) = H_{j,i}(n) + c(n)\left(\frac{\left(\frac{Z^{++}(n)-Z^+(n)}{\delta_2\hat{\Delta}_j(n)}\right)-\left(\frac{Z^{-+}(n)-Z^-(n)}{\delta_2\hat{\Delta}_j(n)}\right)}{2\delta_1\Delta_i(n)} - H_{j,i}(n)\right). \tag{7.7}$$

$$\theta_k(n+1) = \Gamma_k\left(\theta_k(n) + a(n)\sum_{l=1}^{N} M_{k,l}(n)\left(\frac{Z^-(n)-Z^+(n)}{2\delta_1\Delta_l(n)}\right)\right). \tag{7.8}$$

In the above, we set $H_{j,i}(n+1) = H_{i,j}(n+1)$ for $j > i$. Further, $M(n) = [[M_{i,j}(n)]]_{i,j=1}^{N}$ denotes the inverse of the matrix $H(n) = P([[H_{i,j}(n)]]_{i,j=1}^{N})$

Remark 7.1. In practice, an additional averaging over a certain prescribed number L of the faster iterates (7.3)-(7.6) in between two successive updates of the other recursions viz., (7.7) and (7.8), respectively, is seen to improve performance. The same is true for the other algorithms as well.

7.3.2 Three-Simulation Newton SPSA (N-SPSA3)

N-SPSA3 algorithm, which we will discuss now, is based on three simulations. The gradient estimate here is again the same as that of regular two-simulation SPSA. We explain below the Hessian estimate used.

7.3.2.1 The Hessian Estimate

The three-simulation estimate of the Hessian of $J(\theta)$ with respect to θ is based on the following relationship.

$$\nabla^2_\theta J(\theta) = \lim_{\delta_1,\delta_2 \to 0} E\left[\Delta^{-1}\left(\frac{J(\theta+\delta_1\Delta+\delta_2\hat{\Delta}) - J(\theta+\delta_1\Delta)}{\delta_1\delta_2}\right)(\hat{\Delta}^{-1})^T\right]. \quad (7.9)$$

The expectation above is taken w.r.t. the common distribution of Δ and $\hat{\Delta}$. The relation in (7.9) will be used in the next section to provide the complete algorithm. Also, later in Section 7.5.3, this relation will be proved.

7.3.2.2 The N-SPSA3 Algorithm

Consider three simulated Markov processes $\{X^-(n)\}$, $\{X^+(n)\}$ and $\{X^{++}(n)\}$, that are governed by the parameter sequences $\{\theta(n) - \delta_1\Delta(n)\}$, $\{\theta(n) + \delta_1\Delta(n)\}$, and $\{\theta(n) + \delta_1\Delta(n) + \delta_2\hat{\Delta}(n)\}$, respectively. The aforementioned Markov processes are assumed to be independent of one another. Let $Z^w(n)$, $w \in \{-,+,++\}$, be quantities defined via recursions (7.10)-(7.12) below that are used for averaging the cost function in the three simulations. We initialize $Z^w(0) = 0$, $\forall w \in \{-,+,++\}$.

The algorithm is given as follows: For $n \geq 0$, $j,i = 1,\ldots,N$, $j \leq i$, $k = 1,\ldots,N$,

$$Z^-(n+1) = Z^-(n) + b(n)(h(X^-(n)) - Z^-(n)), \tag{7.10}$$

$$Z^+(n+1) = Z^+(n) + b(n)(h(X^+(n)) - Z^+(n)), \tag{7.11}$$

$$Z^{++}(n+1) = Z^{++}(n) + b(n)(h(X^{++}(n)) - Z^{++}(n)), \tag{7.12}$$

$$H_{j,i}(n+1) = H_{j,i}(n) + c(n)\left(\frac{Z^{++}(n) - Z^+(n)}{\delta_1 \delta_2 \Delta_i(n)\hat{\Delta}_j(n)} - H_{j,i}(n)\right). \tag{7.13}$$

$$\theta_k(n+1) = \Gamma_k\left(\theta_k(n) + a(n)\sum_{l=1}^{N} M_{k,l}(n)\left(\frac{Z^-(n) - Z^+(n)}{2\delta_1\Delta_l(n)}\right)\right). \tag{7.14}$$

In the above, $M(n)$ denotes the inverse of the Hessian $H(n)$, constructed as in the N-SPSA4 algorithm (see Section 7.3.1.2).

7.3.3 *Two-Simulation Newton SPSA (N-SPSA2)*

Now we present the two-simulation Newton SPSA algorithm, N-SPSA2. While the Hessian estimate here is the same as for N-SPSA3, see Section 7.3.2, the gradient estimate is different from regular two-simulation SPSA. We therefore explain below the gradient estimate used.

7.3.3.1 The Gradient Estimate

The two-simulation estimate of the gradient of $J(\theta)$ with respect to θ is based on the following relationship.

$$\nabla_\theta J(\theta) = \lim_{\delta_2 \to 0} E\left[\left(\frac{J(\theta + \delta_1\Delta + \delta_2\hat{\Delta}) - J(\theta + \delta_1\Delta)}{\delta_2}\right)(\hat{\Delta}^{-1})^T\right]. \tag{7.15}$$

The expectation above is taken w.r.t. the common distribution of Δ and $\hat{\Delta}$. The relationship (7.15) will be proven later in Section 7.5.4.

7.3.3.2 The N-SPSA2 Algorithm

Consider two simulated Markov processes $\{X^+(n)\}$ and $\{X^{++}(n)\}$, that are governed by the parameter sequences $\{\theta(n) + \delta_1\Delta(n)\}$, and $\{\theta(n) + \delta_1\Delta(n) + \delta_2\hat{\Delta}(n)\}$, respectively. The aforementioned Markov processes are assumed to be independent of one another. Let $Z^w(n)$, $w \in \{+, ++\}$, be quantities defined via recursions (7.16)-(7.17) below that are used for averaging the cost function in the four simulations. We initialize $Z^w(0) = 0$, $\forall w \in \{+, ++\}$.

The algorithm is given as follows: For $n \geq 0$, $j,i = 1,\ldots,N$, $j \leq i$, $k = 1,\ldots,N$,

$$Z^{+}(n+1) = Z^{+}(n) + b(n)(h(X^{+}(n)) - Z^{+}(n)), \tag{7.16}$$

$$Z^{++}(n+1) = Z^{++}(n) + b(n)(h(X^{++}(n)) - Z^{++}(n)), \tag{7.17}$$

$$H_{j,i}(n+1) = H_{j,i}(n) + c(n)\left(\frac{Z^{++}(n) - Z^{+}(n)}{\delta_1\delta_2\Delta_i(n)\hat{\Delta}_j(n)} - H_{j,i}(n)\right). \tag{7.18}$$

$$\theta_k(n+1) = \Gamma_k\left(\theta_k(n) + a(n)\sum_{l=1}^{N} M_{k,l}(n)\left(\frac{Z^{+}(nL) - Z^{++}(nL)}{\delta_2\hat{\Delta}_l(n)}\right)\right). \tag{7.19}$$

In the above, $H(n)$ and $M(n)$ are as in the earlier N-SPSA algorithms.

7.3.4 One-Simulation Newton SPSA (N-SPSA1)

The N-SPSA1 algorithm which is based on just one simulation will be discussed now. Both the gradient and the Hessian estimates here are different from all the other algorithms. Hence, we first present these two estimates below.

7.3.4.1 The Gradient Estimate

The one-simulation estimate of the gradient of $J(\theta)$ with respect to θ is based on the following relationship.

$$\nabla_\theta J(\theta) = \lim_{\delta_2 \to 0} E\left[\left(\frac{J(\theta + \delta_1\Delta + \delta_2\hat{\Delta})}{\delta_2}\right)(\hat{\Delta}^{-1})^T\right]. \tag{7.20}$$

The expectation above is taken w.r.t. the common distribution of Δ and $\hat{\Delta}$.

7.3.4.2 The Hessian Estimate

The one-simulation estimate of the Hessian of $J(\theta)$ with respect to θ is based on the following relationship.

$$\nabla^2_\theta J(\theta) = \lim_{\delta_1,\delta_2 \to 0} E\left[\Delta^{-1}\left(\frac{J(\theta + \delta_1\Delta + \delta_2\hat{\Delta})}{\delta_1\delta_2}\right)(\hat{\Delta}^{-1})^T\right]. \tag{7.21}$$

The expectation above is taken w.r.t. the common distribution of Δ and $\hat{\Delta}$. Both the gradient estimate (7.20) and the Hessian estimate (7.21) will be validated later in Section 7.5.5.

7.3.4.3 The N-SPSA1 Algorithm

Consider a simulated Markov process $\{X^{++}(n)\}$ governed by the parameter sequence $\{\theta(n)+\delta_1\Delta(n)+\delta_2\hat{\Delta}(n)\}$. Let $Z^{++}(n)$ be quantities defined via the recursion (7.22) below that are used for averaging the cost function in the simulation. We initialize $Z^{++}(0)=0$.

The algorithm is given as follows: For $n \geq 0$, $j,i=1,\ldots,N$, $j \leq i$, $k=1,\ldots,N$,

$$Z^{++}(n+1) = Z^{++}(n)+b(n)(h(X^{++}(n))-Z^{++}(n)). \tag{7.22}$$

$$H_{j,i}(n+1) = H_{j,i}(n)+c(n)\left(\frac{Z^{++}(n)}{\delta_1\delta_2\Delta_i(n)\hat{\Delta}_j(n)}-H_{j,i}(n)\right). \tag{7.23}$$

$$\theta_k(n+1) = \Gamma_k\left(\theta_k(n)-a(n)\sum_{l=1}^{N}M_{k,l}(n)\left(\frac{Z^{++}(nL)}{\delta_2\hat{\Delta}_l(n)}\right)\right). \tag{7.24}$$

In the above, $H(n)$ and $M(n)$ are constructed as in any of the N-SPSA algorithms described previously.

Remark 7.2. An important question that arises is how to decide on the choice of the algorithm (out of the four algorithms described above) for a given problem. As can be seen, the gradient estimates in the N-SPSA4 and N-SPSA3 algorithms are the same, and so are the Hessian estimates in the N-SPSA3 and N-SPSA2 algorithms, respectively. The Hessian estimates for N-SPSA3 (and so also N-SPSA2) have a higher bias in comparison to the ones for N-SPSA4. It has however been observed in [2] that on a low-dimensional parameter setting, N-SPSA4 shows the best computational performance. However, on a high-dimensional setting, N-SPSA3 shows the best results on the whole. The algorithm N-SPSA1 has the largest bias as compared to the other algorithms and does not exhibit good performance in general. We believe that the choice of the algorithm could be guided by various factors, such as (a) the accuracy of the converged parameter value that will in general depend on the overall bias in the estimate as well as (b) the ease of implementation or the simplicity of the procedure. The latter consideration is in particular meaningful in the case when real-time computations have to be performed using resources with limited computational power. In such cases, accuracy of the estimate will have to be balanced against the computational resource available for performing the various computations. For instance, in the case of constrained optimization algorithms considered in Chapter 10, a one-sided version of N-SPSA2 is implemented where one of the simulations is run with the nominal parameter update $\theta(n)$ itself. This is guided by the fact that the said simulation also estimates the Lagrange multiplier, hence to avoid the use of a third simulation, the aforementioned one-sided version is suggested.

7.4 Woodbury's Identity Based Newton SPSA Algorithms

All four algorithms described in the previous section have a high computational complexity, as in each iteration, the estimate of the projected Hessian $H(\cdot)$ needs to be inverted. In this section, we address this problem using a popular identity that goes by the name of Woodbury's identity, for incrementally obtaining the inverse of the projected Hessian. For a matrix of order N, typical inversion techniques require $O(N^3)$ computations and some specialized ones are $O(N^2 log(N))$. We use the Woodbury's identity to provide an incremental inversion scheme here, which is of computational complexity $O(N^2)$. The identity is given below:

$$(A+BCD)^{-1} = A^{-1} - A^{-1}B\left(C^{-1} + DA^{-1}B\right)^{-1} DA^{-1},$$

for matrices A, B, C and D of appropriate dimensions. The iteration scheme for the Hessian matrix in all the four Newton SPSA algorithms can be revised to:

$$H(n+1) = (1-c(n))H(n) + R(n)Y(n)S(n),$$

where,

$$R(n) = \frac{1}{\delta_1}\left[\frac{1}{\Delta_1(n)}\frac{1}{\Delta_2(n)}\cdots\frac{1}{\Delta_{|A|\times|B|}(n)}\right]^T, S(n) = \frac{1}{\delta_2}\left[\frac{1}{\hat{\Delta}_1(n)}\frac{1}{\hat{\Delta}_2(n)}\cdots\frac{1}{\hat{\Delta}_{|A|\times|B|}(n)}\right],$$

and a scalar quantity,

$$Y(n) = \begin{cases} c(n)\left((Z^{++}(n) - Z^{+}(n)) - (Z^{-+}(n) - Z^{-}(n))\right) & \text{for N-SPSA4,} \\ c(n)\left(Z^{++}(n) - Z^{+}(n)\right) & \text{for N-SPSA3 and N-SPSA2,} \\ c(n)\left(Z^{++}(n)\right) & \text{for N-SPSA1.} \end{cases}$$

Now, Woodbury's identity applied to $M(n+1) = H(n+1)^{-1}$ gives

$$M(n+1) = \left(\frac{M(n)}{1-b(n)}\left[I - \frac{Y(n)R(n)S(n)M(n)}{1-c(n)+Y(n)S(n)M(n)R(n)}\right]\right). \quad (7.25)$$

This computationally cost-effective procedure for estimating the inverse can now be used to create four new algorithms, W-SPSA4, W-SPSA3, W-SPSA2 and W-SPSA1 with four, three, two and one simulations respectively. We provide the full algorithm W-SPSA4 below:

The W-SPSA4 Algorithm

$$Z^{-}(n+1) = Z^{-}(n) + b(n)(h(X^{-}(n)) - Z^{-}(n)), \tag{7.26}$$

$$Z^{+}(n+1) = Z^{+}(n) + b(n)(h(X^{+}(n)) - Z^{+}(n)), \tag{7.27}$$

$$Z^{-+}(n+1) = Z^{-+}(n) + b(n)(h(X^{-+}(n)) - Z^{-+}(n)), \tag{7.28}$$

$$Z^{++}(n+1) = Z^{++}(n) + b(n)(h(X^{++}(n)) - Z^{++}(n)), \tag{7.29}$$

$$Y(n) = c(n)\left((Z^{++}(n) - Z^{+}(n)) - (Z^{-+}(n) - Z^{-}(n))\right), \tag{7.30}$$

$$M(n+1) = P\left(\frac{M(n)}{1-b(n)}\left[I - \frac{Y(n)R(n)S(n)M(n)}{1-c(n)+Y(n)S(n)M(n)R(n)}\right]\right), \tag{7.31}$$

$$\theta_k(n+1) = \Gamma_k\left(\theta_k(n) + a(n)\sum_{l=1}^{N} M_{k,l}(n)\left(\frac{Z^{-}(n) - Z^{+}(n)}{2\delta_1\Delta_l(n)}\right)\right). \tag{7.32}$$

As previously with N-SPSA4, the operator $P(\cdot)$ ensures that updates to $M(n+1)$ are symmetric and positive definite. The algorithms W-SPSA3, W-SPSA2 and W-SPSA1 can similarly be derived using the appropriate $Y(n)$ and the update (7.25) in N-SPSA3, N-SPSA2 and N-SPSA1, respectively.

7.5 Convergence Analysis

7.5.1 Assumptions

We make the following assumptions for the analysis of N-SPSA algorithms:

Assumption 7.1. The single-stage cost function $h : \mathbb{R}^d \to \mathbb{R}$ is Lipschitz continuous.

Assumption 7.2. The long-run average cost $J(\theta)$ is twice continuously differentiable in θ with bounded third derivatives.

Assumptions 7.1 and 7.2 are standard requirements. For instance, Assumption 7.2 is a technical requirement that ensures that the Hessian of the objective exists and is used to push through suitable Taylor series arguments in the proof.

Let $\{\theta(n)\}$ be a sequence of random parameters obtained using (say) an iterative scheme on which the process $\{X(n)\}$ depends. Let $\mathscr{H}(n) = \sigma(\theta(m), X(m), m \leq n)$, $n \geq 1$ denote a sequence of associated σ-fields. We call $\{\theta(n)\}$ *non-anticipative* if for all Borel sets $A \subset \mathbb{R}^d$,

$$P(X(n+1) \in A \mid \mathscr{H}(n)) = p(\theta(n), X(n), A).$$

Under a non-anticipative $\{\theta(n)\}$, the process $\{X(n), \theta(n)\}$ is Markov. The sequences $\{\theta(n)\}$ resulting in the algorithms discussed in Section 7.3 and Section 7.4, are, for instance, non-anticipative. We shall assume the existence of a stochastic Lyapunov function (below).

Assumption 7.3. There exist $\varepsilon_0 > 0$, $K \subset \mathbb{R}^d$ compact and $V \in C(\mathbb{R}^d)$ such that $\lim_{\|x\| \to \infty} V(x) = \infty$ and under any non-anticipative $\{\theta(n)\}$,

1. $\sup_n E[V(X(n))^2] < \infty$ and
2. $E[V(X(n+1)) \mid \mathscr{H}(n)] \leq V(X(n)) - \varepsilon_0$, whenever $X(n) \notin K$, $n \geq 0$.

Assumption 7.3 is required to ensure that the system remains stable under a tunable parameter. It will not be required if the cost function $h(\cdot)$ is bounded. As before, we let $\|\cdot\|$ denote the Euclidean norm. Also, for any matrix $A \in \mathbb{R}^{N \times N}$, its norm is defined as the induced matrix norm, also denoted using $\|\cdot\|$ and defined according to $\|A\| = \max_{\{x \in \mathbb{R}^N \mid \|x\| = 1\}} \|Ax\|$.

Like any descent algorithms, the aim here is to find a local minimum. So, one needs to ensure that the Hessian estimate after each iterate is positive definite and symmetric. This is achieved by projecting the Hessian estimate to the space of positive definite and symmetric matrices using the operator P described before.

Assumption 7.4. (i) Let $A(n), B(n), n \geq 0$ be sequences of matrices in $\mathbb{R}^{N \times N}$ such that $\lim_{n \to \infty} \|A(n) - B(n)\| = 0$. Then $\lim_{n \to \infty} \|P(A(n)) - P(B(n))\| = 0$ as well.

(ii) Let $C(n), n \geq 0$ be a sequence of matrices in $\mathbb{R}^{N \times N}$, such that $\sup_n \|C(n)\| < \infty$. Then

$$\sup_n \|P(C(n))\|, \ \sup_n \|P(C(n))^{-1}\| < \infty,$$

as well.

Various operators described for instance via the *modified Choleski factorization procedure*, see [1], or the ones presented in [11] and [13], respectively, can be used for projecting the Hessian updates onto the space of positive definite and symmetric matrices. The continuity requirement in Assumption 7.4 on the operator P can be easily imposed in the *modified Choleski factorization procedure* and the operators in [11]. Also the procedure in [13] has been shown (there) to satisfy this requirement. In fact, since $\|A(n) - B(n)\| \to 0$ as $n \to \infty$, the eigenvalues of $A(n)$ and $B(n)$ asymptotically become equal, since they are themselves uniformly continuous functions of the elements of these matrices. A sufficient condition [1, pp.35] for the other requirements in Assumption 7.4 is that the eigenvalues of each projected Hessian update be both bounded above as well as away from zero. Thus for some scalars c_1, $c_2 > 0$ let

$$c_1\|z\|^2 \leq z^T P(C(n))z \leq c_2\|z\|^2, \ \ \forall z \in \mathbb{R}^N, \ \ n \geq 0. \tag{7.33}$$

Then all the eigenvalues of $P(C(n))$, $\forall n$, lie between c_1 and c_2. The above also ensures that the procedure does not get stuck at a non-stationary point. Now by [1, Propositions A.9 and A.15],

$$\sup_n \|P(C(n))\|, \ \sup_n \|\{P(C(n))\}^{-1}\| < \infty.$$

Most projection operators are seen to satisfy (7.33) either by explicitly projecting eigenvalues to the positive half line as with [13] or via (7.33) getting automatically enforced (such as in the *modified Choleski factorization* procedure). A more general condition than (7.33) is, however, given on [1, pp.36].

We show in Lemma 7.7 that $\sup_n \|H(n)\| < \infty$ w.p. 1, where $H(n)$ is the nth update of the Hessian. Assumption 7.4 is a technical requirement and is needed in the convergence analysis. All the algorithms discussed in this chapter, require (i) estimation of the the long-run average cost objective for various perturbed parameter sequences, (ii) obtain the aforementioned estimates of the long-run average cost, and finally, (iii) obtain a parameter update using the Hessian and gradient estimates. Assumption 7.5 provides conditions on the three step-size schedules, $a(n), b(n), c(n), n \geq 0$ for achieving the necessary timescale separations as discussed above.

Assumption 7.5. The step-sizes $a(n), b(n), c(n), n \geq 0$ satisfy the following requirements:

$$\sum_n a(n) = \sum_n b(n) = \sum_n c(n) = \infty, \tag{7.34}$$

$$\sum_n (a(n)^2 + b(n)^2 + c(n)^2) < \infty, \tag{7.35}$$

$$a(n) = o(c(n)) \text{ and } c(n) = o(b(n)). \tag{7.36}$$

Finally, the algorithms require $2N$ independent parameter perturbation random variables $\Delta_i(n)$, $\hat{\Delta}_i(n)$, $i = 1, \ldots, N$, at each parameter update step n.

Assumption 7.6. The random variables $\Delta_i(n)$, $\hat{\Delta}_i(n)$, $n \geq 0$, $i = 1, \ldots, N$, are mutually independent, mean-zero, have a common distribution and satisfy $E[(\Delta_i(n))^{-2}], E[(\hat{\Delta}_i(n))^{-2}] \leq \bar{K}$, for some $\bar{K} < \infty$.

7.5.2 Convergence Analysis of N-SPSA4

Let for any continuous function $v(\cdot) : C \to \mathbb{R}^N$,

$$\tilde{\Gamma}(v(y)) = \lim_{0<\eta\to 0} \left(\frac{\Gamma(y + \eta v(y)) - y}{\eta} \right). \tag{7.37}$$

Note that if $y \in C^o$ (the interior of C), then $\tilde{\Gamma}(v(y)) = v(y)$. However, if $y \in \partial C$ (the boundary of C) and $v(y)$ is such that $y + \eta v(y) \notin C$ for $\eta > 0$ how-so-ever small, then $\tilde{\Gamma}(v(y))$ is the projection of $v(y)$ to the boundary of C. Also, the limit in (7.37) is well defined as a consequence of the fact that C is compact and convex. If that is not the case, one may alternatively consider the set of all limit points of (7.37).

Now, let $\bar{M}(\theta) = P(\nabla^2 J(\theta))^{-1}$ denote the inverse of the *projected Hessian matrix* corresponding to parameter θ and let $\bar{M}_{k,l}(\theta)$ be its (k,l)'th element. Consider the following ODE:

$$\dot{\theta}(t) = \tilde{\Gamma}(-\bar{M}(\theta(t))\nabla J(\theta(t))). \tag{7.38}$$

Let

$$K \stackrel{\Delta}{=} \{\theta \in C \mid \nabla J(\theta)^T \tilde{\Gamma}(-\bar{M}(\theta)\nabla J(\theta)) = 0\}.$$

Further, given $\eta > 0$, let $K^\eta = \{\theta \in C \mid \|\theta - \theta_0\| \leq \eta,\ \theta_0 \in K\}$ be the set of all points that are within a distance η from the set K. Further, let $\hat{K} = \{\theta \in C \mid \tilde{\Gamma}(-\bar{M}(\theta)\nabla J(\theta)) = -\bar{M}(\theta)\nabla J(\theta)\}$. It is easy to see that $C^o \subseteq \hat{K}$, where C^o is the interior of C.

We first provide the main convergence result below.

Theorem 7.1. *Given $\eta > 0$, there exists $\hat{\delta} > 0$, such that for all $\delta_1, \delta_2 \in (0, \hat{\delta}]$, the algorithm (7.3)-(7.8) converges to K^η with probability one.*

The proof of Theorem 7.1 will be established by the following steps:

(i) Lemmas 7.2-7.5 and Corollary 7.6 are used to show that $Z^{(\cdot)}(n)$ sequences in (7.3)-(7.6), converge almost surely to the long-run cost objective $J(\cdot)$ with corresponding perturbed parameter sequences.
(ii) Lemmas 7.7-7.8 are used to show that the noise terms in the Hessian estimate are bounded and diminish to zero.
(iii) Proposition 7.9 and Lemma 7.10 collectively show that the Hessian estimate $H(n)$ indeed converges almost surely to the Hessian of the long-run average cost objective, $J(\cdot)$.
(iv) Lemma 7.11 shows that the inverse of the projected Hessian estimate also converges almost surely to the inverse of the projected Hessian of $J(\cdot)$.
(v) From this result and the convergence results from Section 5.2.3 related to the basic gradient SPSA algorithm 5.2.2, we conclude the proof for the main result. The formal proof of Theorem 7.1 is given at the end of this section.

Let $\mathscr{G}(l) = \sigma(\theta(p),\Delta(p),\hat{\Delta}(p),H_{j,i}(p),X^-(p),X^+(p),X^{-+}(p),X^{++}(p),p \leq l,$ $i,j = 1,\ldots,N)$, $l \geq 1$, denote σ-fields generated by the quantities above. Note that recursions (7.3)-(7.6) can be rewritten as

$$Z^w(p+1) = Z^w(p) + b(p)(h(X^w(p)) - Z^w(p)), \tag{7.39}$$

$w \in \{-,+,-+,++\}$.

Define sequences $\{M^w(p)\}$, $w \in \{-,+,-+,++\}$, as follows:

$$M^w(p) = \sum_{m=1}^{p} b(m)(h(X^w(m)) - E[h(X^w(m)) \mid \mathscr{G}(m-1)]).$$

Lemma 7.2. *The sequences* $\{M^w(p), \mathscr{G}(p)\}$, $w \in \{-,+,-+,++\}$ *are almost surely convergent martingale sequences.*

Proof. It is easy to see that $\{M^w(p), \mathscr{G}(p)\}$, $w \in \{-,+,-+,++\}$ are martingale sequences. Let $A^w(p), p \geq 0$, $w \in \{-,+,-+,++\}$ denote the quadratic variation processes associated with these martingale sequences. Thus,

$$A^w(p) = E\left[\sum_{m=1}^{p} (M^w(m+1) - M^w(m))^2 \mid \mathscr{G}(m-1)\right] + E\left[(M^w(0))^2\right]$$

$$= E\left[\sum_{m=1}^{p} b^2(m+1)\left(h(X^w(m+1)) - E[h(X^w(m+1))^2 \mid \mathscr{G}(m)]\right)\right] + E\left[(M^w(0))^2\right]$$

$$= \sum_{m=1}^{p} b^2(m+1)\left(E[h^2(X^w(m+1)) \mid \mathscr{G}(m)] - E^2[h(X^w(m+1)) \mid \mathscr{G}(m)]\right) + E\left[(M^w(0))^2\right].$$

Now observe that because $h(\cdot)$ is Lipschitz continuous (cf. Assumption 7.1),

$$|h(X^w(m))| - |h(0)| \leq |h(X^w(m)) - h(0)| \leq K\|X^w(m)\|,$$

where $K > 0$ is the Lipschitz constant. Thus,

$$|h(X^w(m))| \leq C_1(1 + \|X^w(m)\|)$$

for $C_1 = \max(K, |h(0)|) < \infty$. By Assumption 7.3, it follows that $\sup_m E[\|X^w(m)\|^2] < \infty$. Now from Assumption 7.5 (cf. equation (7.35)), it follows that $A^w(p) \to A^w(\infty)$ as $p \to \infty$ almost surely, $\forall w \in \{-,+,-+,++\}$. The claim now follows from the martingale convergence theorem (Theorem B.2). □

Lemma 7.3. *The recursions (7.3)-(7.6) are uniformly bounded with probability one.*

Proof. Recall that the recursions (7.3)-(7.6) are analogously written as (7.39). The latter can be rewritten as

$$Z^w(p+1) = Z^w(p) + b(p)(E[h(X^w(p)) \mid \mathcal{G}(p-1)] - Z^w(p)) + M^w(p) - M^w(p-1), \tag{7.40}$$

$w \in \{-,+,-+,++\}$. Now as a consequence of Lemma 7.2, it is sufficient to show the boundedness of the following recursion:

$$Z^w(p+1) = Z^w(p) + b(p)(E[h(X^w(p)) \mid \mathcal{G}(p-1)] - Z^w(p)). \tag{7.41}$$

As in Lemma 7.2, it can be shown that

$$|h(X^w(m))| \leq C_1(1 + \|X^w(m)\|),$$

for a constant $C_1 > 0$. It again follows from Assumption 7.3 that $\sup_m E[\|X^w(m)\| \mid \mathcal{G}(m-1)] < \infty$ almost surely. The rest now follows easily from the Borkar and Meyn stability theorem (Theorem D.1). □

Now, define a sequence of time points $\{s(n), n \geq 0\}$ as follows: $s(0) = 0$, $s(n) = \sum_{i=0}^{n-1} a(i)$, $n \geq 1$. For $i = 1, \ldots, N$, let $\Delta_i(t) = \Delta_i(n)$ and $\hat{\Delta}_i(t) = \hat{\Delta}_i(n)$ for $t \in [s(n), s(n+1)]$, $n \geq 0$. Further let $\Delta(t) = (\Delta_1(t), \ldots, \Delta_N(t))^T$ and $\hat{\Delta}(t) = (\hat{\Delta}_1(t), \ldots, \hat{\Delta}_N(t))^T$, respectively. Also, define another sequence of time points $\{t(n)\}$ as follows: $t(0) = 0$, $t(n) = \sum_{i=0}^{n-1} b(i)$, $n \geq 1$. Consider the following system of ODEs: For $i, j \in \{1, \ldots, N\}$, $w \in \{-,+,-+,++\}$,

$$\dot{\theta}_i(t) = 0, \tag{7.42}$$

$$\dot{H}_{j,i}(t) = 0, \tag{7.43}$$

$$\dot{Z}^w(t) = J(\theta^w(t)) - Z^w(t). \tag{7.44}$$

Here and in the rest of the chapter, the following notation for the parameters is used: $\theta^-(t) = (\theta(t) - \delta_1\Delta(t))$, $\theta^+(t) = (\theta(t) + \delta_1\Delta(t))$, $\theta^{-+}(t) = (\theta(t) - \delta_1\Delta(t) + \delta_2\hat{\Delta}(t))$ and $\theta^{++}(t) = (\theta(t) + \delta_1\Delta(t) + \delta_2\hat{\Delta}(t))$, respectively.

Define now functions $\bar{z}^w(t)$, $w \in \{-,+,-+,++\}$ according to $\bar{z}^w(t(n)) = Z^w(n)$ with the maps $t \to \bar{z}^w(t)$ corresponding to continuous linear interpolations on intervals $[t(n),t(n+1)]$. Given $T>0$, define $\{T(n)\}$ as follows: $T(0)=0$ and for $n \geq 1$, $T(n) = \min\{t(m) \mid t(m) \geq T(n-1)+T\}$. Let $I(n) = [T(n),T(n+1)]$. Note that there exists some integer $m(n)>0$ such that $T(n)=t(m(n))$. Define also functions $z^{w,n}(t)$, $w \in \{-,+,-+,++\}, t \in I(n)$, $n \geq 0$, according to

$$\dot{z}^{w,n}(t) = J(\theta^w(t)) - z^{w,n}(t), \tag{7.45}$$

with $z^{w,n}(T(n)) = \bar{z}^w(t(m(n))) = Z^w(m(n))$.

Lemma 7.4. $\lim_{n\to\infty} \sup_{t\in I(n)} \|z^{w,n}(t) - \bar{z}^w(t)\| = 0 \; \forall w \in \{-,+,-+,++\}$, *w.p.1.*

Proof. The proof follows as in [3, Chapter 2, Lemma 1] (see Proposition 3.2). □

Lemma 7.5. *Given* $T,\varepsilon>0$, $((\theta_i(t(n)+\cdot),\ H_{j,i}(t(n)+\cdot),\ \bar{z}^w(t(n)+\cdot)),\ i,j \in \{1,\ldots,N\},\ w \in \{-,+,-+,++\}$, *is a bounded* (T,ε)*-perturbation of (7.42)-(7.44) for n sufficiently large.*

Proof. Observe that the iterations (7.7)-(7.8) of the algorithm can be written as

$$H_{j,i}(n+1) = H_{j,i}(n) + b(n)\xi_1(n),$$

$$\theta_i(n+1) = \Gamma_i(\theta_i(n) + b(n)\xi_2(n)),$$

respectively, where $\xi_1(n)$ and $\xi_2(n)$ are both $o(1)$ as a consequence of Assumption 7.5 (cf. equation (7.36)). The claim now follows from Lemma 7.4. □

Corollary 7.6. *For all* $w \in \{-,+,-+,++\}$,

$$\|Z^w(n) - J(\theta^w(n))\| \to 0 \;\; a.s.,$$

as $n \to \infty$.

Proof. The claim follows from the Hirsch lemma (Lemma C.5) applied on (7.44) for every $\varepsilon>0$. □

We now look at the recursion (7.7).

Lemma 7.7. *The iterates* $H_{j,i}(n)$, $n \geq 0$, $j,i \in \{1,\ldots,N\}$, *in (7.7), are uniformly bounded with probability one.*

Proof. The result follows as a consequence of Lemma 7.3 using the Borkar-Meyn theorem (Theorem D.1). □

The next step is to replace $Z^w(n)$ by $J(\theta^w(n))$ in the update of (7.7), in lieu of Corollary 7.6. Thus, let

$$F_{j,i}(\theta(n),\Delta(n),\hat{\Delta}(n)) \stackrel{\Delta}{=} \frac{\left(\frac{J(\theta^{++}(n))-J(\theta^{+}(n))}{\delta_2\hat{\Delta}_j(n)}\right)-\left(\frac{J(\theta^{-+}(n))-J(\theta^{-}(n))}{\delta_2\hat{\Delta}_j(n)}\right)}{2\delta_1\Delta_i(n)}.$$

Also let $\mathscr{F}(n) = \sigma(\theta_i(m), H_{j,i}(m), Z^-(mL), Z^+(mL), Z^{-+}(mL), Z^{++}(mL), m \leq n, i,j = 1,\ldots,N;\ \Delta(m), \hat{\Delta}(m), m < n), n \geq 1$. Let $\{N_{j,i}(n)\}$, $j,i = 1,\ldots,N$ be defined according to

$$N_{j,i}(n) = \sum_{m=0}^{n-1} c(m)(F_{j,i}(\theta(m),\Delta(m),\hat{\Delta}(m)) - E[F_{j,i}(\theta(m),\Delta(m),\hat{\Delta}(m)) \mid \mathscr{F}(m)]).$$

Lemma 7.8. *The sequences $\{N_{j,i}(n),\mathscr{F}(n)\}$, $j,i = 1,\ldots,N$ form almost surely convergent martingale sequences.*

Proof. The claim follows in a similar manner as Lemma 7.2. □

Proposition 7.9. *With probability one, $\forall j,i \in \{1,\ldots,N\}$,*

$$\left| E\left[\frac{\left(\frac{J(\theta^{++}(n))-J(\theta^{+}(n))}{\delta_2\hat{\Delta}_j(n)}\right)-\left(\frac{J(\theta^{-+}(n))-J(\theta^{-}(n))}{\delta_2\hat{\Delta}_j(n)}\right)}{2\delta_1\Delta_i(n)} \middle| \mathscr{F}(n) \right] - \nabla^2_{j,i}J(\theta(n)) \right| \to 0$$

as $\delta_1,\delta_2 \to 0$.

Proof. The proof proceeds using several Taylor series expansions to evaluate the conditional expectation above. Note that

$$J(\theta(n)+\delta_1\Delta(n)+\delta_2\hat{\Delta}(n)) = J(\theta(n)+\delta_1\Delta(n)) + \delta_2\sum_{k=1}^{N}\hat{\Delta}_k(n)\nabla_k J(\theta(n)+\delta_1\Delta(n))$$

$$+\frac{1}{2}\delta_2^2\sum_{k=1}^{N}\sum_{l=1}^{N}\hat{\Delta}_k(n)\nabla^2_{k,l}J(\theta(n)+\delta_1\Delta(n))\hat{\Delta}_l(n) + o(\delta_2^2).$$

Similarly,

$$J(\theta(n)-\delta_1\Delta(n)+\delta_2\hat{\Delta}(n)) = J(\theta(n)-\delta_1\Delta(n)) + \delta_2\sum_{k=1}^{N}\hat{\Delta}_k(n)\nabla_k J(\theta(n)-\delta_1\Delta(n))$$

$$+\frac{1}{2}\delta_2^2\sum_{k=1}^{N}\sum_{l=1}^{N}\hat{\Delta}_k(n)\nabla^2_{k,l}J(\theta(n)-\delta_1\Delta(n))\hat{\Delta}_l(n) + o(\delta_2^2).$$

It can now be seen that

$$E\left[\left(\left(\frac{J(\theta(n)+\delta_1\Delta(n)+\delta_2\hat{\Delta}(n))-J(\theta(n)+\delta_1\Delta(n))}{\delta_2\hat{\Delta}_j(n)}\right)\right.\right.$$

$$-\left(\frac{J(\theta(n)-\delta_1\Delta(n)+\delta_2\hat{\Delta}(n))-J(\theta(n)-\delta_1\Delta(n))}{\delta_2\hat{\Delta}_j(n)}\right)\Bigg)/2\delta_1\Delta_i(n) \mid \mathscr{F}(n)\Bigg]$$

$$= E\Bigg[\frac{\nabla_j J(\theta(n)+\delta_1\Delta(n))-\nabla_j J(\theta(n)-\delta_1\Delta(n))}{2\delta_1\Delta_i(n)}$$

$$+\sum_{k\neq j}\frac{\hat{\Delta}_k(n)}{\hat{\Delta}_j(n)}\frac{\nabla_k J(\theta(n)+\delta_1\Delta(n))-\nabla_k J(\theta(n)-\delta_1\Delta(n))}{2\delta_1\Delta_i(n)}$$

$$+\delta_2\sum_{k=1}^{N}\sum_{l=1}^{N}\frac{\hat{\Delta}_k(n)(\nabla^2_{k,l}J(\theta(n)+\delta_1\Delta(n))-\nabla^2_{k,l}J(\theta(n)-\delta_1\Delta(n)))\hat{\Delta}_l(n)}{4\delta_1\Delta_i(n)\hat{\Delta}_j(n)}$$

$$+o(\delta_2) \mid \mathscr{F}(n)\Bigg] \tag{7.46}$$

Now using Taylor series expansions of $\nabla_j J(\theta(n)+\delta_1\Delta(n))$ and $\nabla_j J(\theta(n)-\delta_1\Delta(n))$ around $\nabla_j J(\theta(n))$ gives

$$\frac{\nabla_j J(\theta(n)+\delta_1\Delta(n))-\nabla_j J(\theta(n)-\delta_1\Delta(n))}{2\delta_1\Delta_i(n)} = \nabla^2_{j,i}J(\theta(n))$$

$$+\sum_{l\neq i}\frac{\Delta_l(n)}{\Delta_i(n)}\nabla^2_{j,l}J(\theta(n))+o(\delta_1^2).$$

A similar expansion can be obtained with index k in place of j in the second term on the RHS of (7.46). Also note that

$$\frac{\nabla^2_{k,l}J(\theta(n)+\delta_1\Delta(n))-\nabla^2_{k,l}J(\theta(n)-\delta_1\Delta(n))}{4\delta_1\Delta_i(n)} = \sum_{m=1}^{N}\frac{\Delta(m)(n)\nabla^3_{k,l,m}J(\theta(n))}{2\Delta_i(n)}+o(\delta_1)$$

Thus,

$$\delta_2\sum_{k=1}^{N}\sum_{l=1}^{N}\frac{\hat{\Delta}_k(n)(\nabla^2_{k,l}J(\theta(n)+\delta_1\Delta(n))-\nabla^2_{k,l}J(\theta(n)-\delta_1\Delta(n)))\hat{\Delta}_l(n)}{4\delta_1\Delta_i(n)\hat{\Delta}_j(n)}$$

$$=\delta_2\sum_{k=1}^{N}\sum_{l=1}^{N}\sum_{m=1}^{N}\frac{\hat{\Delta}_k(n)\Delta(m)(n)\nabla^3_{k,l,m}J(\theta(n))\hat{\Delta}_l(n)}{2\hat{\Delta}_j(n)\Delta_i(n)}+o(\delta_1).$$

Substituting the above in (7.46), one obtains

$$E\Bigg[\Bigg(\left(\frac{J(\theta(n)+\delta_1\Delta(n)+\delta_2\hat{\Delta}(n))-J(\theta(n)+\delta_1\Delta(n))}{2\delta_1\Delta_i(n)\delta_2\hat{\Delta}_j(n)}\right)$$

$$-\left(\frac{J(\theta(n)-\delta_1\Delta(n)+\delta_2\hat{\Delta}(n))-J(\theta(n)-\delta_1\Delta(n))}{2\delta_1\Delta_i(n)\delta_2\hat{\Delta}_j(n)}\right)\Bigg) \Bigg| \mathscr{F}(n)\Bigg]$$

$$= E\left[\nabla^2_{j,i}J(\theta(n)) + \sum_{l\neq i}\frac{\Delta_l(n)}{\Delta_i(n)}\nabla^2_{j,l}J(\theta(n)) + \sum_{k\neq j}\frac{\hat{\Delta}_k(n)}{\hat{\Delta}_j(n)}\nabla^2_{k,i}J(\theta(n))\right.$$

$$+\sum_{k\neq j}\sum_{l\neq i}\frac{\hat{\Delta}_k(n)}{\hat{\Delta}_j(n)}\frac{\Delta_l(n)}{\Delta_i(n)}\nabla^2_{k,l}J(\theta(n)) + \delta_2\sum_{k,l,m=1}^{N}\frac{\hat{\Delta}_k(n)\Delta(m)(n)\nabla^3_{k,l,m}J(\theta(n))\hat{\Delta}_l(n)}{2\hat{\Delta}_j(n)\Delta_i(n)}$$

$$\left. +o(\delta_1)+o(\delta_2) \mid \mathscr{F}(n)\right]$$

$$= \nabla^2_{j,i}J(\theta(n)) + \sum_{l\neq i}E\left[\frac{\Delta_l(n)}{\Delta_i(n)} \mid \mathscr{F}(n)\right]\nabla^2_{j,l}J(\theta(n)) + \sum_{k\neq j}E\left[\frac{\hat{\Delta}_k(n)}{\hat{\Delta}_j(n)} \mid \mathscr{F}(n)\right]\nabla^2_{k,i}J(\theta(n))$$

$$+\sum_{k\neq j}\sum_{l\neq i}E\left[\frac{\hat{\Delta}_k(n)}{\hat{\Delta}_j(n)}\frac{\Delta_l(n)}{\Delta_i(n)} \mid \mathscr{F}(n)\right]\nabla^2_{k,l}J(\theta(n))$$

$$+\delta_2\sum_{k=1}^{N}\sum_{l=1}^{N}\sum_{m=1}^{N}E\left[\frac{\hat{\Delta}_k(n)\hat{\Delta}_l(n)\Delta(m)(n)}{2\hat{\Delta}_j(n)\Delta_i(n)} \mid \mathscr{F}(n)\right]\nabla^3_{k,l,m}J(\theta(n))+o(\delta_1)+o(\delta_2).$$

Now, by Assumption 7.6, it is easy to see that all conditional expectations in the last equality above equal zero. Thus

$$E\left[\left(\left(\frac{J(\theta(n)+\delta_1\Delta(n)+\delta_2\hat{\Delta}(n))-J(\theta(n)+\delta_1\Delta(n))}{2\delta_1\Delta_i(n)\delta_2\hat{\Delta}_j(n)}\right)\right.\right.$$

$$\left.\left.-\left(\frac{J(\theta(n)-\delta_1\Delta(n)+\delta_2\hat{\Delta}(n))-J(\theta(n)-\delta_1\Delta(n))}{2\delta_1\Delta_i(n)\delta_2\hat{\Delta}_j(n)}\right)\right) \mid \mathscr{F}(n)\right]$$

$$= \nabla^2_{j,i}J(\theta(n))+o(\delta_1)+o(\delta_2).$$

The claim follows. □

Consider now the following ODEs: For $j,i = 1,\ldots,N$,

$$\left.\begin{array}{l}\dot{H}_{j,i}(t) = \nabla^2_{j,i}J(\theta(t)) - H_{j,i}(t),\\ \dot{\theta}_i(t) = 0.\end{array}\right\} \tag{7.47}$$

Next define $\{r(n)\}$ as follows: $r(0) = 0$ and for $n > 0$, $r(n) = \sum_{m=0}^{n-1} c(m)$. Define $\bar{H}(t) = [[\bar{H}_{j,i}(t)]]_{j,i=1}^{N}$ and $\bar{x}^w(t)$, $w \in \{-,+,-+,++\}$, as follows: For $j,i = 1,\ldots,N$, $\bar{H}_{j,i}(r(n)) = H_{j,i}(n)$ and $\bar{x}^w(r(n)) = Z^w(nL)$ with linear interpolations on $[r(n),r(n+1)]$.

The following can now be shown in the same way as Corollary 7.6.

Lemma 7.10. $\|H(n) - \nabla^2 J(\theta(n))\| \to 0$ *a.s. as* $\delta_1,\delta_2 \to 0$ *and* $n \to \infty$,

where $H(n) = [[H_{j,i}(n)]]_{j,i=1}^{N}$.

Lemma 7.11. *With probability one,* $\|P(H(n))^{-1} - P(\nabla^2 J(\theta(n)))^{-1}\| \to 0$ *as* δ_1, $\delta_2 \to 0$ *and* $n \to \infty$.

Proof. Note that

$$\begin{aligned}
&\|P(H(n))^{-1} - P(\nabla^2 J(\theta(n)))^{-1}\| \\
&= \|P(\nabla^2 J(\theta(n)))^{-1}(P(\nabla^2 J(\theta(n)))P(H(n))^{-1} - I)\| \\
&= \|P(\nabla^2 J(\theta(n)))^{-1}(P(\nabla^2 J(\theta(n)))P(H(n))^{-1} - P(H(n))P(H(n))^{-1})\| \\
&= \|P(\nabla^2 J(\theta(n)))^{-1}(P(\nabla^2 J(\theta(n))) - P(H(n)))P(H(n))^{-1}\| \\
&\leq \|P(\nabla^2 J(\theta(n)))^{-1}\| \cdot \|P(\nabla^2 J(\theta(n))) - P(H(n))\| \cdot \|P(H(n))^{-1}\| \\
&\leq \sup_n \|P(\nabla^2 J(\theta(n)))^{-1}\| \sup_n \|P(H(n))^{-1}\| \cdot \|P(\nabla^2 J(\theta(n))) - P(H(n))\| \\
&\to 0 \text{ as } n \to \infty,
\end{aligned}$$

by Assumption 7.4. In the above, I denotes the $N \times N$-identity matrix. The claim follows. □

Proof of Theorem 7.1. For $i = 1, \ldots, N$, let $\{R_i(n), n \geq 1\}$ be defined according to

$$R_i(n) = \sum_{m=0}^{n-1} a(m) \sum_{k=1}^{N} \bar{M}_{i,k}(\theta(m)) \left(\frac{J(\theta(m) - \delta_1 \Delta(m)) - J(\theta(m) + \delta_1 \Delta(m))}{2\delta_1 \Delta_k(m)} - E\left[\frac{J(\theta(m) - \delta_1 \Delta(m)) - J(\theta(m) + \delta_1 \Delta(m))}{2\delta_1 \Delta_k(m)} \mid \mathscr{F}(m) \right] \right),$$

$n \geq 1$. Then it is easy to see that $\{R_i(n), \mathscr{F}(n)\}$, $i = 1, \ldots, N$, are almost surely convergent martingale sequences. Now recursion (7.8) of the algorithm can be rewritten as

$$\theta_i(n+1) = \Gamma_i(\theta_i(n) + a(n) \sum_{k=1}^{N} \bar{M}_{i,k}(\theta(n))(E[(J(\theta(n) - \delta_1 \Delta(n)) - J(\theta(n) + \delta_1 \Delta(n)))/2\delta_1 \Delta_k(n) \mid \mathscr{F}(n)] + (R_i(n+1) - R_i(n)) + a(n)\alpha(n)), \tag{7.48}$$

where $(R_i(n+1) - R_i(n))$ is $o(1)$ by the above and $\alpha(n)$ vanishes as $n \to \infty$ and $\delta_1, \delta_2 \to 0$ by Corollary 7.6 and Lemma 7.11.

Using Taylor series expansions of $J(\theta(n) - \delta_1 \Delta(n))$ and $J(\theta(n) + \delta_1 \Delta(n))$, respectively, around $\theta(n)$ and taking the conditional expectation above, it is easy to see that recursion (7.8) can be rewritten as

$$\theta_i(n+1) = \Gamma_i(\theta_i(n) - a(n) \sum_{k=1}^{N} \bar{M}_{i,k}(\theta(n)) \nabla_k J(\theta(n)) + a(n)\xi_{\delta_1}(n) + (R_i(n+1) - R_i(n)) + a(n)\alpha(n)), \tag{7.49}$$

where $\xi_{\delta_1}(n)$ vanishes as $n \to \infty$ and $\delta_1 \to 0$. Note that (7.49) can be viewed, using a standard approximation argument as in [5, pp.191-196] and Proposition 7.9, as a discretization of the ODE (7.38) with certain error terms that, however, vanish asymptotically (as $n \to \infty$) and in the limit as $\delta_1, \delta_2 \to 0$. Now $J(\theta)$ itself serves as an associated Lyapunov function for the ODE (7.38). The claim now follows from Lasalle's invariance theorem [7]. □

Remark 7.3. Note that for $\theta \in \hat{K} \cap K$, $\nabla J(\theta) = 0$ by positive definiteness of $\bar{M}(\theta)$. Further, on the set $K \backslash \hat{K}$, if $\nabla J(\theta) \neq 0$, one has $\tilde{\Gamma}_i(-(\bar{M}(\theta)\nabla J(\theta))_i) = 0$ for all those i $(i = 1, \ldots, N)$ for which $\nabla_i J(\theta) \neq 0$. (Here $-(\bar{M}(\theta)\nabla J(\theta))_i$ corresponds to the ith component of the vector $(\bar{M}(\theta)\nabla J(\theta))$.) The latter correspond to spurious fixed points that, however, can occur only on the projection set boundaries (since $C^o \subseteq \hat{K}$) [6, pp. 79].

7.5.3 Convergence Analysis of N-SPSA3

The analysis proceeds in exactly the same way as for N-SPSA4. Note, however, that the form of the Hessian estimator here is different. Hence, we show that the Hessian estimator is consistent.

Let $\mathscr{F}_1(n) \stackrel{\Delta}{=} \sigma(\theta_i(m), H_{j,i}(m), Z^-(mL), Z^+(m), Z^{++}(m), m \leq n, i, j = 1, \ldots, N;$ $\Delta(m), \hat{\Delta}(m), m < n), n \geq 1$ be a sequence of sigma fields.

Proposition 7.12. *With probability one,* $\forall j, i \in \{1, \ldots, N\}$

$$\left| E\left[\frac{J(\theta(n) + \delta_1\Delta(n) + \delta_2\hat{\Delta}(n)) - J(\theta(n) + \delta_1\Delta(n))}{\delta_1\delta_2\Delta_i(n)\hat{\Delta}_j(n)} \mid \mathscr{F}_1(n)\right] - \nabla^2_{j,i}J(\theta(n))\right|$$

$$\longrightarrow 0 \text{ as } \delta_1, \delta_2 \to 0.$$

Proof. Note as before that

$$\frac{J(\theta(n) + \delta_1\Delta(n) + \delta_2\hat{\Delta}(n)) - J(\theta(n) + \delta_1\Delta(n))}{\delta_1\delta_2\Delta_i(n)\hat{\Delta}_j(n)} = \sum_{l=1}^{N}\frac{\hat{\Delta}_l(n)\nabla_l J(\theta(n) + \delta_1\Delta(n))}{\delta_1\Delta_i(n)\hat{\Delta}_j(n)}$$

$$+ \frac{\delta_2}{2}\sum_{l=1}^{N}\sum_{m=1}^{N}\frac{\hat{\Delta}_l(n)\hat{\Delta}(m)(n)\nabla^2_{l,m}J(\theta(n) + \delta_1\Delta(n))}{\delta_1\Delta_i(n)\hat{\Delta}_j(n)} + o(\delta_2) \qquad (7.50)$$

Taking again appropriate Taylor series expansions of $\nabla_l J(\theta(n) + \delta_1\Delta(n))$ and $\nabla^2_{l,m}J(\theta(n) + \delta_1\Delta(n))$ around $\theta(n)$, substituting in (7.50), taking the conditional expectation w.r.t. $\mathscr{F}_1(n)$, one obtains

$$E\left[\frac{J(\theta(n) + \delta_1\Delta(n) + \delta_2\hat{\Delta}(n)) - J(\theta(n) + \delta_1\Delta(n))}{\delta_1\delta_2\Delta_i(n)\hat{\Delta}_j(n)} \mid \mathscr{F}_1(n)\right]$$

$$=\sum_{l=1}^{N}\frac{1}{\delta_1}E\left[\frac{\hat{\Delta}_l(n)}{\Delta_i(n)\hat{\Delta}_j(n)} \mid \mathscr{F}_1(n)\right]\nabla_l J(\theta(n))+\nabla^2_{j,i}J(\theta(n))$$

$$+\sum_{l=1,l\neq j}^{N}\sum_{k=1,k\neq i}^{N} E\left[\frac{\hat{\Delta}_l(n)\Delta_k(n)}{\hat{\Delta}_j(n)\Delta_i(n)} \mid \mathscr{F}_1(n)\right]\nabla^2_{l,k}J(\theta(n))$$

$$+\frac{\delta_1}{2}\sum_{l=1}^{N}\sum_{k=1}^{N}\sum_{m=1}^{N} E\left[\frac{\hat{\Delta}_l(n)\Delta_k(n)\Delta(m)(n)}{\Delta_i(n)\hat{\Delta}_j(n)} \mid \mathscr{F}_1(n)\right]\nabla^3_{k,m,l}J(\theta(n))$$

$$+\frac{\delta_2}{2\delta_1}\sum_{l=1}^{N}\sum_{m=1}^{N} E\left[\frac{\hat{\Delta}_l(n)\hat{\Delta}(m)(n)}{\Delta_i(n)\hat{\Delta}_j(n)} \mid \mathscr{F}_1(n)\right]\nabla^2_{l,m}J(\theta(n))$$

$$+\frac{\delta_2}{2}\sum_{l=1}^{N}\sum_{m=1}^{N}\sum_{k=1}^{N} E\left[\frac{\hat{\Delta}_l(n)\hat{\Delta}(m)(n)\Delta_k(n)}{\Delta_i(n)\hat{\Delta}_j(n)} \mid \mathscr{F}_1(n)\right]\nabla^3_{l,m,k}J(\theta(n))+o(\delta_1)+o(\delta_2).$$

From Assumption 7.6, all the conditional expectation terms on the RHS above equal zero. Thus

$$E\left[\frac{J(\theta(n)+\delta_1\Delta(n)+\delta_2\hat{\Delta}(n))-J(\theta(n)+\delta_1\Delta(n))}{\delta_1\delta_2\Delta_i(n)\hat{\Delta}_j(n)} \mid \mathscr{F}_1(n)\right]$$

$$=\nabla^2_{j,i}J(\theta(n))+o(\delta_1)+o(\delta_2).$$

The claim follows. □

Theorem 7.13. *Given $\eta>0$, there exists $\hat{\delta}>0$ such that for all $\delta_1,\delta_2\in(0,\hat{\delta}]$, the algorithm (7.10)-(7.14) converges to the set K^η with probability one.*

Proof. Follows in the same manner as Theorem 7.1. □

7.5.4 Convergence Analysis of N-SPSA2

As with N-SPSA3, the analysis proceeds along similar lines as for N-SPSA4. Let $\mathscr{F}_2(n) \stackrel{\Delta}{=} \sigma(\theta_i(m), H_{j,i}(m), Z^+(m), Z^{++}(m), m\leq n, i,j=1,\ldots,N; \Delta(m), \hat{\Delta}(m), m<n), n\geq 1$ denote a sequence of sigma fields. Since the form of the Hessian estimate here is the same as in N-SPSA3, the conclusions of Proposition 7.12 continue to hold with $\mathscr{F}_2(n)$ in place of $\mathscr{F}_1(n)$. Note, however, that the form of the gradient estimate here is different. We have the following result for the gradient estimate.

Proposition 7.14. *For all $k=1,\ldots,N$,*

$$\lim_{\delta_1,\delta_2\to 0} \|E\left[\frac{J(\theta(n)+\delta_1\Delta(n)) - J(\theta(n)+\delta_1\Delta(n)+\delta_2\hat{\Delta}(n))}{\delta_2\hat{\Delta}_k(n)} \mid \mathscr{F}_2(n)\right] + \nabla_k J(\theta(n))\| = 0,$$

with probability one.

Proof. Note that

$$\frac{J(\theta(n)+\delta_1\Delta(n)) - J(\theta(n)+\delta_1\Delta(n)+\delta_2\hat{\Delta}(n))}{\delta_2\hat{\Delta}_k(n)} = -\sum_{l=1}^{N}\frac{\hat{\Delta}_l(n)}{\hat{\Delta}_k(n)}\nabla_l J(\theta(n)+\delta_1\Delta(n))$$

$$-\frac{\delta_2}{2}\sum_{l=1}^{N}\sum_{j=1}^{N}\frac{\hat{\Delta}_l(n)}{\hat{\Delta}_k(n)}\hat{\Delta}_j(n)\nabla^2_{l,j}J(\theta(n)+\delta_1\Delta(n)) + o(\delta_2). \qquad (7.51)$$

Again

$$\nabla_l J(\theta(n)+\delta_1\Delta(n)) = \nabla_l J(\theta(n)) + \delta_1\sum_{j=1}^{N}\Delta_j(n)\nabla^2_{l,j}J(\theta(n)) + o(\delta_1),$$

$$\nabla^2_{l,j}J(\theta(n)+\delta_1\Delta(n)) = \nabla^2_{l,j}J(\theta(n)) + \delta_1\sum_{m=1}^{N}\Delta(m)(n)\nabla^3_{l,j,m}J(\theta(n)) + o(\delta_1).$$

Substituting the above in (7.51) and taking conditional expectations, we have

$$E\left[\frac{J(\theta(n)+\delta_1\Delta(n)) - J(\theta(n)+\delta_1\Delta(n)+\delta_2\hat{\Delta}(n))}{\delta_2\hat{\Delta}_k(n)} \mid \mathscr{F}_2(n)\right]$$

$$= -\nabla_k J(\theta(n)) - \sum_{l=1,l\neq k}^{N} E\left[\frac{\hat{\Delta}_l(n)}{\hat{\Delta}_k(n)} \mid \mathscr{F}_2(n)\right]\nabla_l J(\theta(n))$$

$$-\delta_1\sum_{l=1}^{N}\sum_{j=1}^{N} E\left[\frac{\hat{\Delta}_l(n)\Delta_j(n)}{\hat{\Delta}_k(n)} \mid \mathscr{F}_2(n)\right]\nabla^2_{l,j}J(\theta(n))$$

$$-\frac{\delta_2}{2}\sum_{l=1}^{N}\sum_{j=1}^{N} E\left[\frac{\hat{\Delta}_l(n)\hat{\Delta}_j(n)}{\hat{\Delta}_k(n)} \mid \mathscr{F}_2(n)\right]\nabla^2_{l,j}J(\theta(n))$$

$$-\frac{\delta_1\delta_2}{2}\sum_{l=1}^{N}\sum_{j=1}^{N}\sum_{m=1}^{N} E\left[\frac{\hat{\Delta}_l(n)\hat{\Delta}_j(n)\Delta(m)(n)}{\hat{\Delta}_k(n)} \mid \mathscr{F}_2(n)\right]\nabla^3_{l,j,m}J(\theta(n)) + o(\delta_1) + o(\delta_2).$$

Now it is easy to see using Assumption 7.6 that all conditional expectation terms on the RHS above equal zero. Thus,

$$E\left[\frac{J(\theta(n)+\delta_1\Delta(n)) - J(\theta(n)+\delta_1\Delta(n)+\delta_2\hat{\Delta}(n))}{\delta_2\hat{\Delta}_k(n)} \mid \mathscr{F}_2(n)\right]$$

$$= -\nabla_k J(\theta(n)) + o(\delta_1) + o(\delta_2).$$

The claim follows. □

Theorem 7.15. *Given $\eta > 0$, there exists $\hat{\delta} > 0$, such that for all $\delta_1, \delta_2 \in (0, \hat{\delta}]$, the algorithm (7.16)-(7.19) converges to the set K^η with probability one.*

Proof. Using an appropriate martingale construction, it is easy to see that recursion (7.19) can be rewritten as

$$\theta_i(n+1) = \Gamma_i(\theta_i(n) + a(n)\sum_{k=1}^{N} \bar{M}_{i,k}(\theta(n))(E[(J(\theta(n)+\delta_1\Delta(n))$$

$$-J(\theta(n)+\delta_1\Delta(n)+\delta_2\hat{\Delta}(n)))/\delta_2\hat{\Delta}_k(n) \mid \mathscr{F}_2(n)]) + \alpha_1(n) + a(n)\alpha_2(n)), \quad (7.52)$$

where $\alpha_1(n)$ is $o(1)$ and $\alpha_2(n)$ becomes asymptotically negligible as $\delta_1, \delta_2 \to 0$. The claim now follows in a similar manner as Theorem 7.1 upon using the conclusions of Proposition 7.14. □

7.5.5 Convergence Analysis of N-SPSA1

The analysis in this case also proceeds along similar lines as that of N-SPSA4. Let $\mathscr{F}_3(n) = \sigma(\theta_i(m), H_{j,i}(m), Z^{++}(m), m \leq n, i, j = 1, \ldots, N; \Delta(m), \hat{\Delta}(m), m < n)$, $n \geq 1$ denote a sequence of sigma fields. The forms of the gradient and the Hessian estimators are both different from the other algorithms. Hence, we first show their unbiasedness.

Proposition 7.16. *With probability one, $\forall j, i \in \{1, \ldots, N\}$,*

$$\lim_{\delta_1,\delta_2 \to 0} \left| E\left[\frac{J(\theta(n)+\delta_1\Delta(n)+\delta_2\hat{\Delta}(n))}{\delta_1\delta_2\Delta_i(n)\hat{\Delta}_j(n)} \mid \mathscr{F}_3(n)\right] - \nabla^2_{j,i}J(\theta(n))\right| = 0. \quad (7.53)$$

Proof. The proof here is similar to that of Proposition 7.12, the only difference being the presence of additional bias terms that arise from the Taylor series expansion of the 'extra' term $E\left[\frac{J(\theta(n)+\delta_1\Delta(n))}{\delta_1\delta_2\Delta_i(n)\hat{\Delta}_j(n)} \mid \mathscr{F}_3(n)\right]$ that in turn results from the Taylor's expansion of the first term in (7.53). Now note that

$$E\left[\frac{J(\theta(n)+\delta_1\Delta(n))}{\delta_1\delta_2\Delta_i(n)\hat{\Delta}_j(n)} \mid \mathscr{F}_3(n)\right] = E\left[\frac{1}{\Delta_i(n)\hat{\Delta}_j(n)} \mid \mathscr{F}_3(n)\right]\frac{J(\theta(n))}{\delta_1\delta_2}$$

$$+\sum_{k=1}^{N} E\left[\frac{\Delta_k(n)}{\Delta_i(n)\hat{\Delta}_j(n)} \mid \mathscr{F}_3(n)\right]\frac{\nabla_k J(\theta(n))}{\delta_2}$$

$$+\frac{\delta_1}{2\delta_2}\sum_{k=1}^{N}\sum_{m=1}^{N} E\left[\frac{\Delta_k(n)\Delta(m)(n)}{\Delta_i(n)\hat{\Delta}_j(n)} \mid \mathscr{F}_3(n)\right] \nabla^2_{k,m}J(\theta(n)) + o(\delta_1).$$

It is easy to see from Assumption 7.6 that all the conditional expectation terms on the RHS above equal zero. The rest follows as in Proposition 7.12. □

Proposition 7.17. *For all $k = 1,\dots,N$,*

$$\lim_{\delta_1,\delta_2\to 0}\left| E\left[\frac{J(\theta(n)+\delta_1\Delta(n)+\delta_2\hat{\Delta}(n))}{\delta_2\hat{\Delta}_k(n)} \mid \mathscr{F}_3(n)\right] - \nabla_k J(\theta(n))\right| = 0,$$

with probability one.

Proof. Note that

$$\frac{J(\theta(n)+\delta_1\Delta(n)+\delta_2\hat{\Delta}(n))}{\delta_2\hat{\Delta}_k(n)} = \frac{J(\theta(n)+\delta_1\Delta(n))}{\delta_2\hat{\Delta}_k(n)} + \sum_{l=1}^{N}\frac{\hat{\Delta}_l(n)}{\hat{\Delta}_k(n)}\nabla_l J(\theta(n)+\delta_1\Delta(n))$$

$$+\frac{\delta_2}{2}\sum_{l=1}^{N}\sum_{j=1}^{N}\frac{\hat{\Delta}_l(n)}{\hat{\Delta}_k(n)}\hat{\Delta}_j(n)\nabla^2_{l,j}J(\theta(n)+\delta_1\Delta(n)) + o(\delta_2). \tag{7.54}$$

Upon comparison with (7.51), it is clear that there is an extra term $\dfrac{J(\theta(n)+\delta_1\Delta(n))}{\delta_2\hat{\Delta}(n)}$ on the RHS of (7.54) that is not present in the corresponding expression in (7.51). Again note that

$$E\left[\frac{J(\theta(n)+\delta_1\Delta(n))}{\delta_2\hat{\Delta}_k(n)} \mid \mathscr{F}_3(n)\right] = E\left[\frac{1}{\hat{\Delta}_k(n)} \mid \mathscr{F}_3(n)\right]\frac{J(\theta(n))}{\delta_2}$$

$$+\delta_1\sum_{l=1}^{N} E\left[\frac{\Delta_l(n)}{\hat{\Delta}_k(n)} \mid \mathscr{F}_3(n)\right]\frac{\nabla_l J(\theta(n))}{\delta_2}$$

$$+\frac{\delta_1^2}{2}\sum_{l=1}^{N}\sum_{m=1}^{N} E\left[\frac{\Delta_l(n)\Delta(m)(n)}{\hat{\Delta}_k(n)} \mid \mathscr{F}_3(n)\right]\frac{\nabla^2_{l,m}J(\theta(n))}{\delta_2} + o(\delta_1).$$

It is easy to see from Assumption 7.6 that all the conditional expectation terms on the RHS above equal zero. The rest now follows as in Proposition 7.14. □

Theorem 7.18. *Given $\eta > 0$, there exists $\hat{\delta} > 0$, such that for all $\delta_1, \delta_2 \in (0, \hat{\delta}]$, the algorithm (7.22)-(7.24) converges to the set K^η with probability one.*

Proof. As before, (7.24) can be rewritten using a martingale argument as

$$\theta_i(n+1) = \Gamma_i(\theta_i(n) - a(n)\sum_{k=1}^{N}\bar{M}_{i,k}(\theta(n))E\left[\frac{J(\theta(n)+\delta_1\Delta(n)+\delta_2\hat{\Delta}(n))}{\delta_2\hat{\Delta}_k(n)} \mid \mathscr{F}_3(n)\right]$$

$$+\alpha_3(n) + a(n)\alpha_4(n)),$$

where $\alpha_3(n)$ is $o(1)$ and $\alpha_4(n)$ vanishes asymptotically as $\delta_1, \delta_2 \to 0$. The rest now follows from Propositions 7.16 and 7.17, in a similar manner as Theorem 7.1. □

7.5.6 Convergence Analysis of W-SPSA Algorithms

The convergence of the W-SPSA4 algorithm follows from that of the N-SPSA4 algorithm with the following lemma in place of Lemma 7.11.

Lemma 7.19.

$$\left\| M(n) - P(\nabla^2 J(\theta(n)))^{-1} \right\| \to 0 \text{ w.p. 1},$$

with $\delta_1, \delta_2 \to 0$ *as* $n \to \infty$, $\forall i, j \in \{1, 2, \ldots, |A| \times |B|\}$.

Proof. From Woodbury's identity, since $M(n), n \geq 1$ sequence of W-SPSA4 is identical to the $P(H(n))^{-1}, n \geq 1$ sequence of N-SPSA4, the result follows from Lemma 7.11. □

On similar lines, one can derive convergence results for W-SPSA3, W-SPSA2 and W-SPSA1 algorithms.

7.6 Concluding Remarks

We presented in this chapter four different Newton SPSA algorithms from [2] for the long-run average cost objective. It has been empirically shown in [2] that N-SPSA4 shows the best results on a low-dimensional setting (considered there), while N-SPSA3 shows the same for high-dimensional parameters. This, however, needs to be verified over other settings. The short comings of Newton-based algorithms are the requirements of (a) projection to the set of positive definite and symmetric matrices and (b) the problem of taking the inverse of the projected Hessian update at each step. For the second problem, we proposed variants to the N-SPSA algorithms that directly update the inverse of the Hessian matrix by making use of the Woodbury's identity. The problem of finding the Hessian inverse is altogether avoided by [13] where the inverse of the geometric mean of the 'projected eigen-values' of the Hessian update at each update epoch replaces the inverse of the Hessian. Finally, [12] proposes certain improved Hessian estimates for Newton SPSA.

References

1. Bertsekas, D.P.: Nonlinear Programming. Athena Scientific, Belmont (1999)
2. Bhatnagar, S.: Adaptive multivariate three-timescale stochastic approximation algorithms for simulation based optimization. ACM Transactions on Modeling and Computer Simulation 15(1), 74–107 (2005)
3. Borkar, V.S.: Stochastic Approximation: A Dynamical Systems Viewpoint. Cambridge University Press and Hindustan Book Agency (Jointly Published), Cambridge and New Delhi (2008)
4. Fabian, V.: Stochastic approximation. In: Rustagi, J.J. (ed.) Optimizing Methods in Statistics, pp. 439–470. Academic Press, New York (1971)
5. Kushner, H.J., Clark, D.S.: Stochastic Approximation Methods for Constrained and Unconstrained Systems. Springer, New York (1978)
6. Kushner, H.J., Yin, G.G.: Stochastic Approximation Algorithms and Applications. Springer, New York (1997)
7. Lasalle, J.P., Lefschetz, S.: Stability by Liapunov's Direct Method with Applications. Academic Press, New York (1961)
8. Patro, R.K., Bhatnagar, S.: A probabilistic constrained nonlinear optimization framework to optimize RED parameters. Performance Evaluation 66(2), 81–104 (2009)
9. Ruppert, D.: A Newton-Raphson version of the multivariate Robbins-Monro procedure. Annals of Statistics 13, 236–245 (1985)
10. Spall, J.C.: Multivariate stochastic approximation using a simultaneous perturbation gradient approximation. IEEE Trans. Auto. Cont. 37(3), 332–341 (1992)
11. Spall, J.C.: Adaptive stochastic approximation by the simultaneous perturbation method. IEEE Trans. Autom. Contr. 45, 1839–1853 (2000)
12. Spall, J.C.: Feedback and weighting mechanisms for improving Jacobian estimates in the adaptive simultaneous perturbation algorithm. IEEE Transactions on Automatic Control 54(6), 1216–1229 (2009)
13. Zhu, X., Spall, J.C.: A modified second-order SPSA optimization algorithm for finite samples. Int. J. Adapt. Control Signal Process. 16, 397–409 (2002)

Chapter 8
Newton-Based Smoothed Functional Algorithms

8.1 Introduction

We saw in Chapter 6, the development of gradient estimates using the SF technique. The idea is that there was to convolve the gradient of the objective with a multi-variate smoothing density functional. Using an integration-by-parts argument, the same is obtained as a convolution of the objective with a scaled version of the density. The density functions that can be used for smoothing include Gaussian, Cauchy and uniform pdfs.

We extend the above idea to obtain Hessian estimates using Gaussian pdfs as smoothing functions. By taking the convolution of the Hessian of the objective with a multi-variate Gaussian pdf, and through an integration-by-parts argument applied twice, one obtains in an elegant manner the same as a convolution of the objective function with a transformed density functional. The transformation involves generating N independent $N(0,1)$–distributed random variates at each update step (where N corresponds to the parameter dimension). The same perturbed simulation is also seen (see Chapter 6) to estimate the gradient of the objective function. This results in a one-simulation Newton SF algorithm where one perturbed simulation estimates both the gradient and the Hessian of the objective.

Next in Section 8.2.2, we derive a two-simulation balanced estimate of the Hessian of the objective function that is seen to have a lower bias than its one-simulation counterpart. As discussed in Chapter 6, the same two simulations are also seen to help in obtaining a balanced SF estimate of the gradient.

Prior work on Hessian-based estimation schemes in the literature has been discussed in Chapter 7 including the simultaneous perturbation techniques presented there. As with Chapter 7, we consider here the long-run average cost objective and develop multi-timescale stochastic approximation algorithms. The material in this chapter is entirely based on [2].

S. Bhatnagar et al.: Stochastic Recursive Algorithms for Optimization, LNCIS 434, pp. 133–148.
springerlink.com

The rest of the chapter is organized as follows: The Hessian estimates are derived in Section 8.2. The two Newton SF algorithms that incorporate these Hessian estimators as well as the gradient SF estimators described in Chapter 6 are then presented in Section 8.3. The proofs of convergence of these algorithms are presented in Section 8.4. Finally, Section 8.5 presents the concluding remarks.

The framework that we use is exactly the same as described in Section 7.2 of Chapter 7, hence the same is not repeated here. Further, we let Assumptions 7.1–7.5 hold.

8.2 The Hessian Estimates

We present in this section both the one-simulation and the two-simulation Hessian SF estimates.

8.2.1 One-Simulation Hessian SF Estimate

Let

$$D^2_{\beta,1}J(\theta) \stackrel{\triangle}{=} \int G_\beta(\theta-\eta)\nabla^2_\eta J(\eta)d\eta, \tag{8.1}$$

denote the convolution of the N-dimensional multi-variate Gaussian pdf $G_\beta(\cdot)$ (i.e., the joint pdf of N independent $N(0,\beta^2)$-distributed random variables) with $\nabla^2_\eta J(\eta)$, the Hessian of $J(\eta)$. The precise expression of $G_\beta(\theta-\eta)$ is

$$G_\beta(\theta-\eta) = \frac{1}{(2\pi)^{N/2}\beta^N}\exp\left(-\frac{1}{2}\sum_{i=1}^{N}\frac{(\theta_i-\eta_i)^2}{\beta^2}\right),$$

where $\theta,\eta \in \mathbb{R}^N$ with $\theta \stackrel{\triangle}{=} (\theta_1,\ldots,\theta_N)^T$ and $\eta \stackrel{\triangle}{=} (\eta_1,\ldots,\eta_N)^T$. It can be seen that in the limit as $\beta \to 0$, $D^2_{\beta,1}J(\theta) \to \nabla^2 J(\theta)$. Thus, for $\beta > 0$ small enough, $D^2_{\beta,1}J(\theta)$ will serve as an estimate of the Hessian $\nabla^2 J(\theta)$. This argument will be made precise later.

Now, as with the G-SF schemes, upon integrating by parts in (8.1), one obtains

$$\begin{aligned} D^2_{\beta,1}J(\theta) &= -\int \nabla_\eta G_\beta(\theta-\eta)\nabla_\eta J(\eta)d\eta \\ &= \int \nabla_\eta G_\beta(\eta)\nabla_\eta J(\theta-\eta)d\eta. \end{aligned}$$

It is easy to see that

$$\nabla_\eta G_\beta(\eta) = -\frac{\eta}{\beta^2}G_\beta(\eta).$$

Hence,

$$
\begin{aligned}
D^2_{\beta,1}J(\theta) &= -\frac{1}{\beta^2}\int \eta G_\beta(\eta)\nabla_\eta J(\eta)d\eta \\
&= -\frac{1}{\beta^2}\int \nabla_\eta(\eta G_\beta(\eta))J(\theta-\eta)d\eta. \qquad (8.2)
\end{aligned}
$$

The last equality above is obtained via another operation involving integration by parts. Before we proceed further, we first evaluate $\nabla_\eta(\eta G_\beta(\eta)) = \nabla_\eta((\eta_1 G_\beta(\eta), \ldots, \eta_N G_\beta(\eta))$. Note that

$$
\nabla_\eta(\eta G_\beta(\eta)) = \begin{bmatrix} \nabla_{\eta_1}(\eta_1 G_\beta(\eta)) & \nabla_{\eta_2}(\eta_1 G_\beta(\eta)) & \cdots & \nabla_{\eta_N}(\eta_1 G_\beta(\eta)) \\ \nabla_{\eta_1}(\eta_2 G_\beta(\eta)) & \nabla_{\eta_2}(\eta_2 G_\beta(\eta)) & \cdots & \nabla_{\eta_N}(\eta_2 G_\beta(\eta)) \\ \cdots & \cdots & \cdots & \cdots \\ \nabla_{\eta_1}(\eta_N G_\beta(\eta)) & \nabla_{\eta_2}(\eta_N G_\beta(\eta)) & \cdots & \nabla_{\eta_N}(\eta_N G_\beta(\eta)) \end{bmatrix}
$$

$$
= \begin{bmatrix} \left(1-\frac{\eta_1^2}{\beta^2}\right) & -\frac{\eta_1\eta_2}{\beta^2} & \cdots & -\frac{\eta_1\eta_N}{\beta^2} \\ -\frac{\eta_2\eta_1}{\beta^2} & \left(1-\frac{\eta_2^2}{\beta^2}\right) & \cdots & -\frac{\eta_2\eta_N}{\beta^2} \\ \cdots & \cdots & \cdots & \cdots \\ -\frac{\eta_N\eta_1}{\beta^2} & -\frac{\eta_N\eta_2}{\beta^2} & \cdots & \left(1-\frac{\eta_N^2}{\beta^2}\right) \end{bmatrix} G_\beta(\eta).
$$

Let $\hat{H}(\eta)$ denote the matrix above that multiplies $G_\beta(\eta)$. Then from (8.2), we have

$$
D^2_{\beta,1}J(\theta) = -\frac{1}{\beta^2}\int \hat{H}(\eta)G_\beta(\eta)J(\theta-\eta)d\eta.
$$

Let $\eta' = \eta/\beta$. Then $\eta = \beta\eta' = (\beta\eta'_1, \ldots, \beta\eta'_N)^T$ (written component-wise). Hence, $d\eta = \beta^N d\eta'_1 \cdots d\eta'_N = \beta^N d\eta'$. Hence (8.2) becomes

$$
D^2_{\beta,1}J(\theta) = \frac{1}{\beta^2}\int \bar{H}(\eta')G_1(\eta')J(\theta-\beta\eta')d\eta', \qquad (8.3)
$$

where $G_1(\eta') = G_\beta(\eta')$ with $\beta = 1$, i.e., the joint p.d.f. of N independent, $N(0,1)$-distributed random variables. Also,

$$
\bar{H}(\eta') = \begin{bmatrix} ((\eta'_1)^2-1) & \eta'_1\eta'_2 & \cdots & \eta'_1\eta'_N \\ \eta'_2\eta'_1 & ((\eta'_2)^2-1) & \cdots & \eta'_2\eta'_N \\ \cdots & \cdots & \cdots & \cdots \\ \eta'_N\eta'_1 & \eta'_N\eta'_2 & \cdots & ((\eta'_N)^2-1) \end{bmatrix}. \qquad (8.4)
$$

Note that since $\eta_i' = \eta_i/\beta$, $i = 1,\ldots,N$, they are independent $N(0,1)$-distributed random variables. Now since η_i' and $-\eta_i'$ have the same distribution, one obtains

$$D^2_{\beta,1}J(\theta) = E\left[\frac{1}{\beta^2}\bar{H}(\bar{\eta})J(\theta+\beta\bar{\eta}) \mid \theta\right],$$

where the expectation above is taken w.r.t. the p.d.f. $G_1(\bar{\eta})$. Hence the form of the estimator for $\nabla^2 J(\theta(n))$ suggested by the above is (for a large integer $M > 0$ and a small scalar $\beta > 0$)

$$\nabla^2 J(\theta(n)) \approx \frac{1}{\beta^2}\frac{1}{M}\sum_{n=1}^{M}\bar{H}(\eta(n))J(\theta(n)+\beta\eta(n)). \tag{8.5}$$

Here $\eta(n) = (\eta_1(n),\ldots,\eta_N(n))^T$ is a vector of independent $N(0,1)$–distributed random variables.

8.2.2 Two-Simulation Hessian SF Estimate

We now describe a two-simulation (balanced) SF estimate of the Hessian. Let

$$D^2_{\beta,2}J(\theta) = E\left[\frac{1}{2\beta^2}\bar{H}(\bar{\eta})(J(\theta+\beta\bar{\eta})+J(\theta-\beta\bar{\eta})) \mid \theta\right],$$

with $\bar{\eta} \stackrel{\triangle}{=} (\eta_1,\ldots,\eta_N)^T$, with $\eta_1,\ldots,\eta_N$ being independent, $N(0,1)$-distributed random variables. Then $D^2_{\beta,2}J(\theta)$ will serve as an estimate of the Hessian $\nabla^2 J(\theta)$. Using a Taylor series argument, it will be seen that this estimate has a lower bias than the one-simulation Hessian SF estimate in Section 8.2.1.

Thus, the form of the two-sided Hessian estimator suggested by the above is the following: For a large integer $M > 0$ and a small scalar $\beta > 0$,

$$\nabla^2 J(\theta(n)) \approx \frac{1}{2\beta^2}\frac{1}{M}\sum_{n=1}^{M}\bar{H}(\eta(n))(J(\theta(n)+\beta\eta(n))+J(\theta(n)-\beta\eta(n))), \tag{8.6}$$

where $\eta(n) \stackrel{\triangle}{=} (\eta_1(n),\ldots,\eta_N(n))^T$ is a vector of $N(0,1)$-distributed random variables.

8.3 The Newton SF Algorithms

We now present the two Newton SF algorithms that are based on the aforementioned one- and two-simulation SF estimates of the Hessian. These algorithms also incorporate the one- and two-simulation gradient SF estimates described in Chapter 6. The two combinations of (a) one-simulation gradient and Hessian estimates as well as the (b) two-simulation gradient and Hessian estimates, respectively, result in one-simulation and two-simulation Newton SF algorithms that are respectively referred to as N-SF1 and N-SF2.

As with the algorithms in Chapters 5–7, it is observed that the performance of the SF algorithms improves considerably when the parameter vector is updated once after a given number L of instants when $L > 1$. This happens as a consequence of the additional data averaging (over L instants) on top of the two-timescale averaging. The value of L is, however, totally arbitrary, and in fact, it is observed (see [1, 2, 3]) that a value of L between 50 and 500 works well in many cases.

8.3.1 The One-Simulation Newton SF Algorithm (N-SF1)

We now describe the Newton SF algorithm which requires one simulation with perturbed parameter $\theta + \beta\eta$. Let $\{X(n)\}$ be the underlying Markov process parametrized with $\theta(n) + \beta\eta(n)$. Let $Z_{i,j}(n), i, j = 1,2,\ldots,N$, denote components of the Hessian estimate at update instant n. Also, let $Z_i(n), i = 1,2,\ldots,N$, denote components of the gradient estimate at update instant n. The algorithm is given as follows: For a large integer $M > 0$ and a small $\beta > 0$, and for $i, j, k = 1,\ldots,N$, $j < k$,

$$Z_{i,i}(n+1) = Z_{i,i}(n) + b(n)\left(\frac{\eta_i^2(n) - 1}{\beta^2}h(X(n)) - Z_{i,i}(n)\right), \tag{8.7}$$

$$Z_{j,k}(n+1) = Z_{j,k}(n) + b(n)\left(\frac{\eta_j(n)\eta_k(n)}{\beta^2}h(X(n)) - Z_{j,k}(n)\right). \tag{8.8}$$

For $j > k$, set $Z_{j,k}(n+1) = Z_{k,j}(n+1)$. For $l = 1,\ldots,N$, update

$$Z_l(n+1) = Z_l(n) + c(n)\left(\frac{\eta_l(n)}{\beta}h(X(n)) - Z_l(n)\right). \tag{8.9}$$

Next, form the matrix $H(n) = P([[Z_{j,k}(n)]]_{j,k=1}^{N})$ and compute its inverse $M(n) = [[M_{j,k}(n)]]_{j,k=1}^{N} \triangleq H(n)^{-1}$. For $i = 1,\ldots,N$, update $\theta_i(n)$ according to

$$\theta_i(n+1) = \Gamma_i\left(\theta_i(n) - a(n)\sum_{k=1}^{N} M_{i,k}(n)Z_k(n)\right), \tag{8.10}$$

where for some $x = (x_1,\ldots,x_N)^T \in \mathbb{R}^N$, $\Gamma(x) \triangleq (\Gamma_1(x_1),\ldots,\Gamma_N(x_N))^T$ represents the projection of x onto the constraint set C. This is an L_2 projection (i.e., the Euclidean-norm projection).

Next, we present the two-simulation Newton SF algorithm that has a lower bias in both its gradient and Hessian estimates over its one-simulation counterpart.

8.3.2 *The Two-Simulation Newton SF Algorithm (N-SF2)*

The Newton SF algorithm, N-SF2, requiring two simulations $\{X^1(n)\}$ and $\{X^2(n)\}$ with $\theta + \beta\eta$ and $\theta - \beta\eta$ respectively, is presented below: For a large integer $M > 0$ and a small $\beta > 0$, and for $i, j, k = 1,\ldots,N$, $j < k$,

$$Z_{i,i}(n+1) = Z_{i,i}(n) + b(n)\left(\frac{\eta_i^2(n) - 1}{2\beta^2}(h(X^1(n)) + h(X^2(n))) - Z_{i,i}(n)\right), \tag{8.11}$$

$$Z_{j,k}(n+1) = Z_{j,k}(n) + b(n)\left(\frac{\eta_j(n)\eta_k(n)}{2\beta^2}(h(X^1(n)) + h(X^2(n))) - Z_{j,k}(n)\right). \tag{8.12}$$

For $j > k$, set $Z_{j,k}(n+1) = Z_{k,j}(n+1)$. Now, for $l = 1,\ldots,N$, update

$$Z_l(n+1) = Z_l(n) + c(n)\left(\frac{\eta_l(n)}{2\beta}(h(X^1(n)) - h(X^2(n))) - Z_l(n)\right). \tag{8.13}$$

Next, form the matrix $H(n) = P([[Z_{j,k}(n)]]_{j,k=1}^{N})$ and compute its inverse $M(n) = [[M_{j,k}(n)]]_{j,k=1}^{N} \triangleq H(n)^{-1}$. Finally, for $i = 1,\ldots,N$,

$$\theta_i(n+1) = \Gamma_i\left(\theta_i(n) - a(n)\sum_{k=1}^{N} M_{i,k}(n)Z_k(n)\right), \tag{8.14}$$

where the projection operator Γ is defined as for the N-SF1 algorithm.

Remark 8.1. As noted previously, the performance of the algorithms N-SF1 and N-SF2 is seen to improve considerably if recursions (8.7)–(8.9) (resp. (8.11)–(8.13)) in N-SF1 (resp. N-SF2) are run for some given number $L > 1$ of instants in between two successive updates of the parameter (cf. recursions (8.10) and (8.14) in N-SF1 and N-SF2, respectively). In such a case, the projected Hessian matrix $H(n)$ and its inverse $M(n)$ will also have to be computed only once every L instants using the most recent information on the average cost samples $Z_{j,k}(n)$, $j,k \in \{1,\ldots,N\}$. Recursions incorporating the L-step averaging are given in [2] and have been analyzed there for this case as well.

8.4 Convergence Analysis of Newton SF Algorithms

The detailed analysis of convergence of both the Newton SF algorithms, N-SF1 and N-SF2 is discussed in the following sections.

8.4.1 Convergence of N-SF1

Consider the ODE

$$\dot{\theta}(t) = \tilde{\Gamma}(-P(\nabla^2 J(\theta(t)))^{-1}\nabla J(\theta(t))), \tag{8.15}$$

where for any $y \in \mathbb{R}^N$ and a bounded, continuous function $v(\cdot) : \mathbb{R}^N \to \mathbb{R}^N$,

$$\tilde{\Gamma}(v(y)) = \lim_{0<\eta\to 0}\left(\frac{\Gamma(y+\eta v(y)) - \Gamma(y)}{\eta}\right).$$

Let

$$K \stackrel{\triangle}{=} \{\theta \in C \mid \nabla J(\theta)^T \tilde{\Gamma}(-\{P(\nabla^2 J(\theta))\}^{-1}\nabla J(\theta)) = 0\}.$$

Further, for any set $S \subseteq C$, given $\eta > 0$, $S^\eta \stackrel{\triangle}{=} \{\theta \in C \mid \parallel \theta - \theta_0 \parallel \leq \eta,\ \theta_0 \in S\}$ shall denote the set of all points in C that are in an 'η-neighborhood' of the set S. Let $\hat{K}$ denote the set $\{\theta \in C \mid \tilde{\Gamma}(-P(\nabla^2 J(\theta))^{-1}\nabla J(\theta)) = -P(\nabla^2 J(\theta))^{-1}\nabla J(\theta)\}$. Let C^o denote the interior of C. Then, one can see that $C^o \subseteq \hat{K}$. Now, the main convergence result for N-SF1 is as follows:

Theorem 8.1. *Under Assumptions 7.1–7.5, given $\eta > 0$, there exists $\hat{\beta} > 0$, such that for all $\beta \in (0, \hat{\beta}]$, the parameter updates $\theta(n), n \geq 0$ obtained using N-SF1 converge to K^η with probability one as $M \to \infty$.*

To prove this theorem, we provide a sequence of Lemmas, Propositions and Corollaries in the rest of the section, as explained below:

1. Proposition 8.2, Lemma 8.3, Proposition 8.4 and Corollary 8.5 prove that the Hessian updates $Z_{i,j}(n)$s in N-SF1 converge to the actual Hessian in the limit as $\beta \to 0$.
2. Lemma 8.6 shows that the inverse of the projected Hessian estimate also converges to the inverse of the projected Hessian of the objective.
3. Lemmas 8.7–8.8 and Corollary 8.9 show that the gradient estimates $Z_i(n)$ also converge to the actual gradient of the objective, again in the limit as $\beta \to 0$.
4. With gradient and Hessian estimates converging to those of the objective itself, the main result is proven using Lasalle's invariance theorem. The formal proof of the main result is given at the end of this subsection.

The theory of multi-timescale stochastic approximation (see Chapter 3.3) allows us to treat $\theta(n) \equiv \theta$ (a constant) while analyzing the Hessian and gradient updates. Let $\mathscr{F}(l) = \sigma(\theta(p), X(p), p \leq l; \eta(p), p < l)$, $l \geq 1$, denote a sequence of sigma fields. Now define sequences $\{M_{l,l}(p)\}$, $\{M_{i,j}(p)\}$, $l,i,j \in \{1,\ldots,N\}$, $i \neq j$ as follows: For $l = 1,\ldots,N$,

$$M_{l,l}(p) = \sum_{m=1}^{p} b(m)\left(\frac{\eta_l^2(m)-1}{\beta^2}h(X(m)) - E\left[\frac{\eta_l^2(m)-1}{\beta^2}h(X(m)) \mid \mathscr{F}(m-1)\right]\right).$$

Further, for $i, j \in \{1,\ldots,N\}$, we have

$$M_{i,j}(p) = \sum_{m=1}^{p} b(m)\left(\frac{\eta_i(m)\eta_j(m)}{\beta^2}h(X(m)) - E\left[\frac{\eta_i(m)\eta_j(m)}{\beta^2}h(X(m)) \mid \mathscr{F}(m-1)\right]\right).$$

Proposition 8.2. *The sequences $\{M_{l,l}(p), \mathscr{F}(p)\}$ and $\{M_{i,j}(p), \mathscr{F}(p)\}$, $l,i,j = 1, \ldots, N$, $i \neq j$ are almost surely convergent martingale sequences.*

Proof. We consider first the sequence $\{M_{l,l}(p), \mathscr{F}(p)\}$. It is easy to see that it is a martingale sequence. To see that it is square integrable, note that

$$E[M_{l,l}^2(p)] \leq \frac{C_p}{\beta^4}\sum_{m=1}^{p} b^2(m)(E[(\eta_l^2(m)-1)^2 h^2(X(m)) + E^2[(\eta_l^2(m)-1)h(X(m)) \mid \mathscr{F}(m-1)]])$$

for some constant $C_p > 0$ (that, however, depends on p). For the second term on RHS above, note that almost surely,

$$E^2[\eta_l^2(m)-1)h(X(m)) \mid \mathscr{F}(m-1)] \leq E[(\eta_l^2(m)-1)^2 h^2(X(m)) \mid \mathscr{F}(m-1)],$$

by the conditional Jensen's inequality. Hence,

$$E[M_{l,l}^2(p)] \leq \frac{2C_p}{\beta^4}\sum_{m=1}^{p} b^2(m)E[(\eta_l^2(m)-1)^2 h^2(X(m))]$$

$$\leq \frac{2C_p}{\beta^4} \sum_{m=1}^{p} b^2(m) E[(\eta_l^2(m)-1)^2]^{1/2} E[h^4(X(m))]^{1/2}$$

by the Cauchy–Schwartz inequality. Since, $h(\cdot)$ is a Lipschitz continuous function, we have

$$|h(X(m))| - |h(0)| \leq |h(X(m)) - h(0)| \leq \hat{C} \parallel X(m) \parallel,$$

where $\hat{C} > 0$ is the Lipschitz constant. Thus,

$$|h(X(m))| \leq C_1(1 + \parallel X(m) \parallel)$$

for $C_1 = \max(\hat{C}, |h(0)|) < \infty$. Hence, one gets

$$E[h^4(X(m))] \leq C_2(1 + E[\parallel X(m) \parallel^4])$$

for (constant) $C_2 = 8C_1^4$. As a consequence of Assumption 7.3, $\sup_m E[\parallel X(m) \parallel^4] < \infty$. It now follows from Assumption 7.5 that $E[M_{l,l}^2(p)] < \infty$, for all $p \geq 1$.

Now note that

$$\sum_p E[(M_{l,l}(p+1) - M_{l,l}(p))^2 \mid \mathscr{F}(p)]$$

$$\leq \sum_p b^2(p+1)\left(E\left[\left(\frac{\eta_l^2(p+1)-1}{\beta^2} h(X_{p+1})\right)^2 \mid \mathscr{F}(p)\right]\right.$$

$$\left. + E\left[E^2\left[\frac{\eta_l^2(p+1)-1}{\beta^2} h(X_{p+1}) \mid \mathscr{F}(p)\right] \mid \mathscr{F}(p)\right]\right)$$

$$\leq \sum_p 2b^2(p+1) E\left[\left(\frac{\eta_l^2(p+1)-1}{\beta^2} h(X_{p+1})\right)^2 \mid \mathscr{F}(p)\right],$$

almost surely. The last inequality above again follows from the conditional Jensen's inequality. It can now be easily seen as before, using Assumptions 7.3 that

$$\sup_p \frac{1}{\beta^2} E[((\tilde{\eta}_l^2(p+1) - 1) h(X_{p+1}))^2 \mid \mathscr{F}(p)] < \infty \text{ w.p.1.}$$

Hence, using Assumption 7.5, it can be seen that

$$\sum_p E[(M_{l,l}(p+1) - M_{l,l}(p))^2 \mid \mathscr{F}(p)] < \infty$$

almost surely. Thus, by the martingale convergence theorem (Theorem D.1), $\{M_{l,l}(p)\}$ are almost surely convergent martingale sequences. A similar proof settles the claim for $\{M_{i,j}(p)\}$ as well. □

Let $Y(n) \stackrel{\triangle}{=} [[Z_{j,k}(n)]]_{j,k=1}^N$ and $Z(n) \stackrel{\triangle}{=} (Z_1(n), \ldots, Z_N(n))^T$. Then, the matrix $H(n)$ in the algorithm corresponds to $P(Y(n))$. In vector–matrix notation, the Hessian update recursions (8.7)–(8.8) can be written as

$$Y(n+1) = Y(n) + b(n)\left(\frac{1}{\beta^2}\bar{H}(\eta(n))h(X_n) - Y(n)\right). \tag{8.16}$$

Lemma 8.3. *The Hessian updates $Y(n), n \geq 0$ are uniformly bounded and further*

$$\| Y(n) - D^2_{\beta,1}J(\theta(n)) \| \to 0 \text{ as } n \to \infty,$$

with probability one.

Proof. The ODE associated with (8.16) is

$$\dot{Y}(t) = D^2_{\beta,1}J(\theta(t)) - Y(t), \tag{8.17}$$

that has $Y^* = D^2_{\beta,1}J(\theta)$ as its unique globally asymptotically stable equilibrium (when $\theta(t) \equiv \theta$). Note that (8.16) can be rewritten as

$$Y(n+1) = Y(n) + b(n)\left(\frac{1}{\beta^2}E[\bar{H}(\eta(n))h(X_n) \mid \mathscr{F}(n)] - Y(n)\right)$$

$$+b(n)\frac{1}{\beta^2}(\bar{H}(\eta(n))h(X_n) - E[\bar{H}(\eta(n))h(X_n) \mid \mathscr{F}(n)]).$$

From Proposition 8.2, we have that almost surely,

$$\sum_n b(n)\frac{1}{\beta^2}(\bar{H}(\eta(n))h(X_n) - E[\bar{H}(\eta(n))h(X_n) \mid \mathscr{F}(n)]) < \infty.$$

Also, it can be seen as in the proof of Proposition 8.2 that $\sup_n E[\bar{H}(\eta(n))h(X_n) \mid \mathscr{F}(n)] < \infty$ with probability one. It is now easy to verify Assumptions D.1 and D.2, as a result of which the claim follows from the Borkar and Meyn theorem (Theorem D.1. □

Proposition 8.4. $\| D^2_{\beta,1}J(\theta(n)) - \nabla^2 J(\theta(n)) \| \to 0$ *as* $\beta \to 0$.

Proof. Recall that

$$D^2_{\beta,1}J(\theta(n)) = E\left[\frac{1}{\beta^2}\bar{H}(\eta(n))J(\theta(n) + \beta\eta(n)) \mid \theta(n)\right],$$

where $\eta(n) = (\eta_1(n), \ldots, \eta_N(n))^T$ is a vector of independent $N(0,1)$ random variates and the expectation is taken w.r.t. the density of $\eta(n)$. Using a Taylor series expansion of $J(\theta(n) + \beta\eta(n))$ around $\theta(n)$, one obtains

$$\begin{aligned}
D^2_{\beta,1}J(\theta(n)) &= E\Bigg[\frac{1}{\beta^2}\bar{H}(\eta(n))(J(\theta(n))+\beta\eta(n)^T\nabla J(\theta(n)) \\
&\qquad +\frac{\beta^2}{2}\eta(n)^T\nabla^2 J(\theta(n))\eta(n)+o(\beta^2)\mid\theta(n)\Bigg] \\
&= \frac{1}{\beta^2}E[\bar{H}(\eta(n))J(\theta(n))\mid\theta(n)]+\frac{1}{\beta}E[\bar{H}(\eta(n))\eta(n)^T\nabla J(\theta(n))\mid\theta(n)] \\
&\qquad +\tfrac{1}{2}E[\bar{H}(\eta(n))\eta(n)^T\nabla^2 J(\theta(n))\eta(n)\mid\theta(n)]+O(\beta).
\end{aligned} \tag{8.18}$$

Now observe that $E[\bar{H}(\eta(n))]=0$ (the matrix of all zero elements) with $E[\bar{H}(\eta(n))]$ being the matrix of expectations of individual elements of $\bar{H}(\eta(n))$. Hence the first term on the RHS of (8.18) equals zero. Now consider the second term on the RHS of (8.18). Note that

$$E[\bar{H}(\eta(n))\eta(n)^T\nabla J(\theta(n))\mid\theta(n)]=$$
$$E\left[\begin{array}{cccc}
(\eta_1^2(n)-1)\eta(n)^T\nabla J(\theta(n)) & \eta_1(n)\eta_2(n)\eta(n)^T\nabla J(\theta(n)) & \cdots & \eta_1(n)\eta_N(n)\eta(n)^T\nabla J(\theta(n)) \\
\eta_2(n)\eta_1(n)\eta(n)^T\nabla J(\theta(n)) & (\eta_2^2(n)-1)\eta(n)^T\nabla J(\theta(n)) & \cdots & \eta_2(n)\eta_N(n)\eta(n)^T\nabla J(\theta(n)) \\
\cdots & \cdots & \cdots & \cdots \\
\eta_N(n)\eta_1(n)\eta(n)^T\nabla J(\theta(n)) & \eta_N(n)\eta_2(n)\eta(n)^T\nabla J(\theta(n)) & \cdots & (\eta_N^2(n)-1)\eta(n)^T\nabla J(\theta(n))
\end{array}\;\middle|\;\theta(n)\right]. \tag{8.19}$$

Consider the first term (corresponding to the first row and first column) above. Note that

$$\begin{aligned}
&E[(\eta_1^2(n)-1)\eta(n)^T\nabla J(\theta(n))\mid\theta(n)] \\
&=E[(\eta_1^3(n)-\eta_1(n),\eta_1^2(n)\eta_2(n)-\eta_2(n),\ldots,\eta_1^2(n)\eta_N(n)-\eta_N(n))^T\nabla J(\theta(n))\mid\theta(n)] \\
&=0.
\end{aligned}$$

Similarly all other terms in (8.19) can be seen to be equal to zero as well. We use here the facts that $E[\eta_1(n)]=E[\eta_1^3(n)]=0$ and $E[\eta_1^2(n)]=1$. Also, $\eta_i(n)$ is independent of $\eta_j(n)$ for all $i\neq j$. Hence the second term on the RHS of (8.18) equals zero as well. Consider now the third term on the RHS of (8.18). Note that

$$\frac{1}{2}E[\bar{H}(\eta(n))\eta(n)^T\nabla^2 J(\theta(n))\eta(n)\mid\theta(n)]=$$

$$\frac{1}{2}E\left[\begin{array}{ccc}
(\eta_1^2(n)-1)\sum\limits_{i,j=1}^{N}\nabla_{ij}J(\theta(n))\eta_i(n)\eta_j(n) & \cdots & \eta_1(n)\eta_N(n)\sum\limits_{i,j=1}^{N}\nabla_{ij}J(\theta(n))\eta_i(n)\eta_j(n) \\
\eta_2(n)\eta_1(n)\sum\limits_{i,j=1}^{N}\nabla_{ij}J(\theta)\eta_i(n)\eta_j(n) & \cdots & \eta_2(n)\eta_N(n)\sum\limits_{i,j=1}^{N}\nabla_{ij}J(\theta(n))\eta_i(n)\eta_j(n) \\
\cdots & \cdots & \cdots \\
\eta_N(n)\eta_1(n)\sum\limits_{i,j=1}^{N}\nabla_{ij}J(\theta(n))\eta_i(n)\eta_j(n) & \cdots & (\eta_N^2(n)-1)\sum\limits_{i,j=1}^{N}\nabla_{ij}J(\theta(n))\eta_i(n)\eta_j(n)
\end{array}\;\middle|\;\theta(n)\right]. \tag{8.20}$$

Consider now the term corresponding to the first row and first column above. Note that

$$E[(\eta_1^2(n)-1)\Sigma_{i,j=1}^N \nabla_{ij}J(\theta(n))\eta_i(n)\eta_j(n) \mid \theta(n)]$$
$$= E[\eta_1^2(n)\Sigma_{i,j=1}^N \nabla_{ij}J(\theta(n))\eta_i(n)\eta_j(n) \mid \theta(n)] - E[\Sigma_{i,j=1}^N \nabla_{ij}J(\theta(n))\eta_i(n)\eta_j(n) \mid \theta(n)]. \quad (8.21)$$

The first term on the RHS of (8.21) equals

$$E[\eta_1^4(n)\nabla_{11}J(\theta(n)) \mid \theta(n)] + E[\Sigma_{i=j,i\neq 1}\eta_1^2(n)\eta_i^2(n)\nabla_{ij}J(\theta(n)) \mid \theta(n)]$$
$$+E[\Sigma_{i\neq j,i\neq 1}\eta_1^2(n)\eta_i(n)\eta_j(n)\nabla_{ij}J(\theta(n)) \mid \theta(n)] = 3\nabla_{11}J(\theta(n)) + \Sigma_{i=j,i\neq 1}\nabla_{ij}J(\theta(n)),$$

since $E[\eta_1^4(n)] = 3$. The second term on RHS of (8.21) equals $-\sum_{i=1}^{N}\nabla_{ii}J(\theta(n))$. Adding the above two terms, one obtains

$$E[(\eta_1^2(n)-1)\sum_{i,j=1}^{N}\nabla_{ij}J(\theta(n))\eta_i(n)\eta_j(n) \mid \theta(n)] = 2\nabla_{11}J(\theta(n)).$$

Consider now the term in the first row and second column of the matrix in (8.20). Note that

$$E[\eta_1(n)\eta_2(n)\Sigma_{i,j=1}^N \nabla_{ij}J(\theta(n))\eta_i(n)\eta_j(n) \mid \theta(n)]$$
$$= 2E[\eta_1^2(n)\eta_2^2(n)\nabla_{12}J(\theta(n)) \mid \theta(n)]$$
$$+E[\Sigma_{(i,j)\notin\{(1,2),(2,1)\}}\eta_1(n)\eta_2(n)\eta_i(n)\eta_j(n)\nabla_{ij}J(\theta(n)) \mid \theta(n)]$$
$$= 2\nabla_{12}J(\theta(n)).$$

Proceeding in a similar manner, it is easy to verify that the (i,j)th term $(i,j \in \{1,\ldots,N\})$ in the matrix in (8.20) equals $2\nabla_{ij}J(\theta(n))$. Substituting the above back in (8.20), one obtains

$$\frac{1}{2}E[\bar{H}(\eta(n))\eta(n)^T\nabla^2 J(\theta(n))\eta(n)] = \nabla^2 J(\theta(n)).$$

The claim now follows from (8.18). □

Corollary 8.5. *We have*

$$\| Y(n) - \nabla^2 J(\theta(n)) \| \to 0 \text{ as } n \to \infty \text{ and } \beta \to 0,$$

with probability one.

Proof. Follows from Lemma 8.3 and Proposition 8.4. □

Lemma 8.6. *With probability one, as $n \to \infty$ and $\beta \to 0$,*

$$\| P(Y(n))^{-1} - P(\nabla^2 J(\theta(n)))^{-1} \| \to 0.$$

Proof. Follows as in Lemma 7.11 (cf. Chapter 7). □

Next, we consider the gradient updates in (8.9). The proof of the following result has been shown in Chapter 6 (see Lemma chap6:Lemma23). Let $Z(n) \stackrel{\triangle}{=} (Z_l(n), l = 1,\ldots,N)^T$.

Lemma 8.7.

$$\| Z(n) - D_{\beta,1}J(\theta(n)) \| \to 0 \textit{ as } n \to \infty,$$

with probability one.

The next result shows the unbiasedness of the gradient estimates (in the limit as $\beta \to 0$). The proof of this result is also given in Chapter 6 (see Proposition 6.5).

Lemma 8.8.

$$\| D_{\beta,1}J(\theta(n)) - \nabla J(\theta(n)) \| \to 0 \textit{ as } \beta \to 0.$$

Combining Lemmas 8.7 and 8.8, one obtains

Corollary 8.9. *With probability one, as $n \to \infty$ and $\beta \to 0$,*

$$\| Z(n) - \nabla J(\theta(n)) \| \to 0.$$

We now consider the slowest timescale recursion involving the θ update (equation 8.10) of the algorithm.

Proof of Theorem 8.1. Recall that the parameter update recursion corresponds to

$$\theta(n+1) = \Gamma(\theta(n) - a(n)P(Y(n))^{-1}Z(n)). \tag{8.22}$$

Note that one can rewrite (8.22) as

$$\theta(n+1) = \Gamma(\theta(n) - a(n)(P(\nabla^2 J(\theta(n))))^{-1}\nabla J(\theta(n))$$

$$+(P(\nabla^2 J(\theta(n))))^{-1} - P(Y(n))^{-1})\nabla J(\theta(n)) + P(Y(n))^{-1}(\nabla J(\theta(n)) - Z(n)))) + O(\beta),$$

where the $O(\beta)$ term comes about because $\beta > 0$ is held fixed in the algorithm. Further, results such as Proposition 8.4, Lemma 8.6 and Corollary 8.9 have been shown for the case when $\beta \to 0$.

Now as a consequence of Lemma 8.6, Corollary 8.5 and Assumption 7.4, the second and third terms multiplying $a(n)$ above asymptotically vanish as $n \to \infty$ and $\beta \to 0$. One can then view (8.22) as a noisy Euler discretization of the ODE (8.15) using a standard approximation argument as [4, pp.191–196]. Note that $J(\theta)$ itself serves as an associated Liapunov function for (8.15) since

$$\frac{dJ(\theta)}{dt} = \nabla J(\theta)^T \dot{\theta} = \nabla J(\theta)^T \tilde{\Gamma}(-P(\nabla^2 J(\theta))^{-1}\nabla J(\theta)) \leq 0.$$

In particular for $\theta \in \hat{K}$, $\frac{dJ(\theta)}{dt} < 0$ if $\nabla J(\theta) \neq 0$. Now since $J(\theta)$ satisfies Assumption 7.2, it is in particular continuous and hence uniformly bounded on the compact set $C \subset \mathbb{R}^N$. Let $\lambda = \sup_\theta J(\theta) < \infty$. Then, $\{\theta \mid J(\theta) \leq \lambda\} = C$. The rest follows from the Lasalle's invariance theorem (Theorem C.4) and the Hirsch lemma (Lemma C.5). □

8.4.2 Convergence of N-SF2

The proof of convergence proceeds along similar lines as N-SF1. Hence, we only present the main results.

Proposition 8.10.

$$\| D^2_{\beta,2} J(\theta(n)) - \nabla^2 J(\theta(n)) \| \to 0 \ as \ \beta \to 0.$$

Proof. Recall that

$$D^2_{\beta,2} J(\theta(n)) = E\left[\frac{1}{2\beta^2}\bar{H}(\eta(n))(J(\theta(n)+\beta\eta(n))+J(\theta(n)-\beta\eta(n))) \mid \theta(n)\right],$$

where $\eta(n) = (\eta_1(n),\ldots,\eta_N(n))^T$ is a vector of independent $N(0,1)$ random variables. Using Taylor series expansions of $J(\theta(n)+\beta\eta(n))$ and $J(\theta(n)-\beta\eta(n))$ around $\theta(n)$, one obtains

$$J(\theta(n)+\beta\eta(n)) = J(\theta(n)) + \beta\eta(n)^T \nabla J(\theta(n)) + \frac{\beta^2}{2}\eta(n)^T \nabla^2 J(\theta(n))\eta(n) + o(\beta^2)), \quad (8.23)$$

$$J(\theta(n)-\beta\eta(n)) = J(\theta(n)) - \beta\eta(n)^T \nabla J(\theta(n)) + \frac{\beta^2}{2}\eta(n)^T \nabla^2 J(\theta(n))\eta(n) + o(\beta^2)). \quad (8.24)$$

From the foregoing, one obtains

$$D^2_{\beta,2} J(\theta(n)) = E\left[\frac{1}{2\beta^2}\bar{H}(\eta(n))(2J(\theta(n)) + \beta^2\eta(n)^T\nabla^2 J(\theta(n))\eta(n) + o(\beta^3)) \mid \theta(n)\right].$$

It has been shown in the proof of Proposition 8.4 that $E[\bar{H}(\eta(n))J(\theta(n)) \mid \theta(n)] = 0$ and $\frac{1}{2}E[\bar{H}(\eta(n))\eta(n)^T\nabla^2 J(\theta(n))\eta(n) \mid \theta(n)] = \nabla^2 J(\theta(n))$, respectively. The claim follows. □

Remark 8.2. The bias in the Hessian estimates in N-SF2 is lower as compared to the same in N-SF1. This is because when the terms in the Taylor series expansions in (8.23)–(8.24) are summed, the gradient terms $\beta\eta(n)^T\nabla J(\theta(n))$ get directly cancelled. This is unlike the corresponding estimate in N-SF1 where the gradient terms only average to zero (and do not have a direct cancellation).

Next, the following result shows the unbiasedness of the gradient estimates and has been shown for the G-SF2 algorithm in Chapter 6.

Proposition 8.11.

$$\| D_{\beta,2}J(\theta(n)) - \nabla J(\theta(n)) \| \to 0 \textit{ as } \beta \to 0.$$

Remark 8.3. It has been shown in Chapter 6 that the bias in the gradient estimates of N-SF2 is much less as compared to that in N-SF1 because of a direct cancellation of many of the terms in the Taylor series expansions in the estimates of N-SF2. This is unlike N-SF1 where these terms in fact average to zero and do not cancel off directly.

The proof of the main result below follows along the same lines as that of Theorem 8.1.

Theorem 8.12. *Under Assumptions 7.1–7.5, given $\eta > 0$, there exists $\hat{\beta} > 0$, such that for all $\beta \in (0, \hat{\beta}]$, the sequence $\{\theta(n)\}$ obtained using N-SF2 converges to K^η with probability one as $M \to \infty$.*

8.5 Concluding Remarks

The SF estimators of the Hessian belong to the class of simultaneous perturbation estimators that require only a few system simulations to perform Hessian updates at each instant regardless of the parameter dimension. The two Hessian estimators that were presented required respectively, one and two system simulations. An advantage with these estimators is that the same system simulations can also be used to estimate the gradient. Hence, they give rise to Newton-based algorithms with one or two simulations.

In the experiments studied in [2] as well as the application in service systems (see Chapter 12), these algorithms have been found to be very efficient. The estimators presented here were based on Gaussian perturbation sequences. It would be interesting to develop similar estimators using Cauchy and uniform perturbation sequences as well.

References

1. Bhatnagar, S.: Adaptive multivariate three-timescale stochastic approximation algorithms for simulation based optimization. ACM Transactions on Modeling and Computer Simulation 15(1), 74–107 (2005)
2. Bhatnagar, S.: Adaptive Newton-based smoothed functional algorithms for simulation optimization. ACM Transactions on Modeling and Computer Simulation 18(1), 2:1–2:35 (2007)
3. Bhatnagar, S., Mishra, V., Hemachandra, N.: Stochastic algorithms for discrete parameter simulation optimization. IEEE Transactions on Automation Science and Engineering 9(4), 780–793 (2011)
4. Kushner, H.J., Clark, D.S.: Stochastic Approximation Methods for Constrained and Unconstrained Systems. Springer, New York (1978)

Part IV
Variations to the Basic Scheme

This part deals with certain variations to the basic scheme. In particular, we consider applications of simultaneous perturbation approaches to (a) discrete parameter optimization, (b) optimization under inequality constraints when both the objective and the constraint functions are certain long-run average cost objectives, and (c) reinforcement learning — a class of methods that deal with the problem of stochastic control under lack of precise model information.

Many times, one is interested in optimizing a certain cost objective over a discrete set of alternatives or parameters and the goal is to find the best possible parameter. The problem becomes even more interesting when long-run average cost objectives are used. In situations when the parameter set is small, traditional approaches for this problem involve estimating the objective function value for each parameter using Monte-Carlo simulation in order to judge the best parameter. Bhatnagar, Mishra and Hemachandra, in a paper in 2011, presented adaptations of gradient SPSA and SF algorithms for the problem of discrete parameter search for long-run average cost objectives. They also presented a novel random projection technique. The adapted algorithms are seen to be better in performance than a well-known algorithm in the literature in the case when the parameter set is small. When the set is large, the adapted algorithms are still seen to show good results and require less computation. Chapter 9 presents the adaptations of gradient SPSA and SF algorithms to the case of discrete parameter optimization.

One is often interested in optimizing a given objective function subject to certain functional (inequality) constraints being met. For instance, one might be interested in finding a path over which throughput is maximized for a given stream of packets passing through a communication network given that the mean delays along that path are below a pre-specified threshold. The problem becomes interesting when both the objective and the constraint functions are certain long-run averages as under such scenarios, the constraint region is also not known precisely. In a paper in 2011, Bhatnagar, Hemachandra and Mishra, presented four simultaneous perturbation algorithms for this purpose, that incorporate a Lagrange multiplier approach. Chapter 10 presents the simultaneous perturbation algorithms for this problem.

In problems of stochastic control, one is often confronted with scenarios where model information (i.e., knowledge of transition probabilities) is not known and yet one wants to pick an optimal feedback control policy. More over, in many real-life situations, the cardinalities of the state and/or action spaces could be large as well making schemes based on numerical techniques computationally infeasible. Reinforcement learning broadly refers to a class of algorithms that are based on simulation based approaches. In many papers, Bhatnagar and several coauthors presented a host of algorithms for these problems for various cost settings and also for cases when (a) the state-action space size is manageable as well as (b) when it is not and approximation methods based on function approximation need to be resorted to. Chapter 11 deals with the applications of simultaneous perturbation approaches to reinforcement learning.

Chapter 9
Discrete Parameter Optimization

9.1 Introduction

We begin by recalling the basic optimization problem discussed in Chapter 1.

$$\text{Find } \theta^* \text{ that solves } \min_{\theta \in C} J(\theta), \tag{9.1}$$

for a given objective function $J : \mathbb{R}^N \to \mathbb{R}$ that is a function of a tunable parameter θ taking values in a set $C \subset \mathbb{R}^N$. In all the chapters in this book, except the current, the set C has been considered to be a compact and convex subset of $\mathbb{R}^N$. In this chapter, however, we assume that the set C is discrete-valued and contains a finite number of points. Moreover, $J(\theta), \theta \in C$ is a long-run average cost objective. This chapter is largely based on [3].

The above problem has attracted considerable attention in the case when the objective is an expected value over certain noisy cost function measurements. In such cases, when the cardinality of the constraint set C is small, two of the techniques that have been widely studied go under the names of ranking and selection (R&S) [1], and multiple comparison procedures (MCP) [9]. The observations for given θ in these procedures are assumed i.i.d., often with the normal distribution.

As the name suggests, in the R&S class of procedures, the objective function value corresponding to each parameter is estimated from sample path observations and then the various estimates are ranked to obtain the best parameter value. The above procedures are in general not applicable when the number of parameters is large due to the amount of computational effort involved. Optimal computing budget allocation (OCBA) [8, 6, 7] is an R&S procedure that is widely regarded as being amongst the best procedures for small-scale discrete optimization. The idea in OCBA is to optimally allocate a given computing budget between various alternatives in a way as to maximize the probability of correct selection. Ordinal

S. Bhatnagar et al.: Stochastic Recursive Algorithms for Optimization, LNCIS 434, pp. 151–166.
springerlink.com

optimization [14] and simulated annealing [15] are amongst the popular procedures used for large parameter sets.

As stated before, the afore-mentioned approaches have been proposed for the case when the objective function is an expectation over noisy cost samples and do not carry over in a straightforward manner to long-run average cost objectives. A problem with optimization under steady-state simulation is that it is expensive to obtain multiple-independent simulation trajectories corresponding to any given parameter, see, however, [12] for an R&S procedure for steady-state simulations.

Gradient search approaches for continuous parameter optimization have also been applied to the case of discrete optimization in [11, 10, 2, 4]. The idea in these methods is to first form a closed convex hull of the discrete search space and consider an alternative continuous optimization problem in the (above) closed convex hull except for a difference between the various techniques. In [11, 4], the 'continuous portion' of the optimization procedure is allowed to proceed as in continuous optimization procedures and upon convergence, the parameter in the original discrete domain that is the closest to the converged parameter value (in the convex hull) is identified as the converged parameter in the discrete set. In [10, 2], after each iterate, the parameter value obtained from the continuous optimization step is projected back to the discrete set, resulting in the parameter update in each step of the procedure being precisely over the discrete set. In [3], a somewhat similar approach as [11, 4] is used except that for purposes of projection of the continuous parameter update to the discrete set, a random projection approach is considered instead of deterministically projecting each update to the discrete set. This is seen to smooth the underlying dynamics of the associated process. A discrete form of SPSA is considered in [13] and some convergence results have been shown for the same. A different form of discrete algorithm as compared to [13] that shows better results has recently been presented in [16]. In what follows, we shall consider general smooth mappings of which the random projection procedure of [3] emerges as a special case. The algorithms of [3] are then presented for these mappings.

The rest of the chapter is organized as follows: discrete optimization framework is presented in Section 9.2. The algorithms are presented in Section 9.3. Finally, concluding remarks are presented in Section 9.4.

9.2 The Framework

We consider the following the problem setting. Let $X^{\theta}(n), n \geq 0$ be a discrete-time stochastic process whose evolution depends on a parameter $\theta \in C \subset \mathbb{R}^N$ for some fixed $N \geq 1$. Let C (the set in which θ takes values) be a (discrete) finite set having the form $C = \prod_{i=1}^{N} C^i$, where $C^i = \{c_i^0, \ldots, c_i^{n_i}\}$, $i = 1, \ldots, N$. Let $C_{i,\min} \equiv c_i^0 < c_i^1 < \cdots < c_i^{n_i} \equiv C_{i,\max}$, for each $i = 1, \ldots, N$. By construction, the set C contains (say) p points $\theta^1, \theta^2, \ldots, \theta^p$, with $p < \infty$. Thus, $C = \{\theta^1, \theta^2, \ldots, \theta^p\}$. For any $\theta \in C$, let $X^\theta(n)$, $n \geq 0$ take values in the set $\mathscr{S} = \{0, 1, 2, \ldots, |\mathscr{S}|\}$, where $|\mathscr{S}| < \infty$ could be a large integer.

For fixed $\theta \in C$, we assume $X^\theta(n), n \geq 0$ to be an ergodic Markov chain with transition probabilities $p_\theta(i, j)$, $i, j \in \mathscr{S}$. Note that when the parameter θ is tuned (i.e., with $\theta(n)$ in place of θ at instant n), the process $X^\theta(n), n \geq 0$, in general will not be Markov. Let $h : \mathscr{S} \to \mathbb{R}$ be a given state-dependent, single-stage cost function. Our aim is to find a $\theta^* \in C$ satisfying (9.1) where for any $\theta \in C$, $J(\theta)$ is the long-run average cost

$$J(\theta) = \lim_{n \to \infty} \frac{1}{n} \sum_{i=1}^{n} h(X^\theta(i)). \tag{9.2}$$

Now let $\bar{C}$ denote the closed convex hull of the set C (i.e., the smallest closed and convex set containing C). Let $\theta(n) \stackrel{\triangle}{=} (\theta_1(n), \ldots, \theta_N(n))^T, n \geq 1$ denote the sequence of updates of the parameter θ. As stated before, the idea in the described procedures will be to consider the parameter updates in the set $\bar{C}$ (and not C). However, when needed, the discrete parameter to use corresponding to the continuous update shall be obtained from a certain projection operator. We describe below some of the projection operators.

9.2.1 The Deterministic Projection Operator

This is the most commonly used operator, even though as we will later explain, proving the convergence of the resulting scheme when using this operator is not easy because of lack of smoothness at some points in the (extended) transition dynamics of the Markov process corresponding to parameters in the closed and convex hull. For any $\theta = (\theta_1, \ldots, \theta_N)^T \in \mathbb{R}^N$, let $\Gamma(\theta) = (\Gamma_1(\theta_1), \ldots, \Gamma_N(\theta_N))^T \in C$ denote the (*deterministic*) projection of θ to the set C and is defined as follows: Let θ_i be such that $c_i^j \leq \theta_i \leq c_i^{j+1}$ for some $c_i^j < c_i^{j+1}$, with $c_i^j, c_i^{j+1} \in C^i$. Now set

$$\Gamma_i(\theta_i) = \begin{cases} c_i^j & \text{if } (c_i^j \leq \theta_i < (c_i^j + c_i^{j+1})/2 \\ c_i^{j+1} & \text{if } (c_i^j + c_i^{j+1})/2 < \theta_i \leq c_i^{j+1}. \end{cases}$$

If $\theta_i = (c_i^j + c_i^{j+1})/2$, then θ_i is set to either c_i^j or c_i^{j+1} according to some prescribed rule. Also,

$$\Gamma_i(\theta_i) = \begin{cases} C_{i,\min} & \text{if } \theta_i \leq C_{i,\min} \\ C_{i,\max} & \text{if } \theta_i \geq C_{i,\max}. \end{cases}$$

It is clear from the above that $\Gamma(\theta) \in C$.

9.2.2 The Random Projection Operator

We now describe the *random projection* technique from [3]. Unlike the commonly used deterministic projection operators, this technique is seen to smooth the transition dynamics of (the extended parametrized Markov process) $\{X^{\theta}(n)\}$, when $\theta \in \bar{C}$. For $\theta \in \mathbb{R}^N$, the projection mapping $\Gamma(\theta)$ in this case is defined as follows: Let θ_i be such that $c_i^j \leq \theta_i \leq c_i^{j+1}$ for some $c_i^j < c_i^{j+1}$, with $c_i^j, c_i^{j+1} \in C^i$, $i = 1, \ldots, N$. Now observe that one can represent θ_i in terms of c_i^j, c_i^{j+1} as $\theta_i = \alpha_i c_i^j + (1-\alpha_i) c_i^{j+1}$, where

$$\alpha_i = \frac{c_i^{j+1} - \theta_i}{c_i^{j+1} - c_i^j} \in [0,1].$$

Thus,

$$\Gamma_i(\theta_i) = \begin{cases} c_i^j & \text{w.p. } (c_i^{j+1} - \theta_i)/(c_i^{j+1} - c_i^j) \\ c_i^{j+1} & \text{w.p. } (\theta_i - c_i^j)/(c_i^{j+1} - c_i^j). \end{cases}$$

Also,

$$\Gamma_i(\theta_i) = \begin{cases} C_{i,\min} & \text{w.p.1 if } \theta_i \leq C_{i,\min} \\ C_{i,\max} & \text{w.p.1 if } \theta_i \geq C_{i,\max}. \end{cases}$$

It is easy to see that $\Gamma(\theta) \in C$. Now note that any $\theta \in \bar{C}$ can be written as a convex combination

$$\theta = \sum_{k=1}^{p} \alpha_k(\theta)\theta^k, \tag{9.3}$$

of the elements of C. The weights $\alpha_k(\theta)$ satisfy $0 \leq \alpha_k(\theta) \leq 1$, $\forall k \in \{1, \ldots, p\}$ and $\sum_{k=1}^{p} \alpha_k(\theta) = 1$. Such a representation is useful to show the convergence of the algorithms even though precise knowledge of the weights $\alpha_k(\theta)$ is not required. One possible manner in which the weights $\alpha_k(\theta)$ can be obtained is by projecting $\theta \in \bar{C}$ to its *nearest neighbours* in C using the weights in the Γ-projection. We consider the following example for illustrative purposes.

Example 9.1. Let θ be a vector with three components $\theta = (\theta_1, \theta_2, \theta_3)^T \in \mathbb{R}^3$. Suppose $c_1^j \leq \theta_1 \leq c_1^{j+1}$, $c_2^k \leq \theta_2 \leq c_2^{k+1}$ and $c_3^l \leq \theta_3 \leq c_3^{l+1}$. Then with appropriate $\alpha^1, \alpha^2, \alpha^3 \in [0,1]$, one can write

$$\theta = (\alpha^1 c_1^j + (1-\alpha^1)c_1^{j+1}, \alpha^2 c_2^k + (1-\alpha^2)c_2^{k+1}, \alpha^3 c_3^l + (1-\alpha^3)c_3^{l+1})^T.$$

For this example, Table 9.1 shows the various points in $C \subset \mathbb{R}^3$ with respect to which the above θ can be expressed as a convex combination, as well as the corresponding weights. The remaining points in C are assigned a weight of zero each.

Table 9.1 Example: Parameter $\theta = (\alpha^1 c_1^j + (1-\alpha^1)c_1^{j+1}, \alpha^2 c_2^k + (1-\alpha^2)c_2^{k+1}, \alpha^3 c_3^l + (1-\alpha^3)c_3^{l+1})^T \in \bar{C} \subset \mathbb{R}^3$ as a convex combination of the elements of C

S.N.	Elements in C	Weights
1.	$(c_1^j, c_2^k, c_3^l)^T$	$\alpha^1\alpha^2\alpha^3$
2.	$(c_1^j, c_2^k, c_3^{l+1})^T$	$\alpha^1\alpha^2(1-\alpha^3)$
3.	$(c_1^j, c_2^{k+1}, c_3^l)^T$	$\alpha^1(1-\alpha^2)\alpha^3$
4.	$(c_1^j, c_2^{k+1}, c_3^{l+1})^T$	$\alpha^1(1-\alpha^2)(1-\alpha^3)$
5.	$(c_1^{j+1}, c_2^k, c_3^l)^T$	$(1-\alpha^1)\alpha^2\alpha^3$
6.	$(c_1^{j+1}, c_2^k, c_3^{l+1})^T$	$(1-\alpha^1)\alpha^2(1-\alpha^3)$
7.	$(c_1^{j+1}, c_2^{k+1}, c_3^l)^T$	$(1-\alpha^1)(1-\alpha^2)\alpha^3$
8.	$(c_1^{j+1}, c_2^{k+1}, c_3^{l+1}$	$(1-\alpha^1)(1-\alpha^2)(1-\alpha^3)$

A procedure as described above can similarly be extended to the case of parameters θ with N components. As a consequence of the above, one can alternatively view the (*randomized*) Γ-projection as a probabilistic projection of $\theta \in \bar{C}$ to the set C so that $\Gamma(\theta) = \theta^k \in C$ with probability $\gamma_k(\theta)$, $k = 1, \ldots, p$, for some $\gamma_k(\theta)$ such that $0 \le \gamma_k(\theta) \le 1$, $\forall k = 1, \ldots, p$, and $\sum_{k=1}^{p} \gamma_k(\theta) = 1$. For instance, in Example 9.1, $\Gamma(\theta) = \theta_1 \stackrel{\triangle}{=} (c_1^j, c_2^k, c_3^l)^T$ with probability $\gamma_1(\theta) = (1-\alpha^1)(1-\alpha^2)(1-\alpha^3)$. It is easy to see that $\gamma_k(\theta)$, $k = 1, \ldots, p$, are continuously differentiable functions of θ.

9.2.3 *A Generalized Projection Operator*

We now present a generalized projection operator that can alternatively be used (in place of the deterministic projection scheme as well as the randomized projection scheme, respectively). The manner in which this operator is constructed, it works as a deterministic projection scheme in some portions of the parameter space and as a randomized projection scheme in some other portions. Unlike the deterministic projection scheme, it has the advantage that it results in a smooth transition dynamics for the extended Markov process with parameters in the closed and convex hull. Over the randomized projection scheme, it has the advantage of a lower computational requirement because in a significant portion of the space, a deterministic projection is used and thus one does not require generation of random numbers for the probabilistic projection in these portions.

Let θ_i be such that $c_i^j \le \theta_i \le c_i^{j+1}$ for some $c_i^j < c_i^{j+1}$, with $c_i^j, c_i^{j+1} \in C^i$, $i = 1, \ldots, N$. Let $\varepsilon > 0$ be a given small constant. Now set

$$\Gamma_i(\theta_i) = \begin{cases} c_i^j & \text{if } (c_i^j \le \theta_i < (c_i^j + c_i^{j+1})/2 - \varepsilon \\ c_i^{j+1} & \text{if } (c_i^j + c_i^{j+1})/2 + \varepsilon \le \theta_i \le c_i^{j+1}. \end{cases}$$

Further, for $(c_i^j + c_i^{j+1})/2 - \varepsilon \le \theta_i \le (c_i^j + c_i^{j+1})/2 + \varepsilon, i = 1, \ldots, N$,

$$\Gamma_i(\theta_i) = \begin{cases} c_i^j & \text{w.p. } f_i(\theta_i) \\ c_i^{j+1} & \text{w.p. } 1 - f_i(\theta_i). \end{cases}$$

Also, as before,

$$\Gamma_i(\theta_i) = \begin{cases} C_{i,\min} & \text{if } \theta_i \le C_{i,\min} \\ C_{i,\max} & \text{if } \theta_i \ge C_{i,\max}. \end{cases}$$

In the above, $f_i(\cdot)$ is a decreasing and continuously differentiable function that takes values in $[0,1]$ and is such that

$$f_i(\theta_i) = \begin{cases} 0 & \text{if } \theta_i = (c_i^j + c_i^{j+1})/2 + \varepsilon \\ 1 & \text{if } (c_i^j + c_i^{j+1})/2 - \varepsilon. \end{cases}$$

For the purposes of analysis, one can view the *generalized* projection operator as another *randomized* projection scheme where the deterministic "if" statements are replaced by similar statements with "w.p.1 if" conditions. For instance, the first set of conditions for the generalized scheme can (for simplicity in analysis) be approximately rewritten as

$$\Gamma_i(\theta_i) = \begin{cases} c_i^j & \text{w.p.1 if } (c_i^j \le \theta_i < (c_i^j + c_i^{j+1})/2 - \varepsilon \\ c_i^{j+1} & \text{w.p.1 if } (c_i^j + c_i^{j+1})/2 + \varepsilon \le \theta_i \le c_i^{j+1}. \end{cases}$$

The same applies to the other deterministic conditions as well.

In a similar manner as the *randomized* projection scheme described above, it is easy to see that one can obtain weights $\beta_k(\theta), k = 1, \ldots, p$ such that $\Gamma(\theta) = \theta^k$ w.p. $\beta_k(\theta)$. Here, $\beta_k(\theta) \in [0,1], \forall k = 1, \ldots, p$ and $\sum_{k=1}^{p} \beta_k(\theta) = 1$ for any $\theta \in \bar{C}$. The quantity ε can be chosen arbitrarily in applications. A larger value of ε will allow for greater "exploration" of the discrete parameter space by the algorithm.

Example 9.2. As an example, let $\theta \stackrel{\triangle}{=} (\theta_1, \theta_2)^T$ be such that $\theta_1 \in [c_1^k, (c_1^k + c_1^{k+1})/2 - \varepsilon)$ and $\theta_2 \in [(c_2^l + c_2^l)/2 - \varepsilon, (c_2^l + c_2^{l+1})/2 + \varepsilon]$. Then, from the manner in which the generalized projection operator is defined, we have that

$$\Gamma_1(\theta_1) = \begin{cases} c_1^k & \text{w.p.1} \\ c_1^{k+1} & \text{w.p.0}. \end{cases}$$

Also,

$$\Gamma_2(\theta_2) = \begin{cases} c_2^l & \text{w.p.} f_2(\theta_2) \\ c_2^{l+1} & \text{w.p.} 1 - f_2(\theta_2). \end{cases}$$

Since each parameter component is mapped independently (of the other components) to its discrete set of points, we have that

$$\Gamma(\theta) = \begin{cases} (c_1^k, c_2^l)^T & \text{w.p.} f_2(\theta_2) \\ (c_1^k, c_2^{l+1})^T & \text{w.p.} 1 - f_2(\theta_2). \end{cases}$$

The remaining components in C are then assigned a weight of 0 each. Thus, if θ^j and θ^m, respectively correspond to the points $(c_1^k, c_2^l)^T$ and $(c_1^k, c_2^{l+1})^T$ within the set C, then one may let $\beta_j(\theta) = f_2(\theta_2)$ and $\beta_m(\theta) = 1 - f_2(\theta_2)$. Further, $\beta_k(\theta) = 0, \forall k \in \{1, \ldots, p\}$ with $k \notin \{j, m\}$.

9.2.4 Regular Projection Operator to $\bar{C}$

We denote the regular projection of any $\theta \in \mathbb{R}^N$ to the set $\bar{C} = \prod_{i=1}^{N} [C_{i,\min}, C_{i,\max}]$, which is the closed convex hull of C (from construction) as $\bar{\Gamma}(\theta) \stackrel{\triangle}{=} (\bar{\Gamma}_1(\theta_1), \ldots, \bar{\Gamma}_N(\theta_N))^T$. Here $\bar{\Gamma}_i(\theta_i) = \min(C_{i,\max}, \max(\theta_i, C_{i,\min}))$. It is easy to obtain this projection by simply comparing each component θ_j of the parameter θ $(j = 1, \ldots, N)$ with the corresponding boundary points $C_{j,\min}$ and $C_{j,\max}$ in the sets C^j and resetting (component-wise) θ_j to $C_{j,\min}$ or $C_{j,\max}$ depending on whether θ_j is below $C_{j,\min}$ or above $C_{j,\max}$. Further, if $C_{j,\min} \leq \theta_j \leq C_{j,\max}$, then $\bar{\Gamma}_j(\theta_j) = \theta_j$.

9.2.5 Basic Results for the Generalized Projection Operator Case

We consider here the generalized projection operator. Similar results with the randomized projection operator have been presented in [3]. Define the transition probabilities $p_\theta(i, j), i, j \in \mathscr{S}$ with $\theta \in \bar{C}$ according to

$$p_\theta(i, j) = \sum_{k=1}^{p} \beta_k(\theta) p_{\theta^k}(i, j), \tag{9.4}$$

with $\beta_k(\theta)$ obtained as described in the generalized projection scheme. It is easy to see that $0 \leq p_\theta(i, j) \leq 1, \forall i, j \in \mathscr{S}, \theta \in \bar{C}$ and $\sum_{j \in \mathscr{S}} p_\theta(i, j) = 1, \forall i \in \mathscr{S}, \theta \in \bar{C}$. Also, $p_\theta(i, j)$ are continuously differentiable in $\theta \in \bar{C}$, because of the fact that $f_k(\theta_k), k = 1, \ldots, p$ are chosen to be continuously differentiable functions. For any $n \geq 1$, let

$p^n_\theta(i,j)$ represent the probability of going from state i to state j in n steps when the underlying parameter is θ.

Lemma 9.1. *For any $\theta \in \bar{C}$, $i,j \in \mathscr{S}$ and any $n \geq 1$*

$$p^n_\theta(i,j) \geq \sum_{l=1}^{p} \beta_l^n(\theta) p^n_{\theta^l}(i,j).$$

Proof. We prove the claim by induction. Note that the claim is true for $n = 1$ from (9.4). Assume that the claim is valid for some $n = K > 1$. We now show that it is true for $n = K+1$. Note that

$$\begin{aligned} p^{K+1}_\theta(i,j) &= \sum_{l\in\mathscr{S}} p_\theta(i,l) p^K_\theta(l,j) \\ &\geq \sum_{l\in\mathscr{S}} \sum_{r=1}^{p} \beta_r(\theta) p_{\theta^r}(i,l) \sum_{m=1}^{p} \beta_m^K(\theta) p^K_{\theta^m}(l,j) \\ &= \sum_{r=1}^{p} \sum_{m=1}^{p} \beta_r(\theta)\beta_m^K(\theta) \sum_{l\in\mathscr{S}} p_{\theta^r}(i,l) p^K_{\theta^m}(l,j) \\ &\geq \sum_{r=1}^{p} \beta_r^{K+1}(\theta) \sum_{l\in\mathscr{S}} p_{\theta^r}(i,l) p^K_{\theta^r}(l,j) \\ &= \sum_{r=1}^{p} \beta_r^{K+1}(\theta) p^{K+1}_{\theta^r}(i,j). \end{aligned}$$

The first inequality above follows from the induction hypothesis while the second inequality follows by considering only values of $m = r$ in the second summation in its preceding expression. The claim follows. □

Lemma 9.2. *For any $\theta \in \bar{C}$, $\{X^\theta(n), n \geq 1\}$ is an ergodic Markov chain.*

Proof. It is easy to see that the process $X^\theta(n), n \geq 1$, $\theta \in \bar{C}$ governed by the transition probabilities $p_\theta(i,j)$, $i,j \in \mathscr{S}$ (defined as in (9.4)) is Markov. Now each of the processes $X^{\theta^k}(n), n \geq 1$, $k = 1,\ldots,p$ is ergodic Markov. Consider now $X^{\theta^k}(n), n \geq 1$ for some $\theta^k \in C$. Since $X^{\theta^k}(n), n \geq 1$ is irreducible, for any $i,j,l \in \mathscr{S}$, there exist integers $n_{1,k}$, $n_{2,k} > 0$ such that $p^{n_{1,k}}_{\theta^k}(i,l) > 0$ and $p^{n_{2,k}}_{\theta^k}(l,j) > 0$. (Here $n_{1,k}$ in general depends on i and l, and $n_{2,k}$ on l and j, respectively.) Now since $X^{\theta^k}(n), n \geq 1$ is aperiodic, there exists $M_k > 0$ (that depends on l) such that $p^n_{\theta^k}(l,l) > 0$ for all $l \in \mathscr{S}$ and $n \geq M_k$, see for instance [5, Lemma 5.3.2, pp.99]. Now

$$\begin{aligned} p^{n_{1,k}+n+n_{2,k}}_{\theta^k}(i,j) &\geq p^{n_{1,k}}_{\theta^k}(i,l) p^n_{\theta^k}(l,l) p^{n_{2,k}}_{\theta^k}(l,j) \\ &> 0 \ \forall n \geq M_k. \end{aligned}$$

Thus $p^n_{\theta k}(i,j) > 0$ for all $n > N_k \stackrel{\triangle}{=} n_{1,k} + n_{2,k} + M_k$, $k = 1,\ldots,p$. Now let $\hat{N} = \max(N_1,\ldots,N_p) < \infty$. Hence, $p^n_{\theta k}(i,j) > 0$ for all $n \geq \hat{N}$ and $k = 1,\ldots,p$. From Lemma 9.1, it follows that $p^n_{\theta}(i,j) > 0$, $\forall n \geq \hat{N}$. The above is true for all $i,j \in \mathscr{S}$ with, however, a possibly different value of $\hat{N}$ for different (i,j)–tuples. Thus $\{X^{\theta}(n)\}$ is irreducible. It can also be seen to be aperiodic by letting $j = i$. Finally, the chain $\{X^{\theta}(n)\}$, $\theta \in \bar{C}$ is positive recurrent since it is irreducible and finite state. □

Now let $\bar{J}(\theta)$ be defined as in (9.2) for $\theta \in \bar{C}$, i.e.,

$$\bar{J}(\theta) = \lim_{n\to\infty} \frac{1}{n} \sum_{m=0}^{n-1} h(X^{\theta}(m)),\ \theta \in \bar{C}.$$

Note that the single-stage cost function $h(\cdot)$ is the same as before. The only difference is in the parameter θ that now takes values in $\bar{C}$. By Lemma 9.2, the above limit is well defined for all $\theta \in \bar{C}$. By definition, $\bar{J}(\theta) = J(\theta)$ for all $\theta \in C$.

Lemma 9.3. *$\bar{J}(\theta)$ is continuously differentiable in $\theta \in \bar{C}$.*

Proof. Let $\gamma(\theta) \stackrel{\triangle}{=} (\gamma_i(\theta), i \in \mathscr{S})^T$ denote the stationary distribution of the Markov chain $X^{\theta}(n), n \geq 1$, $\theta \in \bar{C}$. Then, $\bar{J}(\theta)$ can be written as

$$\bar{J}(\theta) = \sum_{i\in S} h(i)\gamma_i(\theta).$$

Thus it is sufficient to show that $\gamma(\theta)$ is continuously differentiable in θ. Let us denote $P(\theta) \stackrel{\triangle}{=} [[p_{\theta}(i,j), i,j \in \mathscr{S}]]$ as the transition probability matrix when the parameter is $\theta \in \bar{C}$ is held fixed.

The claim will follow using a result from [17, Theorem 2 on pp.402–403]. Let $P^{\infty}(\theta) = \lim_{m\to\infty} \frac{1}{m} \sum_{n=1}^{m} P^n(\theta)$ and $Z(\theta) \stackrel{\triangle}{=} [I - P(\theta) - P^{\infty}(\theta)]^{-1}$, respectively, where I denotes the $(|\mathscr{S}| \times |\mathscr{S}|)$-identity matrix and $P^m(\theta)$ is the matrix of m-step transition probabilities $p^m_{\theta}(i,j)$, $i,j \in \mathscr{S}$. From [17, Theorem 2], one can write

$$\gamma(\theta + \delta e_i) = \gamma(\theta)(I + (P(\theta + \delta e_i) - P(\theta))Z(\theta) + o(h)),$$

where $\delta > 0$ is a small quantity and e_i, $i \in \{1,\ldots,N\}$ is a unit vector with 1 as its ith entry and 0s elsewhere. Hence we get

$$\nabla_i \gamma(\theta) = \gamma(\theta)\nabla_i P(\theta) Z(\theta),\ i = 1,\ldots,N.$$

Thus,

$$\nabla\gamma(\theta) = \gamma(\theta)\nabla P(\theta) Z(\theta). \tag{9.5}$$

By construction (as mentioned before), $\nabla P(\theta)$ exists and is continuous. Hence $\nabla\gamma(\theta)$ exists. Next, we verify that $\nabla\gamma(\theta)$ is continuous as well. Note that $\gamma(\theta)$ is continuous because it is differentiable. The claim will follow if we show that $Z(\theta)$ is also a continuous function.

Let $H(\theta,\theta+\theta_0) = [I-(P(\theta+\theta_0)-P(\theta))]^{-1}$. Then $H(\theta,\theta+\theta_0) \to I$ as $\|\theta_0\| \to 0$. Also, let $U(\theta,\theta+\theta_0) = (P(\theta+\theta_0)-P(\theta))Z(\theta)$. Then $U(\theta,\theta+\theta_0) \to \bar{0}$ as $\|\theta_0\| \to 0$. Here $\bar{0}$ is a matrix with all elements being zero. From [17, Theorem 2], we have

$$Z(\theta+\theta_0) = Z(\theta)H(\theta,\theta+\theta_0)$$
$$-P^{\infty}(\theta)H(\theta,\theta+\theta_0)U(\theta,\theta+\theta_0)Z(\theta)H(\theta,\theta+\theta_0).$$

Hence we get

$$\|Z(\theta+\theta_0)-Z(\theta)\| \le \|Z(\theta)\|\|H(\theta,\theta+\theta_0)-I\|$$
$$+\|P^{\infty}(\theta)\|\|H(\theta,\theta+\theta_0)\|\|U(\theta,\theta+\theta_0)\|\|Z(\theta)\|\|H(\theta,\theta+\theta_0)\|.$$

It can thus be seen that

$$\|Z(\theta+\theta_0)-Z(\theta)\| \to 0 \text{ as } \|\theta_0\| \to 0.$$

The claim follows. □

9.3 The Algorithms

The operator $\bar{\Gamma}$ will be used to project the continuous-valued iterates in the algorithms (below) to the closed convex hull $\bar{C}$ of the set C while Γ will be used to identify the actual parameter value used in the simulation. We present two algorithms: one based on SPSA gradient estimates and the other based on SF estimates. We refer to these as simply SPSA and SFA, respectively. We consider the mapping Γ to be defined via the *generalized* projection scheme. These algorithms but with the *randomized* projection operator are described in [3]. Since we use a long-run average cost objective, both algorithms incorporate two step-size sequences $a(n)$ and $b(n), n \ge 0$ that satisfy Assumption 3.6. Thus, recursions governed by $a(n), n \ge 0$ are slower while those governed by $b(n), n \ge 0$ are faster. In either of the algorithms below, $\theta(m) \stackrel{\triangle}{=} (\theta_1(m),\ldots,\theta_N(m))^T$ shall denote the parameter vector at the end of the mth iteration.

9.3.1 The SPSA Algorithm

Let $\Delta_1(m),\ldots,\Delta_N(m)$ denote independent random variables having the distribution $\Delta_i(m) = \pm 1$ w.p. 1/2, $\forall i = 1,\ldots,N$, $m \ge 0$. Set $\theta_i^1(m) = \Gamma_i(\theta_i(m) + \delta\Delta_i(m))$ and

$\theta_i^2(m) = \Gamma_i(\theta_i(m) - \delta\Delta_i(m))$, respectively, for $i = 1,\ldots,N$, where $\delta > 0$ is a given (small) constant. Other distributions for the perturbation sequences $\Delta_i(m), m \geq 0$, $i = 1,\ldots,N$ may also be used (see Chapter 5). Let $\theta^j(m) \triangleq (\theta_1^j(m),\ldots,\theta_N^j(m))$, $j = 1,2$. Set $Z(0) = 0$.

Generate two parallel simulations $\{X^{\theta^1(m)}(m)\}$ and $\{X^{\theta^2(m)}(m)\}$ governed by parameter sequences $\{\theta^1(m)\}$ and $\{\theta^2(m)\}$, respectively.

Fix a large integer $M > 0$. For $i = 1,\ldots,N$, $m = 0,1,\ldots,M-1$, we have

$$\theta_i(m+1) = \bar{\Gamma}_i\left(\theta_i(m) + a(m)\frac{Z(m+1)}{2\delta\Delta_i(m)}\right), \tag{9.6}$$

$$Z(m+1) = Z(m) + b(m)(h(X^{\theta^2(m)}(m)) - h(X^{\theta^1(m)}(m)) - Z(m)). \tag{9.7}$$

Output $\Gamma(\theta(M))$ as the final parameter.

9.3.2 The SFA Algorithm

Let $\eta_1(m),\ldots,\eta_N(m)$ be independent $N(0,1)$-distributed random variables. Let $\beta > 0$ be a given (small) constant. Let $\theta^j(m) = (\theta_1^j(m),\ldots,\theta_N^j(m))^T$, $j = 1,2$, where $\theta_i^1(m) = \Gamma_i(\theta_i(m) + \beta\eta_i(m))$ and $\theta_i^2(m) = \Gamma_i(\theta_i(m) - \beta\eta_i(m))$, $i = 1,\ldots,N$, respectively.

Generate two parallel simulations $\{X^{\theta^1(m)}(m)\}$ and $\{X^{\theta^2(m)}(m)\}$ governed by parameter sequences $\{\theta^1(m)\}$ and $\{\theta^2(m)\}$, respectively.

Fix a large integer $M > 0$. For $i = 1,\ldots,N$, $m = 0,1,\ldots,M-1$, we have

$$\theta_i(m+1) = \bar{\Gamma}_i(\theta_i(m) + a(m)Z_i(m+1)), \tag{9.8}$$

$$Z_i(m+1) = Z_i(m) + b(m)\left(\frac{\eta_i(m)}{2\beta}(h(X^{\theta^2(m)}(m)) - h(X^{\theta^1(m)}(m))) - Z_i(m)\right). \tag{9.9}$$

Output $\Gamma(\theta(M))$ as the final parameter.

It is important to note that while the updates of the parameter θ in the SPSA and SFA algorithms are performed in the set $\bar{C}$, the actual parameters used in the two simulations at the mth instant, $m \geq 0$, in these algorithms (i.e., $\theta^1(m)$ and $\theta^2(m)$, respectively) are C-valued as a consequence of the Γ-projection.

9.3.3 Convergence Analysis

We show here the convergence of both algorithms. The first result below shows that under the extended dynamics of the Markov chain $X^\theta(n), n \geq 0$, $\theta \in \bar{C}$, each algorithm is analogous to its continuous parameter counterpart where the random projection operator Γ is replaced with $\bar{\Gamma}$.

Lemma 9.4. *Under the extended dynamics of the Markov process $\{X^\theta(n)\}$ defined over all $\theta \in \bar{C}$,*

(i) SPSA is analogous to a similar algorithm in which $\theta^i(m)$ in SPSA is replaced by $\bar{\theta}^i(m) \stackrel{\triangle}{=} (\bar{\theta}^i_j(m), j = 1,\ldots,N)^T$, $i = 1,2$, where $\bar{\theta}^1_j(m) = \bar{\Gamma}_j(\theta_j(m) + \delta\Delta_j(m))$ and $\bar{\theta}^2_j(m) = \bar{\Gamma}_j(\theta_j(m) - \delta\Delta_j(m))$, respectively, $j = 1,\ldots,N$.

(ii) SFA is analogous to a similar algorithm in which $\theta^i(m)$ in SFA is replaced by $\bar{\theta}^i(m) \stackrel{\triangle}{=} (\bar{\theta}^i_j(m), j = 1,\ldots,N)^T$, $i = 1,2$, where $\bar{\theta}^1_j(m) = \bar{\Gamma}_j(\theta_j(m) + \beta\eta_j(m))$ and $\bar{\theta}^2_j(m) = \bar{\Gamma}_j(\theta_j(m) - \beta\eta_j(m))$, respectively, $j = 1,\ldots,N$.

Proof. We prove here the claim in part (i) for the case of SPSA. The same for SFA (in part (ii)) follows in a similar manner. Consider the SPSA algorithm (9.6)–(9.7). Let $\theta(m)$ be a given parameter update that lies in $\bar{C}^o$ (where $\bar{C}^o$ denotes the interior of the set $\bar{C}$). Let $\delta > 0$ be sufficiently small so that $\bar{\theta}^1(m) = (\bar{\Gamma}_j(\theta_j(m) + \delta\Delta_j(m)), j = 1,\ldots,N)^T = (\theta_j(m) + \delta\Delta_j(m)), j = 1,\ldots,N)^T$ and $\bar{\theta}^2(m) = (\bar{\Gamma}_j(\theta_j(m) - \delta\Delta_j(m)), j = 1,\ldots,N)^T = (\theta_j(m) - \delta\Delta_j(m)), j = 1,\ldots,N)^T$. Thus, the perturbed parameters $\bar{\theta}^1(m)$ and $\bar{\theta}^2(m)$ lie in $\bar{C}^o$ as well.

Consider now the Γ-projected parameters $\theta^1(m) = (\Gamma_j(\theta_j(m) + \delta\Delta_j(m)), j = 1,\ldots,N)^T$ and $\theta^2(m) = (\Gamma_j(\theta_j(m) - \delta\Delta_j(m)), j = 1,\ldots,N)^T$, respectively. By the construction of the generalized projection operator, these parameters are equal to $\theta^k \in C$ with probabilities $\beta_k((\theta_j(m) + \delta\Delta_j(m), j = 1,\ldots,N)^T)$ and $\beta_k((\theta_j(m) - \delta\Delta_j(m), j = 1,\ldots,N)^T)$, respectively. When the operative parameter is θ^k, the transition probabilities are $p_{\theta^k}(i,l)$, $i,l \in S$. Thus with probabilities $\beta_k((\theta_j(m) + \delta\Delta_j(m), j = 1,\ldots,N)^T)$ and $\beta_k((\theta_j(m) - \delta\Delta_j(m), j = 1,\ldots,N)^T)$, respectively, the transition probabilities in the two simulations equal $p_{\theta^k}(i,l)$, $i,l \in S$.

Next, consider the alternative (extended) system with parameters $\bar{\theta}^1(m)$ and $\bar{\theta}^2(m)$, respectively. The transition probabilities are now given by

$$p_{\bar{\theta}^i(m)}(j,l) = \sum_{k=1}^{p} \beta_k(\bar{\theta}^i(m)) p_{\theta^k}(j,l),$$

$i = 1,2$, $j,l \in S$. Thus, with probability $\beta_k(\bar{\theta}^i(m))$, a transition probability of $p_{\theta^k}(j,l)$ is obtained in the ith system. Thus, the two systems (original and the one with extended dynamics) are analogous.

Now consider the case when $\theta(m) \in \partial\bar{C}$, i.e., is a point on the boundary of $\bar{C}$). Then, one or more components of $\theta(m)$ are extreme points. For simplicity, assume that only one component (say the ith component) is an extreme point as the same argument carries over if there are more parameter components that are extreme points.

By the ith component of $\theta(m)$ being an extreme point, we mean that $\theta_i(m)$ is either $C_{i,\min}$ or $C_{i,\max}$. The other components $j = 1,\ldots,N, j \neq i$ are not extreme, i.e., $c_j^l \leq \theta_j(m) \leq c_j^{l+1}$ for some $l, l+1 \in \{1,\ldots,p\}$. Thus, one of $\theta_i(m) + \delta\Delta_i(m)$ or $\theta_i(m) - \delta\Delta_i(m)$ will lie outside of the interval $[C_{i,\min}, C_{i,\max}]$ while the other will lie inside of it for $\delta > 0$ small enough. For instance, suppose that $\theta_i(m) = C_{i,\max}$ and that $\theta_i(m) + \delta\Delta_i(m) > C_{i,\max}$ (which will happen if $\Delta_i(m) = +1$). In such a case, $\theta_i^1(m) = \Gamma_i(\theta_i(m) + \delta\Delta_i(m)) = C_{i,\max}$ with probability one. Then, as before, $\theta^1(m)$ can be written as the convex combination $\theta^1(m) = \sum_{k=1}^{p} \beta_k(\theta^1(m))\theta^k$ and the rest follows as before. □

As a consequence of Lemma 9.4, it is sufficient to analyze the convergence of the SPSA and SFA algorithms for the new system with extended transition probabilities and where $\bar{\theta}^i(m)$ is used in place of $\theta^i(m)$, $i = 1,2$. Let for any bounded and continuous function $v(\cdot) : \mathbb{R} \to \mathbb{R}$,

$$\hat{\Gamma}_i(v(y)) = \lim_{0<\eta\to 0} \left(\frac{\bar{\Gamma}_i(y + \eta v(y)) - \bar{\Gamma}_i(y)}{\eta} \right).$$

For $x = (x_1,\ldots,x_N)^T$, let $\hat{\Gamma}(x) = (\hat{\Gamma}_1(x_1), \ldots, \hat{\Gamma}_N(x_N))^T$. Let

$$\hat{K} = \{\theta \in \bar{C} \mid \hat{\Gamma}(-\nabla \bar{J}(\theta)) = 0\}.$$

Given $\varepsilon > 0$, let $\hat{K}^\varepsilon$ be the ε-neighborhood of $\hat{K}$, i.e., the set of points that are within a distance of ε from the set $\hat{K}$. Let P be the set

$$P = \{\hat{\theta} \in C \mid \hat{\theta} = \Gamma(\theta), \theta \in \hat{K}\},$$

and P^ε be its ε-neighborhood.

Theorem 9.5. *Given $\varepsilon > 0$, $\exists \delta_0 > 0$ such that $\forall \delta \in (0, \delta_0)$, $\{\theta(M)\}$ obtained according to the SPSA algorithm satisfies $\theta(M) \to \theta^* \in P^\varepsilon$ almost surely as $M \to \infty$.*

Proof. As a consequence of Lemma 9.4(i), we consider the alternative system with $\bar{\theta}^1(m)$, $\bar{\theta}^2(m)$ in place of $\theta^1(m)$, $\theta^2(m)$, respectively. For this system, it can be shown in a similar manner as Chapter 5 that given any $\varepsilon > 0$, there exists a $\delta_0 > 0$ such that for all $\delta \in (0, \delta_0)$, $\theta(M) \to \theta^*$ for some $\theta^* \in \hat{K}^\varepsilon$ almost surely as $M \to \infty$. Thus $\Gamma(\theta(M)) \to \Gamma(\theta^*)$ as $M \to \infty$ almost surely. Note that since $\theta^* \in \hat{K}^\varepsilon$, we have $\Gamma(\theta^*) \in P^\varepsilon$. The claim follows. □

The following result holds for the SFA algorithm.

Theorem 9.6. *Given* $\varepsilon > 0$, $\exists \beta_0 > 0$ *such that* $\forall \beta \in (0, \beta_0)$, $\{\theta(M)\}$ *obtained according to the SFA algorithm satisfies* $\theta(M) \to \theta^*$ *for some* $\theta^* \in P^{\varepsilon}$ *almost surely as* $M \to \infty$.

Proof. As with Theorem 9.5, as a consequence of Lemma 9.4(ii), we consider the alternative system with $\bar{\theta}^1(m)$, $\bar{\theta}^2(m)$ in place of $\theta^1(m)$, $\theta^2(m)$, respectively. It can now be shown in a similar manner as Chapter 6 that given any $\varepsilon > 0$, there exists a $\beta_0 > 0$ such that for all $\beta \in (0, \beta_0)$, $\theta(M) \to \theta^* \in \hat{K}^{\varepsilon}$ almost surely as $M \to \infty$. The rest follows in a similar manner as Theorem 9.5. □

Remark 9.1. The equivalence between the original system and its (alternate) continuous analog (cf. Lemma 9.4) critically depends on the quantities $\beta_k(\theta)$, $k = 1, \ldots, p$ (that describe the Γ-projection) being continuously differentiable. This is the case when either the *generalized* or the *randomized* projection operator is used in the algorithms. (In the latter case, in fact we have $\alpha_k(\theta)$ in place of $\beta_k(\theta)$ that are seen to be continuously differentiable.) The proofs of Theorems 9.5 and 9.6 are based on this equivalence. On the other hand, if deterministic projections are used (in place of randomized), one can proceed by splitting the convex hull $\bar{C}$ into disjoint regions such that a point $\bar{\theta}$ in any such region will project to a unique $\theta \in C$. This however will result in the transition probabilities (as function of θ) being non-smooth at the boundaries of these regions. The analysis in Lemma 9.3 as well as Theorems 9.5 and 9.6 will not carry through in such a case unless the (deterministic) projection scheme at the boundaries of the afore-mentioned regions is modified in a way that the transition probabilities become smooth. In fact, the *generalized* projection scheme is designed to achieve precisely this.

9.4 Concluding Remarks

We presented in this chapter adaptations of the SPSA and SFA algorithms for discrete parameter optimization for the optimization of a long-run average cost criterion associated with an underlying parametrized Markov chain. The idea was to update the algorithms in a closed and convex hull of the parameter region while the two parallel systems used in these algorithms are run at any instant using a suitable projection of the continuous-valued running parameter update to the underlying discrete set. Three different operators, deterministic, randomized and generalized, respectively, were presented for this purpose. The transition dynamics was first extended to include the case of continuously-valued parameters in the closed and convex hull of the discrete parameter space. It was observed that the transition probabilities and hence also the stationary distribution of the parametrized

Markov process become continuously differentiable in the (continuously-valued) parameter in the extended space. This was, however, the case when either randomized or generalized projection operators (but not deterministic) are used. The generalized projection operator is a hybrid between deterministic and randomized schemes. The regular convergence analysis of the SPSA and SFA algorithms can then be carried over to the discrete parameter setting when randomized or generalized projection operators are used.

In [3], the performance of the SPSA and SFA algorithms has been empirically tested, in the case when randomized projection operators are used, over several experiments on two different settings of admission control. In one of these settings, the parameter set is small and contains around 100 elements (parameters) while in another, it is large and has about 10^8 parameters. Performance comparisons with the equal allocation algorithm as well as the optimal computing budget allocation (OCBA) procedure [8, 6, 7] have also been shown ([3]) in the case when the parameter set has size 100. Over small-sized parameter sets, OCBA is widely regarded in the literature as being amongst the best algorithms for discrete parameter search. While the original OCBA scheme has been proposed for the case when the objective function is an expectation over noisy cost samples, an adaptation of the same for the long-run average cost criterion is described in [3]. It is observed in [3] that for low computing budgets, the performance of SPSA and OCBA is similar and better than SPSA and equal allocation. On the other hand, as the computing budget is increased, SFA shows the best results and is clearly better than OCBA. The performance of SPSA is also better than OCBA in this regime. In the case when the parameter set is large (for instance in the setting with 10^8 parameters in [3], R&S procedures such as OCBA and equal allocation are no longer implementable. It is observed that even in such cases, SPSA and SFA are easily implementable and show good results. An advantage in adapting efficient continuous optimization procedures such as SPSA and SFA to the case of discrete parameter optimization is that the search proceeds along the direction determined by the procedure that makes it computationally more efficient as one does not require storage of cost estimates corresponding to each parameter. On the other hand, in most other R&S procedures such as OCBA, the cost estimate corresponding to each parameter needs to be obtained first (using a given number of simulation samples) and stored before comparisons are drawn. This can result in such procedures being computationally less efficient particularly for large parameter spaces. A potential disadvantage, on the other hand, with adopting SPSA and SFA-based techniques for discrete parameter search is that in some cases, they may get caught in some bad local minima. This is unlike R&S type procedures. A possible future direction is to extensively study the empirical performance of SPSA and SFA with the generalized projection operator on various settings as well as the performance of similar adaptations of the other algorithms described in earlier chapters such as SPSA with Hadamard matrix perturbations as well as the Newton-based algorithms for the discrete parameter optimization setting.

References

1. Bechhofer, R.: A single-sample multiple decision procedure for ranking means of normal populations with known variances. The Annals of Mathematical Statistics 25(1), 16–39 (1954)
2. Bhatnagar, S., Kowshik, H.J.: A discrete parameter stochastic approximation algorithm for simulation optimization. Simulation 81(11), 757–772 (2005)
3. Bhatnagar, S., Mishra, V., Hemachandra, N.: Stochastic algorithms for discrete parameter simulation optimization. IEEE Transactions on Automation Science and Engineering 9(4), 780–793 (2011)
4. Bhatnagar, S., Reddy, I.B.B.: Optimal threshold policies in communication networks via discrete parameter stochastic approximation. Telecommunication Systems 29, 9–31 (2005)
5. Borkar, V.S.: Probability Theory: An Advanced Course. Springer, New York (1995)
6. Branke, J., Chick, S., Schmidt, C.: Selecting a selection procedure. Management Science 53(12), 1916–1932 (2007)
7. Chen, C., Lee, L.: Stochastic simulation optimization: an optimal computing budget allocation, vol. 1. World Scientific Pub. Co. Inc. (2010)
8. Chen, C., Lin, J., Yücesan, E., Chick, S.: Simulation budget allocation for further enhancing the efficiency of ordinal optimization. Discrete Event Dynamic Systems 10(3), 251–270 (2000)
9. Dunnett, C.: A multiple comparison procedure for comparing several treatments with a control. Journal of the American Statistical Association 50(272), 1096–1121 (1955)
10. Gerencsér, L.: Convergence rate of moments in stochastic approximation with simultaneous perturbation gradient approximation and resetting. IEEE Trans. Auto. Cont. 44(5), 894–906 (1999)
11. Gokbayrak, K., Cassandras, C.: Generalized surrogate problem methodology for online stochastic discrete optimization. Journal of Optimization Theory and Applications 114(1), 97–132 (2002)
12. Goldsman, D., Kim, S., Marshall, W., Nelson, B.: Ranking and selection for steady-state simulation: Procedures and perspectives. INFORMS Journal on Computing 14(1), 2–19 (2002)
13. Hill, S.D., Gerencsér, L., Vágó, Z.: Stochastic approximation on discrete sets using simultaneous difference approximations. In: Proceedings of the American Control Conference, Boston, MA, pp. 2795–2798 (2004)
14. Ho, Y., Sreenivas, R., Vakili, P.: Ordinal optimization of deds. Discrete Event Dynamic Systems 2(1), 61–88 (1992)
15. Kirkpatrick, S., Gelatt, C., Vecchi, M.: Optimization by simulated annealing. Science 220(4598), 671 (1983)
16. Wang, Q., Spall, J.C.: Discrete simultaneous perturbation stochastic approximation on loss function with noisy measurements. In: Proceedings of the American Control Conference, San Francisco, CA, pp. 4520–4525 (2011)
17. Schweitzer, P.J.: Perturbation theory and finite Markov chains. Journal of Applied Probability 5, 401–413 (1968)

Chapter 10
Algorithms for Constrained Optimization

10.1 Introduction

The optimization problem that we have considered so far has the form

$$\text{Find } \theta^* \in C \text{ such that } J(\theta^*) = \min_{\theta \in C} J(\theta), \tag{10.1}$$

for a given objective function $J : \mathbb{R}^N \to \mathbb{R}$ and where $C \subset \mathbb{R}^N$ is a given set in which θ takes values.

In many applications, the problem of optimizing the objective needs to be carried out keeping in view that certain functional constraints are satisfied. Many times, these (functional) constraints are specified via some other cost functions being below certain thresholds. For example, in the case of communication networks, a problem of interest could be to find a path from the source to the destination for a user over which the throughput is maximum, subject to the constraint that the mean delay-per-packet is below some threshold (of say one second). Another constraint could similarly be on the probability of packet loss being below some other threshold (say 0.01).

The problem that we are interested in this chapter has the form:

$$\text{Find } \theta^* \text{ such that } J(\theta^*) = \min_{\theta \in C} \{J(\theta) \mid G_i(\theta) \leq \alpha_i,\ i = 1, \ldots, p\}, \tag{10.2}$$

where $G_i(\cdot)$ and α_i, $i = 1, \ldots, p$ are additionally prescribed cost functions and constants that together constitute the functional constraints. The constraint region in which optimization needs to be performed in such a case becomes

$$C \cap \left(\cap_{i=1}^{p} \{\theta \mid G_i(\theta) \leq \alpha_i\}\right).$$

S. Bhatnagar et al.: Stochastic Recursive Algorithms for Optimization, LNCIS 434, pp. 167–186.
springerlink.com

Note that the constraint region here is parameter-dependent since the constraint functions (like the objective) are also parameter-dependent. We specifically consider here the case when the objective and the constraint functions are all long-run averages of certain given sample cost functions whose values at each instant can be estimated through simulation. We incorporate the Lagrange multiplier approach to deal with the inequality constraints.

Since neither the objective nor the constraints are known analytically, information on their gradients and/or Hessians is usually not available (even when they exist). As we shall see, a combination of multi-timescale stochastic approximation and simultaneous perturbation methods proves useful here. Two of these methods are based on the SPSA technique while the other two incorporate the SF approach. The material presented in this chapter is largely based on [2].

Section 10.2 describes the constrained optimization problem framework and the simulation optimization methods are presented subsequently in Section 10.3. A sketch of the convergence analysis of these algorithms is given in Section 10.4. Finally, Section 10.5 presents the concluding remarks. Application of methods similar to the ones described in this chapter as well as those given in Chapter 9, to the context of service systems, has been explored in Chapter 12. Somewhat similar techniques have also been applied in the context of the random early detection (RED) scheme for flow control in Chapter 14.2.

10.2 The Framework

Let $\{X(n), n \geq 1\}$ be an $\mathbb{R}^d$-valued parametrized Markov process with a tunable parameter θ. We assume that $\theta \in C \subset \mathbb{R}^N$, where C is a compact and convex set. Let $p(\theta, x, dy)$ and $\nu_\theta(dx)$, respectively, denote the transition kernel and stationary distribution of $\{X(n)\}$ when θ is the operative parameter.

Let $h, g_1, g_2, \ldots, g_p : \mathbb{R}^d \to \mathbb{R}^+ \cup \{0\}$ be given functions (for some $p \geq 1$). The function h is the single-stage cost while $g_1, \ldots, g_p$ are associated maps that determine the constraints. The aim here is to find a parameter $\theta \in C$ that minimizes the long-run average cost

$$J(\theta) = \lim_{l \to \infty} \frac{1}{l} \sum_{j=0}^{l-1} h(X(j)), \tag{10.3}$$

subject to

$$G_i(\theta) = \lim_{l \to \infty} \frac{1}{l} \sum_{j=0}^{l-1} g_i(X(j)) \leq \alpha_i, \; i = 1, 2, \ldots, p. \tag{10.4}$$

Here $\alpha_1, \ldots, \alpha_p > 0$ are given constants that specify the threshold levels.

We will assume that there exists at least one $\theta \in C$ for which all the inequality constraints (10.4) are satisfied. Note, however, that there may not be a unique (constrained) minimizer θ^*. Thus, it is enough to find a θ^* that minimizes $J(\theta)$ while satisfying all the functional constraints. Note also that, in general, it is very difficult to achieve a global minimum and optimization methods such as simulated annealing that aim at finding a global minimum can be computationally inefficient. Hence, many times, one has to be content with finding a local minimum. We apply the Lagrange relaxation procedure to account for the constraints and provide algorithms for finding a locally optimum parameter.

Lagrangian Relaxation

The constrained long-run average cost optimization problem described above can be expressed using the standard Lagrange multiplier theory as an unconstrained optimization problem. Let $L(\theta, \lambda_1, \lambda_2 \ldots, \lambda_p)$ denote the Lagrangian described by

$$\begin{aligned} L(\theta, \lambda_1, \lambda_2 \ldots, \lambda_p) &= J(\theta) + \sum_{i=1}^{p} \lambda_i (G_i(\theta) - \alpha_i) \\ &= \int \left(h(x) + \sum_{i=1}^{p} \lambda_i (g_i(x) - \alpha_i) \right) \nu_\theta(dx), \end{aligned} \tag{10.5}$$

where $\lambda_1, \lambda_2, \ldots, \lambda_p \in \mathbb{R}^+ \cup \{0\}$ denote the Lagrange multipliers corresponding to the p functional constraints. In the following, we denote by Λ, the vector $\Lambda = (\lambda_1, \ldots, \lambda_p)^T$.

An optimal (θ^*, Λ^*) is a saddle point for the Lagrangian, i.e., $L(\theta, \Lambda^*) \geq L(\theta^*, \Lambda^*) \geq L(\theta^*, \Lambda)$. Thus, it is necessary to design an algorithm which descends in θ and ascends in Λ in order to find the optimum point. An iterative local search procedure would update θ and Λ in descent and ascent directions, respectively. Neither the objective nor the constraint functions have analytical expressions as a consequence of being long-run averages. So, the Lagrangian $L(\theta, \Lambda)$ also does not possess an analytical expression. Thus, any optimization algorithm in this setting must rely on outcomes either from a real system or those obtained using simulation.

Assumptions

Let $\{\theta(n)\}$ be a sequence of random parameters obtained using an iterative scheme on which the process $\{X(n)\}$ depends. Let $\mathcal{H}(n) = \sigma(\theta(m), X(m), m \leq n)$, $n \geq 0$

denote a sequence of associated σ-fields. We call $\{\theta(n)\}$ non-anticipative if for all Borel sets $A \subset \mathbb{R}^d$,

$$P(X(n+1) \in A \mid \mathscr{H}(n)) = p(\theta(n), X(n), A).$$

It is easy to see that sequences $\{\theta(n)\}$ obtained using the algorithms in the next section are non-anticipative. Under non-anticipative $\{\theta(n)\}$, the joint process $\{(X(n), \theta(n))\}$ is Markov.

Assumption 10.1. The process $\{X(n)\}$ is ergodic Markov for any given $\theta \in C$.

Assumption 10.2. The single-stage cost and constraint functions $h, g_1, g_2, \ldots, g_p : \mathbb{R}^d \to \mathbb{R}^+ \cup \{0\}$ are all Lipschitz continuous.

Assumption 10.3. The functions $J(\cdot)$ and $G_i(\cdot)$, $i = 1, 2, \ldots, p$ are twice continuously differentiable functions with bounded third derivatives.

Assumption 10.4. There exist $\varepsilon_0 > 0$, $K \subset \mathbb{R}^d$ compact and $V \in C(\mathbb{R}^d)$ such that $\lim_{\|x\| \to \infty} V(x) = \infty$ and under any non-anticipative $\{\theta(n)\}$,

(i) $\sup_n E[V(X(n))^2] < \infty$, for any given X_0, and
(ii) $E[V(X(n+1)) \mid \mathscr{H}(n)] \leq V(X(n)) - \varepsilon_0$ a.s., whenever $X(n) \notin K$, $n \geq 0$.

Here $C(\mathbb{R}^d)$ is the set of all real-valued continuous functions on $\mathbb{R}^d$. Also, $\|\cdot\|$ denotes the Euclidean vector norm. The same norm also denotes (by an abuse of notation) the matrix norm induced by the Euclidean vector norm (i.e., $\|A\| = \sup_{\|x\|=1} \|Ax\|$, $A \in \mathbb{R}^{N \times N}$).

Let $P : \mathbb{R}^{N \times N} \to \{$positive definite and symmetric matrices$\}$ denote an operator that projects any $N \times N$-matrix to the space of positive definite and symmetric matrices. We let $P(A) = A$, if A is positive definite and symmetric. For a matrix A, let $\{P(A)\}^{-1}$ denote the inverse of the matrix $P(A)$.

Assumption 10.5. If $\{A(n)\}$ and $\{B(n)\}$ are sequences of matrices in $\mathbb{R}^{N \times N}$ such that $\lim_{n \to \infty} \|A(n) - B(n)\| = 0$, then $\lim_{n \to \infty} \|P(A(n)) - P(B(n))\| = 0$ as well. Further, for any sequence $\{C(n)\}$ of matrices in $\mathbb{R}^{N \times N}$, if $\sup_n \|C(n)\| < \infty$, then $\sup_n \|P(C(n))\|$, $\sup_n \|\{P(C(n))\}^{-1}\| < \infty$ as well.

Assumption 10.6. Let $a(n)$, $b(n)$, $c(n)$ and $d(n)$, $n \geq 0$ be sequences of positive step-sizes that satisfy the requirements

$$\sum_{n=0}^{\infty} a(n) = \sum_{n=0}^{\infty} b(n) = \sum_{n=0}^{\infty} c(n) = \sum_{n=0}^{\infty} d(n) = \infty, \tag{10.6}$$

$$\sum_{n=0}^{\infty} a(n)^2, \sum_{n=0}^{\infty} b(n)^2, \sum_{n=0}^{\infty} c(n)^2, \sum_{n=0}^{\infty} d(n)^2 < \infty, \tag{10.7}$$

$$\lim_{n\to\infty} \frac{a(n)}{b(n)} = \lim_{n\to\infty} \frac{b(n)}{c(n)} = \lim_{n\to\infty} \frac{c(n)}{d(n)} = 0. \tag{10.8}$$

Note that Assumption 10.1 ensures, in particular, that the long-run average cost (10.3) and the constraint functions (10.4) are well defined for any θ. Assumption 10.2 ensures that the single-stage cost functions $h, g_1, \ldots, g_p$ exhibit an at most linear growth (as a function of the state). Assumption 10.3 is a technical requirement used to push through a Taylor's argument (See Section 10.4).

Assumption 10.4 concerns the existence of a stochastic Lyapunov function $V(\cdot)$. This ensures that the system remains stable under a tunable parameter. Note that Assumption 10.4 will not be required if the functions $h(\cdot)$ and $g_i(\cdot)$, $i = 1, \ldots, p$ are bounded in addition. Assumption 10.5 is required for Newton-based algorithms where one projects the Hessian estimate after each iteration onto the space of positive definite and symmetric matrices. This ensures that the algorithm progresses along the negative gradient direction at each update epoch. Finally, Assumption 10.6 ensures a difference in timescales in recursions governed with the various step sizes as explained in earlier chapters.

10.3 Algorithms

As explained below, an algorithm for the constrained optimization problem would require three or four nested loops depending on whether the algorithm is a gradient-based scheme or is Newton based.

1. The inner-most loop in any of these schemes would aggregate data over various simulation runs for given Λ and θ.
2. The next outer loop would update θ for a given Λ-update, so that corresponding to that update, the optimum θ is attained. The update in this loop, in turn, may depend on the outcome of another (nested) inner loop depending on whether the scheme works only with gradients or requires both gradient as well as Hessian computations.
3. Finally, the outer-most loop would update the Lagrange multipliers Λ using outcomes of the aforementioned loop updates.

Note that a regular procedure as described above will take a very long time to converge because any of the outer-loop updates would have to wait for convergence of the corresponding inner-loop procedures. Multi-timescale stochastic approximation again comes to our rescue. By having coupled stochastic updates with different (diminishing) step-size schedules, with each converging to zero at a different rate, one can have recursions that proceed simultaneously and yet converge to an equilibrium solution.

We describe four stochastic approximation algorithms for this purpose. Two of these are gradient-based algorithms while the other two are Newton-based schemes. These algorithms incorporate SPSA and SF gradient/Hessian estimates that are, however, different from the balanced estimates presented in Chapters 7 and 8, respectively. All four algorithms use two simulations each, one of which corresponds to the running parameter update in each algorithm, while the other is from a perturbed parameter sequence that in turn depends on the particular scheme (gradient/Hessian as well as SPSA/SF) used. The running parameter (θ) update in each of these algorithms is also used to aggregate data for the Lagrange multiplier updates. Hence, for reducing the simulation load, we incorporate the estimates from the running parameter into the gradient/Hessian estimators as well.

All four step-size schedules (cf. Assumption 10.6) are used for the Newton algorithms while the gradient algorithms rely on the sequences $a(n)$, $b(n)$, and $c(n)$, $n \geq 0$, respectively. For $x = (x_1, \ldots, x_N)^T \in \mathbb{R}^N$, let $\Gamma(x) = (\Gamma_1(x_1), \ldots, \Gamma_N(x_N))^T$ represent the projection of x onto the set C. Further, let $\hat{\Gamma} : \mathbb{R} \to [0, \bar{L}]$ denote the projection $\hat{\Gamma}(y) = \min(\bar{L}, \max(y, 0))$ for any $y \in \mathbb{R}$, where $\bar{L} > 0$ is a large constant. $\bar{L} < \infty$ ensures that the stochastic recursions that use $\hat{\Gamma}$ stay uniformly bounded. In the following, we let $\theta(n) \stackrel{\triangle}{=} (\theta_1(n), \ldots, \theta_N(n))^T$, denote the nth update of the parameter θ and $\lambda_i(n)$ (resp. $\Lambda(n)$), the nth update of λ_i (resp. Λ), $i = 1, \ldots, p$. In what follows, before presenting the four algorithms, we first present the gradient or gradient/Hessian estimates used in each scheme depending on whether the same is a gradient or Newton-based algorithm, since as mentioned before, the forms of the gradient/Hessian estimators used here are different from those proposed in Chapters 7 and 8.

10.3.1 Constrained Gradient-Based SPSA Algorithm (CG-SPSA)

The Gradient Estimate

Let $\Delta_1, \ldots, \Delta_N$ be independent random variables satisfying Assumption 5.4. Let $\Delta = (\Delta_1, \ldots, \Delta_N)^T$ and $\Delta^{-1} = (1/\Delta_1, \ldots, 1/\Delta_N)^T$, respectively. Then, the form of the estimate of the gradient of $L(\theta, \Lambda)$ w.r.t. θ is obtained from the following relationship:

$$\nabla_\theta L(\theta, \Lambda) = \lim_{\delta \downarrow 0} E\left[\left(\frac{L(\theta + \delta\Delta, \Lambda) - L(\theta, \Lambda)}{\delta}\right)\Delta^{-1}\right],$$

where the expectation is w.r.t. the distribution of Δ. This is essentially same as the one-sided SPSA gradient estimation scheme discussed in Chapter 5.3.2.

The Algorithm

Generate two parallel simulations $\{X(n)\}$ and $\{X'(n)\}$ such that at any instant n, $X(n)$ is governed by $\theta(n)$ while $X'(n)$ is governed by $\theta(n)+\delta\Delta(n)$, where $\delta>0$ is a given small constant. Also, $\Delta(n)$ is the vector $\Delta(n)=(\Delta_1(n),\ldots,\Delta_N(n))^T$. Here $\Delta_l(n)$, $l=1,\ldots,N$, $n\geq 0$ being independent random variables satisfying Assumption 5.4. We have for $l=1,\ldots,N$, $i=1,\ldots,p$,

$$\begin{aligned}
Z(n+1)=&Z(n)\\
&+c(n)\Bigg(h(X'(n))+\sum_{i=1}^{p}\lambda_i(n)g_i(X'(n))\\
&-h(X(n))-\sum_{i=1}^{p}\lambda_i(n)g_i(X(n))-Z(n)\Bigg), \qquad (10.9)\\
\theta_l(n+1)=&\Gamma_l\left(\theta_l(n)-b(n)\frac{Z(n)}{\delta\Delta_l(n)}\right), \qquad (10.10)\\
Y_i(n+1)=&Y_i(n)+c(n)(g_i(X(n))-Y_i(n)), \qquad (10.11)\\
\lambda_i(n+1)=&\hat{\Gamma}(\lambda_i(n)+a(n)(Y_i(n)-\alpha_i)). \qquad (10.12)
\end{aligned}$$

In the above, $Z(n)$ is an estimate of $(L(\theta(n)+\delta\Delta(n),\Lambda(n))-L(\theta(n),\Lambda(n)))$. Also, $Z(n)/(\delta\Delta_l(n))$ is an estimate of $\nabla_{\theta_l}L(\theta(n),\Lambda(n))$, the partial derivative corresponding to the lth component of θ, $l=1,\ldots,N$. Further, $Y_i(n)$ is the nth estimate of the constraint function $G_i(\theta(n))$ in (10.4).

10.3.2 Constrained Newton-Based SPSA Algorithm (CN-SPSA)

The Gradient and Hessian Estimates

The gradient and Hessian estimates in CN-SPSA depend on two independent sequences of perturbation random variables. The form of the gradient estimates here is different from the one used in CG-SPSA. Let $\Delta_1,\ldots,\Delta_N$, $\hat{\Delta}_1,\ldots,\hat{\Delta}_N$ be mutually independent random variables satisfying Assumption 7.6. Let Δ and Δ^{-1} be as before. Also, let $\hat{\Delta}=(\hat{\Delta}_1,\ldots,\hat{\Delta}_N)^T$ and $\hat{\Delta}^{-1}=(1/\hat{\Delta}_1,\ldots,1/\hat{\Delta}_N)^T$, respectively. Then, the estimates of the gradient and Hessian of $L(\theta,\Lambda)$ w.r.t. θ are based on the following relationships that are prove later in Propositions 10.11 and 10.10, respectively.

$$\nabla_\theta L(\theta,\Lambda)=\lim_{\delta_1,\delta_2\downarrow 0}E\left[\left(\frac{L(\theta+\delta_1\Delta+\delta_2\hat{\Delta},\Lambda)-L(\theta,\Lambda)}{\delta_2}\right)\hat{\Delta}^{-1}\right],$$

$$\nabla^2_\theta L(\theta,\Lambda) = \lim_{\delta_1,\delta_2\downarrow 0} E\left[\triangle^{-1}\left(\frac{L(\theta+\delta_1\Delta+\delta_2\hat{\Delta},\Lambda)-L(\theta,\Lambda)}{\delta_1\delta_2}\right)(\hat{\Delta}^{-1})^T\right].$$

The expectations above are taken w.r.t. the joint distribution of Δ and $\hat{\Delta}$.

The Algorithm

Let $\Delta(n), \hat{\Delta}(n), n \geq 0$ be two sequences of independent perturbation vectors $\Delta(n) \stackrel{\triangle}{=} (\Delta_1(n),\ldots,\Delta_N(n))^T$ and $\hat{\Delta}(n) \stackrel{\triangle}{=} (\hat{\Delta}_1(n),\ldots,\hat{\Delta}_N(n))^T$, respectively, with $\Delta_l(n)$, $\hat{\Delta}_l(n)$, $l = 1,\ldots,N, n \geq 0$ satisfying Assumption 7.6. Let $\delta_1, \delta_2 > 0$ be two small constants. Generate two parallel simulations $\{X(n)\}$ and $\{X'(n)\}$ such that at any instant n, $X(n)$ is governed by the parameter $\theta(n)$ while $X'(n)$ is governed by $\theta(n)+\delta_1\Delta(n)+\delta_2\hat{\Delta}(n)$. Then, the update rule of CN-SPSA algorithm is as follows:

$$\begin{aligned} Z(n+1) = &Z(n)+d(n)\Big(h(X'(n))+\sum_{i=1}^{p}\lambda_i(n)g_i(X'(n)) \\ &- h(X(n))-\sum_{i=1}^{p}\lambda_i(n)g_i(X(n))-Z(n)\Big). \end{aligned} \tag{10.13}$$

For $j,l \in \{1,\ldots,N\}, j \leq l$,

$$H_{j,l}(n+1) = H_{j,l}(n)+c(n)\left(\frac{Z(n)}{\delta_1\delta_2\Delta_l(n)\hat{\Delta}_j(n)} - H_{j,l}(n)\right), \tag{10.14}$$

For $l = 1,\ldots,N, i = 1,\ldots,p$,

$$\theta_l(n+1) = \Gamma_l\left(\theta_l(n)-b(n)\sum_{k=1}^{N}M_{l,k}(n)\left(\frac{Z(n)}{\delta_2\hat{\Delta}_k(n)}\right)\right), \tag{10.15}$$

$$Y_i(n+1) = Y_i(n)+c(n)(g_i(X(n))-Y_i(n)), \tag{10.16}$$

$$\lambda_i(n+1) = \hat{\Gamma}(\lambda_i(n)+a(n)(Y_i(n)-\alpha_i)). \tag{10.17}$$

In the above,

- We set $H_{j,l}(n+1) = H_{l,j}(n+1)$ for $j > l$.
- $M(n) = [[M_{k,l}(n)]]_{k,l=1}^N$ denotes the inverse of the matrix $H(n) \stackrel{\triangle}{=} P([[H_{k,l}(n)]]_{k,l=1}^N)$.
- $Z(n)$ is an estimate of $(L(\theta(n)+\delta_1\Delta(n)+\delta_2\hat{\Delta}(n),\Lambda(n))-L(\theta(n),\Lambda(n)))$.

- $Z(n)/(\delta_2\hat{\Delta}_k(n))$ is an estimate of the partial derivative $\nabla_{\theta_k}L(\theta(n),\Lambda(n))$ and $Z(n)/(\delta_1\delta_2\Delta_l(n)\hat{\Delta}_j(n))$ is an estimate of the (j,l)th component of the Hessian matrix $H(n)$.

10.3.3 Constrained Gradient-Based SF Algorithm (CG-SF)

The Gradient Estimate

The gradient estimate here will involve a one-sided form described below. Let $\eta = (\eta_1,\ldots,\eta_N)^T$ be a vector of independent $N(0,1)$-distributed random variates. The gradient estimates for CG-SF are based on the following relationship whose proof is given later in Proposition 10.2.

$$\nabla_\theta L(\theta,\Lambda) = \lim_{\beta\downarrow 0} E\left[\frac{\eta}{\beta}(L(\theta+\beta\eta,\Lambda)-L(\theta,\Lambda))\right],$$

where the expectation is taken w.r.t. the distribution of η.

The Algorithm

Let $\beta > 0$ be a given small constant. Let $\eta(n) \stackrel{\triangle}{=} (\eta_1(n),\ldots,\eta_N(n))^T$, where $\eta_l(n)$, $l=1,\ldots,N$, $n\geq 0$ are independent $N(0,1)$-distributed random variables. Generate two parallel simulations $\{X(n)\}$ and $\{X'(n)\}$ such that at any instant n, $X(n)$ is governed by the parameter $\theta(n)$ while $X'(n)$ is governed by $\theta(n)+\beta\eta(n)$. Then for $l=1,\ldots,N$, $i=1,\ldots,p$, we have

$$\begin{aligned} Z_l(n+1) =& Z_l(n)+c(n)\left(\frac{\eta_l(n)}{\beta}\left(h(X'(n))+\sum_{i=1}^{p}\lambda_i(n)g_i(X'(n))\right.\right.\\ &\left.\left.-h(X(n))-\sum_{i=1}^{p}\lambda_i(n)g_i(X(n))\right)-Z_l(n)\right), \quad (10.18)\\ \theta_l(n+1) =& \Gamma_l(\theta_l(n)-b(n)Z_l(n)), \quad (10.19)\\ Y_i(n+1) =& Y_i(n)+c(n)(g_i(X(n))-Y_i(n)), \quad (10.20)\\ \lambda_i(n+1) =& \hat{\Gamma}(\lambda_i(n)+a(n)(Y_i(n)-\alpha_i)). \quad (10.21) \end{aligned}$$

Here $Z_l(n)$ is an estimate of $\nabla_{\theta_l}L(\theta(n),\Lambda(n))$.

10.3.4 *Constrained Newton-Based SF Algorithm (CN-SF)*

The Gradient and Hessian Estimates

The gradient estimate in CN-SF is the same as the one in CG-SF. As in the case of the N-SF algorithms of Chapter 8, where there are no functional constraints, (see (8.4)), let

$$\bar{H}(\eta) \stackrel{\triangle}{=} \begin{bmatrix} (\eta_1^2-1) & \eta_1\eta_2 & \cdots & \eta_1\eta_N \\ \eta_2\eta_1 & (\eta_2^2-1) & \cdots & \eta_2\eta_N \\ \cdots & \cdots & \cdots & \cdots \\ \eta_N\eta_1 & \eta_N\eta_2 & \cdots & (\eta_N^2-1) \end{bmatrix}, \tag{10.22}$$

where $\eta = (\eta_1,\ldots,\eta_N)^T$ is a vector of mutually independent $N(0,1)$ random variables. The estimate for the Hessian in CN-SF is obtained from the following relationship:

$$\nabla_\theta^2 L(\theta,\Lambda) = \lim_{\beta\to 0} \frac{1}{\beta^2} E\left[\bar{H}(\eta)(L(\theta+\beta\eta,\Lambda) - L(\theta,\Lambda))\right], \tag{10.23}$$

where the expectation is taken w.r.t. the distribution of η.

The Algorithm

Let $\beta > 0$ be a given small constant. Let $\eta(n) \stackrel{\triangle}{=} (\eta_1(n),\ldots,\eta_N(n))^T$, where $\eta_l(n)$, $l = 1,\ldots,N$, $n \geq 0$ are mutually independent $N(0,1)$-distributed random variables. Generate two parallel simulations $\{X(n)\}$ and $\{X'(n)\}$ such that at any instant n, $X(n)$ is governed by the parameter $\theta(n)$ while $X'(n)$ is governed by $\theta(n)+\beta\eta(n)$. For $i,j,k = 1,\ldots,N$, $j<k$, update

$$\begin{aligned} Z_{i,i}(n+1) =& (1-d(n))Z_{i,i}(n) \\ &+ d(n)\left(\frac{\eta_i^2(n)-1}{\beta^2}\left(-h(X(n)) - \sum_{i=1}^{p} \lambda_i(n) g_i(X(n))\right.\right. \\ &\left.\left. + h(X'(n)) + \sum_{i=1}^{p} \lambda_i(n) g_i(X'(n))\right)\right), \end{aligned} \tag{10.24}$$

$$Z_{j,k}(n+1) = (1-d(n))Z_{j,k}(n) + d(n)\left(\frac{\eta_j(n)\eta_k(n)}{\beta^2}\left(-h(X(n)) - \sum_{i=1}^{p} \lambda_i(n) g_i(X(n)) + h(X'(n)) + \sum_{i=1}^{p} \lambda_i(n) g_i(X'(n))\right)\right). \quad (10.25)$$

For $j > k$, set $Z_{j,k}(n+1) = Z_{k,j}(n+1)$. Next for $l = 1,\ldots,N$, $i = 1,\ldots,p$, update

$$Z_l(n+1) = Z_l(n) + c(n)\left(\frac{\eta_l(n)}{\beta}\left(h(X'(n)) + \sum_{i=1}^{p} \lambda_i(n) g_i(X'(n)) - h(X(n)) - \sum_{i=1}^{p} \lambda_i(n) g_i(X(n))\right) - Z_l(n)\right), \quad (10.26)$$

$$\theta_l(n+1) = \Gamma_l\left(\theta_l(n) - b(n)\sum_{k=1}^{N} M_{l,k}(n) Z_k(n)\right), \quad (10.27)$$

$$Y_i(n+1) = Y_i(n) + c(n)(g_i(X(n)) - Y_i(n)), \quad (10.28)$$

$$\lambda_i(n+1) = \hat{\Gamma}(\lambda_i(n) + a(n)(Y_i(n) - \alpha_i)). \quad (10.29)$$

In the above, $M(n) = [[M_{i,j}(n)]]_{i,j=1}^{N} \stackrel{\triangle}{=} H(n)^{-1}$ denotes the inverse of the Hessian matrix $H(n) = P([[Z_{i,j}(n)]]_{i,j=1}^{N})$. Also, $Z_{i,i}(n)$ (resp. $Z_{j,k}(n)$) is the nth estimate of the (i,i)th (resp. (j,k)th) element of the Hessian matrix $\nabla_\theta^2 L(\theta,\lambda)$. Further, as with CG-SF, $Z_l(n)$ is an estimate of $\nabla_{\theta_l} L(\theta(n),\Lambda(n))$.

Remark 10.1. The quantities $Y_i(n)$ in each of these algorithms are used in the updates of the Lagrange multipliers $\lambda_i(n), i = 1,\ldots,p, n \geq 0$, for which one requires the nominal parameter updates $\theta(n), n \geq 0$. For simulation efficiency, the gradient/Hessian estimators in these algorithms have been designed in a way as to make use of the simulations with the nominal parameters as well. On the other hand, the two-simulation-balanced estimators of the gradient/Hessian described in Chapters 7 and 8 could be slightly more efficient as compared to the one-sided (unbalanced) estimators used here because of less higher-order biases in the former. The resulting algorithms with such (balanced) estimators would however require three parallel simulations. Nevertheless, it would be interesting to empirically study the comparisons of such algorithms (with balanced estimators) with the algorithms presented here both in terms of accuracy as well as computational effort.

10.4 A Sketch of the Convergence

We first provide a sketch of the convergence proof for the algorithm CN-SF. Later, we describe the changes necessary in the analysis of the other algorithms.

Convergence Analysis of CN-SF

The analysis of (10.24)–(10.27) works along the lines of SF schemes discussed in Chapter 8. Since $a(n) = o(b(n))$, a multi-timescale stochastic approximation analysis allows us to treat $\Lambda(n)$ as a constant while analyzing (10.24)–(10.27), under which condition these updates reduce to a scheme similar to the one used in Chapter 8 except with different gradient and Hessian estimators. In the following, we, therefore, first show that the gradient and Hessian estimators that we use are strongly consistent.

Let $\bar{H}(\eta(n))$ be defined as in (10.22) with $\eta(n)$ in place of η, where $\eta(n)$ are the random variables described in the algorithm.

Proposition 10.1.

$$\left\| E\left[\frac{1}{\beta^2}\bar{H}(\eta(n))(L(\theta(n)+\beta\eta(n),\Lambda(n)) - L(\theta(n),\Lambda(n))) \mid \theta(n),\Lambda(n)\right] - \nabla^2_{\theta(n)}L(\theta(n),\Lambda(n))\right\| \to 0, \ as\ \beta \to 0.$$

Proof. Note that

$$E\left[\frac{1}{\beta^2}\bar{H}(\eta(n))(L(\theta(n)+\beta\eta(n),\Lambda(n)) - L(\theta(n),\Lambda(n))) \mid \theta(n),\Lambda(n)\right]$$
$$= E[\frac{1}{\beta^2}\bar{H}(\eta(n))(J(\theta(n)+\beta\eta(n)) + \sum_{i=1}^{p}\lambda_i(n)G_i(\theta(n)+\beta\eta(n))$$
$$-J(\theta(n)) - \sum_{i=1}^{p}\lambda_i(n)G_i(\theta(n))) \mid \theta(n),\Lambda(n)].$$

Using Taylor series expansions of $J(\theta(n)+\beta\eta(n))$ and $G_i(\theta(n)+\beta\eta(n))$, respectively, around $\theta(n)$, one obtains

$$L(\theta(n)+\beta\eta(n),\Lambda(n)) - L(\theta(n),\Lambda(n))$$
$$= \beta\eta(n)^T(\nabla J(\theta(n)) + \sum_{i=1}^{p}\lambda_i(n)\nabla G_i(\theta(n)))$$
$$+\frac{\beta^2}{2}\eta(n)^T(\nabla^2 J(\theta(n)) + \sum_{i=1}^{p}\lambda_i(n)\nabla^2 G_i(\theta(n)))\eta(n) + o(\beta^2). \quad (10.30)$$

Hence,

$$E\left[\frac{1}{\beta^2}\bar{H}(\eta(n))(L(\theta(n)+\beta\eta(n),\Lambda(n))-L(\theta(n),\Lambda(n)))\mid\theta(n),\Lambda(n)\right]$$

$$=E\left[\frac{1}{\beta}\bar{H}(\eta(n))\eta(n)^T\left(\nabla J(\theta(n))+\sum_{i=1}^{p}\lambda_i(n)\nabla G_i(\theta(n))\right)\mid\theta(n),\Lambda(n)\right]$$

$$+\frac{1}{2}E\left[\bar{H}(\eta(n))\eta(n)^T\left(\nabla^2 J(\theta(n))+\sum_{i=1}^{p}\lambda_i(n)\nabla^2 G_i(\theta(n))\right)\eta(n)\mid\theta(n),\Lambda(n)\right]+O(\beta).$$

The first term on the RHS above equals zero, while the second term equals $\nabla^2_{\theta(n)} L(\theta(n),\Lambda(n))$. □

Proposition 10.2.

$$\|E\left[\frac{1}{\beta}\eta(n)(L(\theta(n)+\beta\eta(n),\Lambda(n))-L(\theta(n),\Lambda(n)))\mid\theta(n),\Lambda(n)\right]$$
$$-\nabla_{\theta(n)}L(\theta(n),\Lambda(n))\|\to 0$$

as $\beta\to 0$.

Proof. Note that

$$E\left[\frac{1}{\beta}\eta(n)(L(\theta(n)+\beta\eta(n),\Lambda(n))-L(\theta(n),\Lambda(n)))\mid\theta(n),\Lambda(n)\right]$$

$$=E[\eta(n)\eta(n)^T\nabla_{\theta(n)}L(\theta(n),\Lambda(n))\mid\theta(n),\Lambda(n)]$$

$$+\frac{\beta}{2}E[\eta(n)\eta(n)^T\nabla^2_{\theta(n)}L(\theta(n),\Lambda(n))\eta(n)\mid\theta(n),\Lambda(n)]+o(\beta)$$

$$=\nabla_{\theta(n)}L(\theta(n),\Lambda(n))+o(\beta).$$

The last equality follows since the second term on the RHS of the first equality above equals zero and $E[\eta(n)\eta(n)^T]=I$, the identity matrix. The claim follows. □

Consider now the recursion (10.27). Since $a(n)=o(b(n))$, we treat $\Lambda(n)\equiv\Lambda$, a constant, in the analysis of (10.27). The ODE associated with (10.27) is thus

$$\dot{\theta}(t)=\tilde{\Gamma}(-\{P(\nabla^2_\theta L(\theta(t),\Lambda))\}^{-1}\nabla_\theta L(\theta(t),\Lambda)), \tag{10.31}$$

where for any $y\in\mathbb{R}^N$ and a bounded, continuous function $v(\cdot):\mathbb{R}^N\to\mathbb{R}^N$,

$$\tilde{\Gamma}(v(y))=\lim_{0<\eta\to 0}\left(\frac{\Gamma(y+\eta v(y))-\Gamma(y)}{\eta}\right).$$

Let

$$K_\Lambda \stackrel{\triangle}{=} \{\theta \in C \mid \nabla_\theta L(\theta,\Lambda)^T \tilde{\Gamma}(-\{P(\nabla_\theta^2 L(\theta,\Lambda))\}^{-1}\nabla_\theta L(\theta,\Lambda)) = 0\}.$$

Further, let

$$\hat{K}_\Lambda \stackrel{\triangle}{=} \{\theta \in C \mid \tilde{\Gamma}(-\{P(\nabla_\theta^2 L(\theta,\Lambda))\}^{-1}\nabla_\theta L(\theta,\Lambda)) = -\{P(\nabla_\theta^2 L(\theta,\Lambda))\}^{-1}\nabla_\theta L(\theta,\Lambda)\}.$$

Let C^o denote the interior of C. Then, one can see that $C^o \subseteq \hat{K}_\Lambda$ for any $\Lambda = (\lambda_1,\ldots,\lambda_p)^T$, with $\lambda_i \geq 0$, $i = 1,\ldots,p$. In the light of Propositions 10.1 and 10.2, we have the following result whose proof follows in a similar manner as Theorem 8.12 (Chapter 8).

Theorem 10.3. *Let $\lambda_i(n) \equiv \lambda_i \; \forall n \geq 0$, for some $\lambda_i \geq 0$, $i = 1,\ldots,p$. Then the sequence $\{\theta(n)\}$ converges as $\beta \to 0$ to a point $\theta^\Lambda \in K_\Lambda$ with probability one, where $\Lambda = (\lambda_1,\ldots,\lambda_p)^T$.*

Consider now the recursion (10.28) and consider the following sequence of ODEs: For $l = 1,\ldots,N, i = 1,\ldots,p$,

$$\dot{\lambda}_i(t) = 0, \tag{10.32}$$
$$\dot{\theta}_l(t) = 0, \tag{10.33}$$
$$\dot{Y}_i(t) = G_i(\theta(t)) - Y_i(t). \tag{10.34}$$

In lieu of (10.32)–(10.33), $\theta(t) \equiv \theta$ for some $\theta \in \mathbb{R}^N$ and (10.34) can be rewritten as

$$\dot{Y}_i(t) = G_i(\theta) - Y_i(t). \tag{10.35}$$

We now have the following result:

Proposition 10.4. $\|Y_i(n) - G_i(\theta(n))\| \to 0$ *w.p. 1, as $n \to \infty$, for all $i = 1,\ldots,p$.*

Proof. Since $a(n) = o(c(n))$ and $b(n) = o(c(n))$, one can treat in a similar manner as the foregoing, $\Lambda(n)$ and $\theta(n)$ to be constants when analyzing (10.28). Rewrite (10.28) as

$$Y_i(n+1) = Y_i(n) + c(n)(G_i(\theta(n)) + \xi_i(n) + M(n+1) - Y_i(n)),$$

where $\xi_i(n) = E[g_i(X(n)) \mid \mathscr{F}(n-1)] - G_i(\theta(n))$ with $\mathscr{F}(n) = \sigma(X(m),\theta(m), m \leq n)$, $n \geq 0$ being the associated σ-fields and $M(n+1) = (g_i(X(n)) - E[g_i(X(n)) \mid \mathscr{F}(n-1)])$. Let $N(n) = \sum_{m=0}^{n} c(m)M_{m+1}, n \geq 0$. We will first verify that $(N(n), \mathscr{F}(n))$, $n \geq 0$ is a square-integrable martingale. Note that

$$E[N(n)^2] \leq K_n \sum_{m=0}^{n} c(m)^2 E[g_i^2(X(n)) + E^2[g_i(X(n)) \mid \mathscr{F}(n-1)]],$$

for some $K_n > 0$ that depends on n. By the conditional Jensen's inequality, we have $E^2[g_i(X(n)) \mid \mathscr{F}(n-1)] \leq E[g_i^2(X(n)) \mid \mathscr{F}(n-1)]$. Thus, $E[N(n)^2] \leq 2K_n \sum_{m=0}^{n} c(m)^2$ $E[g_i^2(X(n))]$. Now since $g_i(\cdot)$ is Lipschitz continuous,

$$|g_i(X(n))| \leq |g_i(0)| + |g_i(X(n)) - g_i(0)| \leq |g_i(0)| + \bar{K}_i \|X(n)\|,$$

where $\bar{K}_i > 0$ is the Lipschitz constant for the function $g_i(\cdot)$. Thus, $|g_i(X(n))| \leq K_i(1 + \|X(n)\|)$, where $K_i = \max(|g_i(0)|, \bar{K}_i)$. Hence $E[g_i^2(X(n))] \leq 2K_i^2$ $(1 + E[\|X(n)\|^2])$. Now from Assumption 10.4, we have that $\sup_n E[\|X(n)\|^2] < \infty$. Hence, $E[N(n)^2] < \infty$ for all $n \geq 0$. Further, $E[N(n+1) \mid \mathscr{F}(n)] = N(n)$ w.p.1. Thus, $N(n), n \geq 0$ is a square-integrable martingale sequence. Now,

$$\sum_n E[(N(n+1) - N(n))^2 \mid \mathscr{F}(n)]$$

$$= \sum_n c(n+1)^2 E[(g_i(X(n+1)) - E[g_i(X(n+1)) \mid \mathscr{F}(n)])^2 \mid \mathscr{F}(n)]$$

$$\leq 2 \sum_n c(n+1)^2 E[g_i^2(X(n+1)) + E^2[g_i(X(n+1)) \mid \mathscr{F}(n)] \mid \mathscr{F}(n)]$$

$$\leq 4 \sum_n c(n+1)^2 E[g_i^2(X(n+1)) \mid \mathscr{F}(n)].$$

From Assumption 10.4 and the Lipschitz continuity of $g_i(\cdot)$, it can again be seen that $\sup_n E[g_i^2(X(n+1)) \mid \mathscr{F}(n)] < \infty$ almost surely. Since $\sum_n c(n)^2 < \infty$, we have that

$$\sum_n E[(N(n+1) - N(n))^2 \mid \mathscr{F}(n)] < \infty \text{ w.p.1.}$$

Hence, from the martingale convergence theorem (Theorem B.2), $\{N(n)\}$ is an almost surely convergent sequence. Now, $\xi_i(n), n \geq 0$ constitutes the Markov noise. Note, however, that as a consequence of Assumption 10.1, along the 'natural timescale', $\xi_i(n) \to 0$ w.p. 1 as $n \to \infty$, $\forall i = 1, \ldots, p$. Since the natural timescale is faster than the timescale of the stochastic recursion, the latter sees the quantity $\xi_i(n)$ as having converged to zero, see [6, Chapter 6.2] for a detailed analysis of natural timescale recursions. The rest follows in a straightforward manner from the Hirsch lemma (Lemma C.5) applied to the ODE (10.35) for every $\varepsilon > 0$. □

Finally, we consider the slowest timescale recursion (10.29). In the light of Proposition 10.4, one may consider the following alternate recursion: For $i = 1, \ldots, p$,

$$\lambda_i(n+1) = \hat{\Gamma}(\lambda_i(n) + a(n)(G_i(\theta(n)) - \alpha_i)). \tag{10.36}$$

Let for any $\lambda \in \mathbb{R}$ and a bounded, continuous function $w(\cdot) : \mathbb{R} \to \mathbb{R}$,

$$\bar{\hat{\Gamma}}(w(\lambda)) = \lim_{0 < \eta \to 0} \left(\frac{\hat{\Gamma}(\lambda + \eta w(\lambda)) - \hat{\Gamma}(\lambda)}{\eta} \right).$$

Then (10.36) is an Euler discretization with (nonuniform) step sizes $a(n)$ of the ODE

$$\dot{\lambda}_i(t) = \bar{\hat{\Gamma}}(G_i(\theta(t)) - \alpha_i), \tag{10.37}$$

$i = 1, \ldots, p$. Let

$$F \stackrel{\triangle}{=} \{\Lambda = (\lambda_1, \ldots, \lambda_p)^T \mid \lambda_i \in [0, \bar{L}], \bar{\hat{\Gamma}}(G_i(\theta^\Lambda) - \alpha_i) = 0,\ \forall i = 1, \ldots, p,\ \theta^\Lambda \in K_\Lambda\}.$$

Also, let

$$F^- \stackrel{\triangle}{=} \{\Lambda = (\lambda_1, \ldots, \lambda_p)^T \mid \lambda_i \in [0, \bar{L}), \bar{\hat{\Gamma}}(G_i(\theta^\Lambda) - \alpha_i) = 0,\ \forall i = 1, \ldots, p,\ \theta^\Lambda \in K_\Lambda\}.$$

A standard stochastic approximation argument using the Hirsch lemma and Theorem 10.3 also shows the following:

Theorem 10.5. *$\Lambda(n) \to \Lambda^*$ for some $\Lambda^* \stackrel{\triangle}{=} (\lambda_1^*, \ldots, \lambda_p^*)^T \in F$ as $n \to \infty$ and $\beta \to 0$ with probability one.*

Let Λ^* be as in Theorem 10.5. The next proposition shows that the limiting point θ^{Λ^*} corresponding to Λ^* satisfies all the inequality constraints viz., $G_i(\theta^{\Lambda^*}) \leq \alpha_i$, $\forall i = 1, \ldots, p$. In other words, the limiting point θ^{Λ^*} is a feasible point of the constrained optimization problem (10.3)-(10.4).

Proposition 10.6. *For any $\Lambda^* \in F^-$, the corresponding parameter $\theta^{\Lambda^*} \in K_{\Lambda^*}$ satisfies all inequality constraints $G_i(\theta^{\Lambda^*}) \leq \alpha_i$, $\forall i = 1, \ldots, p$.*

Proof. Suppose not. Then for some $i \in \{1, \ldots, p\}$, $G_i(\theta^{\Lambda^*}) > \alpha_i$. Hence,

$$\hat{\Gamma}(G_i(\theta^{\Lambda^*}) - \alpha_i) = \lim_{\eta \to 0} \frac{\hat{\Gamma}(\lambda_i^* + \eta(G_i(\theta^{\Lambda^*}) - \alpha_i)) - \lambda_i^*}{\eta}$$

$$= \lim_{\eta \to 0} \frac{\lambda_i^* + \eta(G_i(\theta^{\Lambda^*}) - \alpha_i) - \lambda_i^*}{\eta} = G_i(\theta^{\Lambda^*}) - \alpha_i > 0,$$

which is a contradiction since $\Lambda^* \in F^-$. The second equality above follows because $\lambda_i^* \geq 0$ and $G_i(\theta^{\Lambda^*}) > \alpha_i$. Hence, for sufficiently small $\eta > 0$, $\lambda_i^* + \eta(G_i(\theta^{\Lambda^*}) - \alpha_i) \in F^-$ as well and hence

$$\hat{\Gamma}(\lambda_i^* + \eta(G_i(\theta^{\Lambda^*}) - \alpha_i)) = \lambda_i^* + \eta(G_i(\theta^{\Lambda^*}) - \alpha_i).$$

The claim follows. □

We call $\hat{\Lambda} \in F$ a spurious fixed point of the ODE (10.37) if the non-projected version of the same, i.e.,

$$\dot{\lambda}_i(t) = G_i(\theta(t)) - \alpha_i,$$

does not have $\hat{\Lambda}$ as a fixed point. Such a fixed point is introduced by the projection operator in the ODE and would lie on the boundary of the constraint set (cf. [7]).

Corollary 10.7. *For any $\Lambda^* \in F$ for which $\lambda_i^* = \bar{L}$ for some $i = 1,\ldots,p$, and $G_i(\theta^{\Lambda^*}) > \alpha_i$, is a spurious fixed point of (10.37).*

Proof. Observe that

$$\hat{\Gamma}(\lambda_i^* + \eta(G_i(\theta^{\Lambda^*}) - \alpha_i)) = \lambda_i^* = \bar{L},$$

since $\lambda_i^* + \eta(G_i(\theta^{\Lambda^*}) - \alpha_i) > \bar{L}$ for any $\eta > 0$. Hence, $\bar{\hat{\Gamma}}(G_i(\theta^{\Lambda}) - \alpha_i) = 0$. The claim follows. □

Proposition 10.8. *For $\Lambda^* \in F$, if $G_i(\theta^{\Lambda^*}) < \alpha_i$, for some $i \in \{1,\ldots,p\}$, then $\lambda_i^* = 0$.*

Proof. We consider both possibilities viz., (a) $\lambda_i^* = 0$ and (b) $\lambda_i^* > 0$, respectively. Consider (a) first. It is easy to see that for $\lambda_i^* = 0$ and $G_i(\theta^{\Lambda^*}) < \alpha_i$, we have that $\hat{\Gamma}(\lambda_i^* + \eta(G_i(\theta^{\Lambda^*}) - \alpha_i)) = 0$ as well for all $\eta > 0$.

Now consider (b). Note that for $\lambda_i^* > 0$, one can find $\eta_0 > 0$, such that for all $0 < \eta \le \eta_0$,

$$\hat{\Gamma}(\lambda_i^* + \eta(G_i(\theta^{\Lambda^*}) - \alpha_i)) = \lambda_i^* + \eta(G_i(\theta^{\Lambda^*}) - \alpha_i) > 0.$$

Thus

$$\bar{\hat{\Gamma}}(G_i(\theta^{\Lambda^*}) - \alpha_i) = G_i(\theta^{\Lambda^*}) - \alpha_i < 0,$$

which is a contradiction since $\Lambda^* \in F$. Thus, for $\Lambda^* \in F$, $G_i(\theta^{\Lambda^*}) < \alpha_i$ for some $i \in \{1,\ldots,p\}$ is only possible provided $\lambda_i^* = 0$. □

Remark 10.2. From Theorem 10.5, $\Lambda(n) \to \Lambda^*$ for some $\Lambda^* \stackrel{\triangle}{=} (\lambda_1^*,\ldots,\lambda_p^*)^T$ with $\lambda_i^* \in [0,\bar{L}]$, $\forall i = 1,\ldots,p$ such that $\theta^{\Lambda^*} \in K_{\Lambda^*}$ and $\bar{\hat{\Gamma}}(G_i(\theta^{\Lambda^*}) - \alpha_i) = 0$, $\forall i = 1,\ldots,p$. Note that for given Λ, the condition $\bar{\hat{\Gamma}}(G_i(\theta^{\Lambda}) - \alpha_i) = 0$, $i = 1,\ldots,p$ is the same as $\bar{\hat{\Gamma}}(\nabla_{\lambda_i} L(\theta^{\Lambda},\Lambda)) = 0$. Using the envelope theorem of mathematical economics [8, pp.964-966], one may conclude that (10.37) corresponds to

$$\dot{\lambda}_i(t) = \bar{\hat{\Gamma}}(\nabla_{\lambda_i} L(\theta^{\Lambda^*},\Lambda^*)), \tag{10.38}$$

$i = 1,\ldots,p$, interpreted in the 'Caratheodory' sense, see [5, Lemma 4.3]. The parameter tuple $(\theta(n),\Lambda(n))$ can then be seen to converge to a local minimum – local maximum tuple for the Lagrangian $L(\cdot,\cdot)$.

Convergence Analysis of CG-SF

Again in view of $a(n) = o(b(n))$, we let $\Lambda(n) \equiv \Lambda$, a constant, when analyzing (10.18)–(10.19). Proposition 10.2 shows that the gradient estimator is strongly consistent. (Note that the form of the gradient estimates in both CN-SF and CG-SF are the same.) Now let $K'_\Lambda \stackrel{\triangle}{=} \{\theta \in C \mid \tilde{\Gamma}(-\nabla_\theta L(\theta,\Lambda)) = 0\}$. One can see that K'_Λ is analogous to K_Λ, except for some spurious fixed points on the boundary (in addition to the regular fixed points). The conclusions of Theorem 10.3 now continue to hold with K'_Λ in place of K_Λ.

Finally, let the set F be now defined as

$$F \stackrel{\triangle}{=} \{\Lambda = (\lambda_1,\ldots,\lambda_p)^T \mid \lambda_i \in [0,\bar{L}], \bar{\tilde{\Gamma}}(G_i(\theta^\Lambda) - \alpha_i) = 0,\ \forall i = 1,\ldots,p,\ \theta^\Lambda \in K'_\Lambda\}.$$

The conclusions of Theorem 10.5 as well as Propositions 10.6 and 10.8 continue to hold with the set F defined above.

Convergence Analysis of CG-SPSA

As before, since $a(n) = o(b(n))$, let $\Lambda(n) \equiv \Lambda$ (a constant) $\forall n$, while analyzing (10.18)–(10.19).

Proposition 10.9. *With probability one,* $\forall l \in \{1,\ldots,N\}$, *as* $\delta \to 0$,

$$\left| E\left[\frac{L(\theta(n)+\delta\Delta(n),\Lambda(n)) - L(\theta(n),\Lambda(n))}{\delta\Delta_l(n)} \mid \theta(n),\Lambda(n)\right] - \nabla_{\theta_l} L(\theta(n),\Lambda(n))\right| \to 0.$$

Proof. Follows from a routine Taylor series based argument and the properties of the perturbations $\Delta_i(n), i = 1,\ldots,N, n \geq 0$ (cf. Assumption 5.4). □

The rest of the analysis is now identical to that of CG-SF.

Convergence Analysis of CN-SPSA

Again since $a(n) = o(b(n))$, let $\Lambda(n) \equiv \Lambda$ (a constant) $\forall n$, when analyzing (10.13)-(10.15). The analysis of these recursions follows in a similar manner as that of the recursions in Chapter 7. The Hessian and gradient estimators used are seen to be strongly consistent. The gradient estimator here is defined from the same two simulations that are used to estimate the Hessian in this scheme. Hence, the gradient estimator here is significantly different when compared with the estimator used in CG-SPSA.

Proposition 10.10. *With probability one, $\forall j, i \in \{1,\ldots,N\}$,*

$$\left| E\left[\frac{L(\theta(n)+\delta_1\Delta(n)+\delta_2\hat{\Delta}(n),\Lambda(n)) - L(\theta(n),\Lambda(n))}{\delta_1\delta_2\Delta_i(n)\hat{\Delta}_j(n)} \mid \theta(n),\Lambda(n)\right] - \nabla^2_{\theta_j,\theta_i}L(\theta(n),\Lambda(n))\right| \to 0,$$

as $\delta_1, \delta_2 \to 0$.

Proof. Follows again by using an argument based on Taylor series expansion, see for instance, Chapter 7 for a proof of unbiasedness of a similar estimator. □

Proposition 10.11. *With probability one, $\forall k \in \{1,\ldots,N\}$,*

$$\left| E\left[\frac{L(\theta(n)+\delta_1\Delta(n)+\delta_2\hat{\Delta}(n),\Lambda(n)) - L(\theta(n),\Lambda(n))}{\delta_2\hat{\Delta}_k(n)} \mid \theta(n),\Lambda(n)\right] - \nabla_{\theta_k}L(\theta(n),\Lambda(n))\right| \to 0,$$

as $\delta_1, \delta_2 \to 0$.

Proof. As before, follows using an argument based on an appropriate Taylor series expansion, see Chapter 5 for a proof of unbiasedness of a similar estimator. □

The rest of the analysis now follows in a similar manner as that of CN-SF.

10.5 Concluding Remarks

We presented in this chapter simulation-based algorithms for optimizing an objective function under inequality constraints that are in turn obtained from some other related objective functions. Both the objective and the constraint functions were considered to have a long-run average form. Hence neither the objective nor the constraints are analytically known functions of the parameter. The Lagrange relaxation approach was used to handle the inequality constraints and the algorithms were based on multi-timescale stochastic approximation for performing parameter search with the long-run average objective and under similar constraint functions. Similar approaches have also been used in the context of the random early detection (RED) scheme for flow control in TCP/IP networks [9, 4] (see Chapter 14.2). Such techniques have also been applied in the context of reinforcement learning for the problem of controlling a stochastic dynamic system under functional constraints, see [5, 1, 3]. An application, in the context of service systems, of similar methods as explored in this chapter for the problem of discrete parameter constrained optimization is considered in Chapter 12.

References

1. Bhatnagar, S.: An actor-critic algorithm with function approximation for discounted cost constrained markov decision processes. Systems and Control Letters 59, 760–766 (2010)
2. Bhatnagar, S., Hemachandra, N., Mishra, V.: Stochastic approximation algorithms for constrained optimization via simulation. ACM Transactions on Modeling and Computer Simulation 21, 15:1–15:22 (2011)
3. Bhatnagar, S., Lakshmanan, K.: An online actorcritic algorithm with function approximation for constrained Markov decision processes. Journal of Optimization Theory and Applications 153(3), 688–708 (2012)
4. Bhatnagar, S., Patro, R.K.: A proof of convergence of the B-RED and P-RED algorithms in random early detection. IEEE Communication Letters 13(10), 809–811 (2009)
5. Borkar, V.S.: An actor-critic algorithm for constrained Markov decision processes. Systems and Control Letters 54, 207–213 (2005)
6. Borkar, V.S.: Stochastic Approximation: A Dynamical Systems Viewpoint. Cambridge University Press and Hindustan Book Agency (Jointly Published), Cambridge and New Delhi (2008)
7. Kushner, H.J., Yin, G.G.: Stochastic Approximation Algorithms and Applications. Springer, New York (1997)
8. Mas-Colell, A., Whinston, M.D., Green, J.R.: Microeconomic Theory. Oxford University Press, Oxford (1995)
9. Patro, R.K., Bhatnagar, S.: A probabilistic constrained nonlinear optimization framework to optimize RED parameters. Performance Evaluation 66(2), 81–104 (2009)

Chapter 11
Reinforcement Learning

11.1 Introduction

Reinforcement learning (RL) [22], [5] is one of the most active research areas in machine learning and artificial intelligence (AI). While it has its roots in AI, RL has found tremendous applications in problems involving sequential decision making under uncertainty or stochastic control [3, 4, 19].

The basic framework in RL involves interactions between an "agent", i.e., the learner/controller and the "environment". The task of the agent is to observe the state of the environment and select an action. On its part, the environment reacts (to the agent's selection of an action) by probabilistically changing its state. In addition, the environment hands over the agent a certain reward that could be positive or negative — a negative reward implies a penalty or cost. The agent next observes the new state of the environment, again selects an action and the process is repeated. The goal of the agent is to select an action at each time instant (upon observing the state of the environment) in a way as to maximize a long-term reward. The reward that the agent receives from the environment when it selects an action plays the role of a "reinforcement" signal. The agent uses this signal to update its strategy to select actions based on the environment state. Many times, these updates are incremental in nature resulting in algorithms that gradually converge to the optimal strategies.

One of the challenges unique to RL is the tradeoff between "exploration" and "exploitation". To obtain a large reward, an RL agent might select actions it has previously tried that were found to yield a high reward. However, in doing so, it would not be exploring other (unexplored) actions that could potentially result in an even higher reward. Thus, it needs to strike a balance between exploiting actions known to give high rewards and exploring newer actions.

RL algorithms are largely classified under two broad categories: those that deal with the problem of prediction and those that deal with control. Usually, it is convenient to assume that the controller selects actions according to a *policy* that is

S. Bhatnagar et al.: Stochastic Recursive Algorithms for Optimization, LNCIS 434, pp. 187–220.
springerlink.com

a decision rule that suggests which action to pick in which state. The policy can also be a function of time. In the problem of prediction, one is often interested in evaluating, in terms of the long-term reward or cost, the *value* of a given policy. When the state space is large or high-dimensional, estimating the value itself of a given policy may be difficult and one requires function approximation-based approaches. An efficient RL algorithm for the problem of prediction is temporal difference (TD) learning. In the problem of control, on the other hand, the aim is to find an optimal control policy. In actor-critic RL algorithms, for instance, the problem of prediction forms a subtask of the control problem. In particular, for performing an update on the policy, the 'value' of the current update of the policy is first estimated. Multi-timescale stochastic approximation helps in such scenarios. Another important RL algorithm for the problem of control is Q-learning. One of the problems that algorithms such as TD and Q-learning with function approximation suffer from is the off-policy problem. In particular, Q-learning with function approximation can in fact diverge. We discuss this in more detail in the later sections.

The area of RL has seen several impressive applications. Examples include the development of a world class computer backgammon player [24] as well as the control of an inverted autonomous helicopter flight [27]. RL is applicable in situations when information about the system model is not precisely known, however, states can be simulated via a simulator or else directly observed from a real system.

In this chapter, we shall study recently developed algorithms for RL that are based on simultaneous perturbation techniques. These algorithms are largely from [9, 1, 7, 8, 6]. We discuss algorithms that are based on full-state representations as well as those that incorporate function approximation. A distinguishing feature of these algorithms is that they are based on simultaneous perturbation ideas. Apart from being easily implementable, some of these algorithms also exhibit significant improvements over well-known algorithms in the literature. For instance, we show that the multi-timescale simultaneous perturbation variant of Q-learning with function approximation is convergent and does not suffer from the off-policy problem. We shall be concerned here with the problem of minimizing costs rather than maximizing rewards. The two problems can be seen to be analogous if one defines costs as negative rewards.

11.2 Markov Decision Processes

Consider a discrete time stochastic process $\{X(n), n \geq 0\}$ that takes values at each instant in a set S called the state space. Suppose that the evolution of $\{X(n)\}$ depends on a control-valued sequence $\{Z(n)\}$. Let $A(X(n))$ denote the set of all controls (also called actions) that are available to the controller when the state of the MDP is $X(n) \in S$. Then, $A \triangleq \cup_{X \in S} A(X)$ is the set of all possible actions, also called the action space. We let S and A be both finite sets.

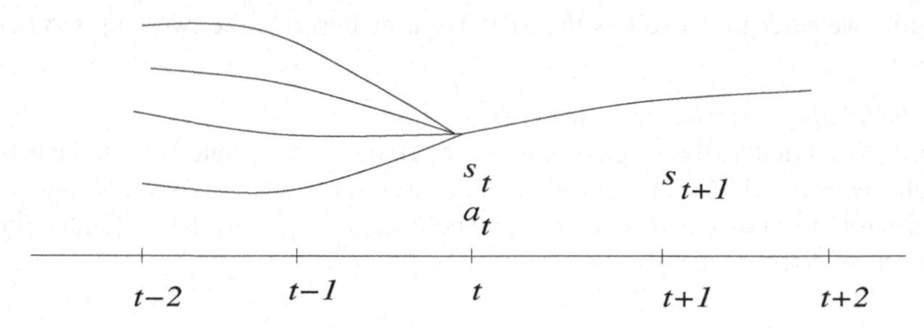

Fig. 11.1 The Controlled Markov Behaviour

As illustrated in Fig. 11.1, the process $\{X(n)\}$ is said to be a Markov decision process (MDP) if it satisfies the following (controlled Markov) property: For any states $i_0, i_1, \ldots, i_{n-1}, i, j \in S$ and actions $a_j \in A(i_j)$, $j = 0, 1, \ldots, n-1$, $a \in A(i)$, we have

$$Pr(X(n+1) = j \mid X(n) = i, Z(n) = a, X(n-1) = i_{n-1}, Z(n-1) = a_{n-1}, \ldots, X_0 = i_0, Z_0 = a_0) \triangleq p(i, j, a). \quad (11.1)$$

Here, $p(i, j, a)$, $i, j \in S$, $a \in A(i)$ are called controlled transition probabilities and satisfy the properties $0 \leq p(i, j, a) \leq 1$ and $\sum_{j \in S} p(i, j, a) = 1$.

Problems in MDPs fall into two main categories: finite horizon and infinite horizon. In finite horizon MDPs, the process is observed and controlled only over a fixed number N of time instants while in infinite horizon problems, the same is done over an infinite number of instants. We shall be concerned here with infinite horizon problems. The text [3] deals primarily with finite horizon problems.

It is often convenient for an agent to select actions according to a policy, i.e., a rule for selecting actions. By an admissible policy π, we mean a sequence of functions $\pi = \{\mu_0, \mu_1, \mu_2, \ldots,\}$ with each $\mu_k : S \to A$, such that $\mu_k(i) \in A(i)$, $\forall i \in S$, $k \in \{0, 1, \ldots\}$. Let Π be the set of all admissible policies. If $\mu_k = \mu$, $\forall k = 0, 1, \ldots$, for some μ independent of k, then we call the policy π (or many times μ itself) a stationary deterministic policy (SDP). By a randomized policy

(RP) ψ, we mean a sequence $\psi = \{\phi_0, \phi_1, \phi_2, \ldots\}$ with each $\phi(n) : S \to \mathscr{P}(A)$, $n = 0, 1, \ldots$, where $\mathscr{P}(A)$ is the set of all probability vectors on A. The above map is defined so that for each $i \in S$, $\phi(n)(i) \in \mathscr{P}(A(i))$, $n = 0, 1, 2, \ldots$, with $\mathscr{P}(A(i))$ being the set of all probability vectors on $A(i)$. A stationary randomized policy (SRP) is a RP ψ for which $\phi(n) = \phi \; \forall n = 0, 1, \ldots$. By an abuse of notation, we refer to ϕ itself as the SRP. We now describe the two long-term cost criteria.

- *The infinite horizon discounted cost criterion:*
  Let $r(i,a)$ denote the single-stage cost when the current state is i and the action chosen is $a \in A(i)$. Let $\alpha \in (0,1)$ be the discount factor – a given constant. The aim is to minimize over all admissible policies, $\pi = \{\mu_0, \mu_1, \mu_2, \ldots\}$, the infinite horizon discounted cost:

$$V^{\pi}(i) = E\left[\sum_{k=0}^{\infty} \alpha^k r(X_k, \mu_k(X_k)) \mid X_0 = i\right], \tag{11.2}$$

  starting from a given initial state $i \in S$. For a given policy π, the function $V^{\pi}(\cdot)$ is called the value function corresponding to policy π. The optimal cost or value function $V^*(\cdot)$ is now defined by

$$V^*(i) = \min_{\pi \in \Pi} V^{\pi}(i), \qquad i \in S. \tag{11.3}$$

  One can show that an optimal SDP exists in this case and the optimal cost V^* satisfies the Bellman equation

$$V^*(i) = \min_{u \in A(i)} \left(r(i,a) + \sum_{j \in S} \alpha p(i,j,a) V^*(j) \right). \tag{11.4}$$

  From the form of the cost function (11.2), it is evident that the single-stage costs over the first few stages only matter since costs for subsequent stages get beaten down due to the exponential weighting via the discount factor. Hence, this form of the cost function is useful in cases when one is primarily interested in controlling the transient performance of the system.
- *The long-run average cost criterion:*
  Let $r(i,a)$ denote the single-stage cost in this case as before. The aim here is to find an admissible policy $\pi \in \Pi$ that minimizes the associated infinite horizon average cost $J(\pi)$ defined as

$$J(\pi) = \lim_{N \to \infty} \frac{1}{N} E\left[\sum_{k=0}^{N-1} r(X_k, \mu_k(X_k)) \mid X_0 = i\right], \tag{11.5}$$

starting from any initial state i. In case, the limit in (11.5) does not exist, one may replace the same with 'limsup'. However, we shall assume that the resulting Markov chain under any SDP π is ergodic, i.e., aperiodic, irreducible (and hence also positive recurrent since it is finite state). Under this assumption, the limit above can be shown to exist. Further, for any SDP π, a function $h^\pi : S \rightarrow \mathbb{R}$ exists that satisfies the Bellman equation corresponding to the policy π (also called the Poisson equation):

$$J(\pi) + h^\pi(i) = r(i,\mu(i)) + \sum_{j\in S} p(i,j,\mu(i))h^\pi(j),$$

for all $i \in S$.

Let J^* denote the optimal cost. Then,

$$J^* = \min_{\pi} J(\pi).$$

Note that $J(\pi)$ (and so also J^*) does not depend on the initial state i (see (11.5)) because of the ergodicity assumption. Let $h^*(\cdot)$ now denote the differential cost function. It can again be seen that an optimal SDP exists and the resulting Bellman equation for optimality corresponds to

$$J^* + h^*(i) = \min_{a\in A(i)} \left(r(i,a) + \sum_{j\in S} p(i,j,a)h^*(j) \right), \ \forall i \in S. \tag{11.6}$$

Note that the cost function (11.5) here is a long-run average of single-stage cost functions. This form of the cost function is useful in scenarios where one is interested in controlling the steady-state system performance but not so much the transient system behaviour.

11.3 Numerical Procedures for MDPs

We now describe two of the important numerical procedures for MDPs assuming that information on transition probabilities is completely known. Moreover, the controller can fully observe the states of the system. In the case of the RL algorithms that we discuss in the next section, information on transition probabilities is usually assumed to be unavailable. In some instances, even the states of the MDP may not be fully observable. Such systems are then modelled in the setting of partially observed MDPs (POMDPs). We shall, however, not be concerned with POMDPs in this article.

The numerical schemes that we describe below go by the names of value iteration and policy iteration [4], [19], respectively. A few other computational procedures largely based on these two aproaches are also available in the literature. We describe both procedures below for the two different cost criteria.

11.3.1 Numerical Procedures for Discounted Cost MDPs

Value Iteration

- **Step 1:** Set an initial estimate $V_0(\cdot)$ of the value function.
- **Step 2:** For $n = 0, 1, 2, \ldots$, for all $i \in S$, iterate the following until convergence:

$$V_{n+1}(i) = \min_{a \in A(i)} \left(r(i,a) + \alpha \sum_{j \in S} p(i,j,a) V(n)(j) \right). \tag{11.7}$$

The iteration (11.7) is based on the Bellman equation for optimality (11.4) and can be seen to be a fixed point iteration corresponding to a given contraction operator. Hence, it can be seen to converge ([4], [19]). In practice, the above iteration can be stopped after a certain large number (N) of iterates. The value $V_N(\cdot)$ would then be an estimate of the true value function $V^*(\cdot)$.

Policy Iteration

- **Step 1:** Set an initial estimate μ_0 of the optimal SDP.
- **Step 2 (Policy Evaluation):** Given the SDP update μ_n at iteration n, solve the Bellman equation (11.4) with actions obtained only according to the policy μ_n. Thus the iteration there does not involve the 'min' operation. In vector-matrix notation, this corresponds to solving

$$V^{\mu_n} = R^{\mu_n} + \alpha P^{\mu_n} V^{\mu_n}, \tag{11.8}$$

where V^{μ_n} is the value function under the stationary policy μ_n. Further, R^{μ_n} and P^{μ_n} are the cost vector and transition probability matrix, respectively, under policy μ_n. The unique solution to (11.8) corresponds to

$$V^{\mu_n} = (I - \alpha P^{\mu_n})^{-1} R^{\mu_n}.$$

Computing the inverse above could be problematic in many cases particularly when the number of states is large. One can in such a case solve (11.8) by running a value iteration scheme (for the given policy μ_n) as follows: For $m = 0, 1, 2, \ldots$,

$$V_{m+1}^{\mu_n}(i) = r(i, \mu_n(i)) + \alpha \sum_{j \in S} p(i, j, \mu_n(i)) V_m^{\mu_n}(j),$$

starting from any $V_0^{\mu_n}(i)$, $i \in S$. Upon convergence of the *inner loop*, viz., $V_m^{\mu_n}(i) \to V^{\mu_n}(i)$, as $m \to \infty$, a policy update is performed as below.

- **Step 3 (Policy Improvement):**

$$\mu_{n+1}(i) \in \arg\min \left(r(i, \cdot) + \alpha \sum_{j \in S} p(i, j, \cdot) V^{\mu_n}(j) \right).$$

Even though we have described the above procedures considering SDP updates, one can also describe these with SRP updates.

11.3.2 Numerical Procedures for Long-Run Average Cost MDPs

The Relative Value Iteration Scheme

Note that (11.6) constitutes a system of $|S|$ equations with $|S|+1$ unknowns. One way to solve this system could be to set $h^*(i_0) = 0$ for some $i_0 \in S$ and solve the system for $J^*, h^*(i)$, $i \in S \backslash \{i_0\}$. Another procedure (that we consider here), that is commonly used and is called the relative value iteration scheme, is to arbitrarily select a state $i_0 \in S$ as a reference state and estimate J^* from the estimates of $h^*(i_0)$ and perform a value iteration type recursion on the Bellman equation (11.6) in a similar manner as for the discounted cost setting. Thus, arbitrarily initialize $h_0(i)$, $\forall i \in S$. Further for all $n \geq 1$, $i \in S$,

$$h_{n+1}(i) = \min_{a \in A(i)} \left(r(i, a) + \sum_{j \in S} p(i, j, a) h_n(j) \right) - h_n(i_0). \tag{11.9}$$

This procedure is slightly different from the value iteration scheme for discounted cost MDPs for the following reason: If one ignores the minimization operation in (11.3), then obtaining a solution to (11.3) would involve inverting the full-rank matrix $(I - \alpha P)$. On the other hand, if one ignores the minimization step in (11.6), then a solution would involve inverting the matrix $(I - P)$ that, however, is singular as one of its eigenvalues is zero. Note that the solution h^* to (11.9) is not unique as addition of a constant function $\bar{h}$ (say $\bar{h}(i) = k$, $\forall i$, for some $k \in \mathbb{R}$) to h^* results in the equation (11.6) being satisfied as well. This is not so in the discounted cost setting because of the presence of the discount factor α multiplying the expected value of the next state $\sum_{j\in S} p(i,j,a)V^*(j))$, see (11.4).

Policy Iteration

- **Step 1:** Set an initial estimate μ_0 of the optimal SDP. Arbitarily select a state $i_0 \in S$ as a reference state.
- **Step 2 (Policy Evaluation):** Given the SDP update μ_n, perform the following iteration: For $m = 0,1,2,\ldots$, $i \in S$,

$$h^{\mu_n}_{m+1}(i) = r(i,\mu_n(i)) + \alpha \sum_{j\in S} p(i,j,\mu_n(i))h^{\mu_n}_m(j) - h^{\mu_n}_m(i_0), \quad (11.10)$$

starting from any $h^{\mu_n}_0(i)$, $i \in S$. Upon convergence of the *inner loop*, viz., $h^{\mu_n}_m(i) \to h^{\mu_n}(i)$, as $m \to \infty$, for all $i \in S$, a policy update is performed as below.
- **Step 3 (Policy Improvement):**

$$\mu_{n+1}(i) \in \arg\min \left(r(i,\cdot) + \alpha \sum_{j\in S} p(i,j,\cdot)h^{\mu_n}(j) \right). \quad (11.11)$$

Proofs of convergence of the above procedures in the cases of both discounted cost and long-run average cost criteria can be found, for instance, in [4].

In the next section, we present RL algorithms for both discounted and long-run average cost criteria in the case of full-state representations. The basic underlying assumption in this setting is that the sizes of the state and action spaces are manageable and as such do not result in significant computational challenges for the given algorithms. Subsequently, we shall also present algorithms with function approximation that are, in particular, useful when the state and/or action spaces are large.

11.4 Reinforcement Learning Algorithms for Look-up Table Case

In all our algorithms, we let $|S| < \infty$, i.e., the state space is finite. The action sets $A(i)$ corresponding to any state i will be chosen to be either finite or else compact and convex subsets of a given Euclidean space.

11.4.1 An Actor-Critic Algorithm for Infinite Horizon Discounted Cost MDPs

We first present, from [9], an actor-critic algorithm based on the policy iteration technique. We let all sets $A(i)$ be compact (non-discrete) and convex subsets of $\mathbb{R}^N$. The algorithms we present here fall under the broad class of actor-critic algorithms that are based on the policy iteration approach. Recall that policy iteration involves updates using two nested loops: an outer-loop update occurring only on convergence of the inner-loop procedure corresponding to the given outer-loop update. Using two-timescale stochastic approximation, one is able to run recursions for both loops simultaneously.

We make the following assumption on the cost function and transition probabilities.

Assumption 11.1. For all $i, j \in S$, $a \in A(i)$, both $r(i,a)$ and $p(i,j,a)$ are continuously differentiable w.r.t. a.

Under Assumption 11.1, one can show that an optimal stationary policy exists for this problem and the optimal cost V^* satisfies the Bellman equation. Since any action $a_i \stackrel{\triangle}{=} (a_i^1,\ldots,a_i^N)^T \in A(i) \subset \mathbb{R}^N$, $i \in S$, one can identify a SDP π directly with the vector $(a_1^1,\ldots,a_1^N, a_2^1,\ldots,a_2^N, \ldots, a_s^1,\ldots,a_s^N)^T$ or simply with the block vector $(a_1,\ldots,a_s)^T$ of actions ordered lexicographically according to states, i.e., the jth component $(j = 1,\ldots,s)$ of this vector would correspond to the action taken in state j. Let $V_\pi(i)$ be the stationary value or cost-to-go function corresponding to the SDP π starting from $i \in S$.

Lemma 11.1. *Under Assumption 11.1, $V_\pi(i)$, $\forall i \in S$ are continuously differentiable functions of π.*

Proof. It follows from the Bellman equation for the given SDP π that $V_\pi \stackrel{\triangle}{=} (V_\pi(i), i \in S)^T$ satisfies

$$V_\pi = (I - \alpha P^\pi)^{-1} R^\pi,$$

where $P^\pi = [[p(i,j,\pi(i))]]_{i,j\in S}$ is the transition probability matrix of the Markov chain under SDP π and $R^\pi = (r(i,\pi(i)), i \in S)^T$ is the single-stage cost vector. The claim follows from an application of the Cramer's rule. □

Let $\pi(n) \stackrel{\triangle}{=} (a_1(n),\ldots,a_s(n))^T$ with each $a_i(n) \stackrel{\triangle}{=} (a_i^1(n),\ldots,a_i^N(n))^T$ denote the nth update of policy π. For simplicity assume that for $n \geq 0$, $\triangle(n) \in \mathbb{R}^{N|S|}$ is a vector of mutually independent, ± 1-valued, mean zero random variables $\triangle_i^j(n)$, $j = 1,\ldots,N$, $i \in S$ (viz., $\triangle(n) = (\triangle_1^1(n), \ldots, \triangle_1^N(n), \triangle_2^1(n), \ldots, \triangle_2^N(n), \ldots, \triangle_s^1(n),$

..., $\triangle_s^N(n))^T$). More general $\Delta_i^j(n)$ can be considered. In fact, any $\Delta_i^j(n)$ that satisfy Assumption 5.4 may be considered.

Let $\Gamma_i : \mathbb{R}^N \rightarrow A(i)$ be the projection operator that projects any N-dimensional vector x to the action set $A(i)$ (i.e., $\Gamma_i(x) \in A(i)$), $i \in S$. Further, let $\delta > 0$ be a given (small) constant. Consider two step-size sequences $\{b(n)\}$ and $\{c(n)\}$ that satisfy

$$\sum_n b(n) = \sum_n c(n) = \infty, \quad \sum_n b(n)^2, \sum_n c(n)^2 < \infty \quad \text{and} \quad c(n) = o(b(n)). \tag{11.12}$$

The Actor-Critic Algorithm

Let $\{Y_n^1(i,a)\}$ and $\{Y_n^2(i,a)\}$ be two independent families of i.i.d. random variables, each having the distribution $p(i,\cdot,a)$. The algorithm is as follows: For all $i \in S$, $r = 1,2$, we initialize $V_0^r(i) = 0$. Then, $\forall i \in S$, $j = 1,\ldots,N$, we have

$$a_i^j(n+1) = \Gamma_i^j\left(a_i^j(n) + c(n)\left(\frac{V_n^1(i) - V_n^2(i)}{2\delta\triangle_i^j(n)}\right)\right), \tag{11.13}$$

$$V_{n+1}^r(i) = V_n^r(i) + b(n)(r(i,\pi_i^r(n)) + \alpha V_n^r(Y_n^r(i,\pi_i^r(n))) - V_n^r(i)). \tag{11.14}$$

This algorithm is an actor-critic algorithm based on policy iteration and performs both the value function updates ($V_n^r(i)$) and the policy updates ($a_i^j(n)$). As with other multi-timescale algorithms, an additional averaging (on top of the two timescale averaging) over L epochs (with $L > 1$ set arbitrarily) of the value corresponding to a given policy update is seen to improve algorithmic behaviour.

Convergence Analysis

Let $\mathscr{F}_l = \sigma(a_i(p), \triangle_i(p), V_p^1(i), V_p^2(i), p \leq l, i \in S; Y_p^1(i,\pi_i^1(p)), Y_p^2(i,\pi_i^2(p)), p < l, i \in S)$, $l \geq 1$ be a sequence of associated sigma fields. We first analyze the faster recursion. Define a sequence $\{t(n)\}$ according to $t(0) = 0$ and for $n \geq 1$, $t(n) = \sum_{m=0}^{n-1} c(m)$. Let $a_i^j(t), t \geq 0$ be defined according to $a_i^j(t(n)) = a_i^j(n)$ with continuous linear interpolation in between. Also, let $\triangle_i^j(t(n)) = \triangle_i^j(n)$ $\forall n \geq 0$ with $\triangle_i^j(t) = \triangle_i^j(t(n)) = \triangle_i^j(n)$ $\forall t \in [t(n), t(n+1))$.

Consider the following system of ODEs: $\forall j = 1,\ldots,N$, $i \in S$, $r = 1,2$,

$$\dot{a}_i^j(t) = 0, \tag{11.15}$$

$$\dot{V}_t^l(i) = R^{\pi^l(t)} + (\alpha P^{\pi^l(t)} - I)V_t^l(i) \tag{11.16}$$

Here $R^{\pi^l(t)} = (r(i,\pi^l(t)), i \in S)^T$ is the vector (over all states) of the single-stage costs under SDP $\pi^l(t)$ and $P^{\pi^l(t)} = [[p(i,j,\pi^l(t))]]_{i,j\in S}$ is the transition probability matrix of the Markov process $\{X(n)\}$ when actions are chosen according to the above SDP. Note that the iterates (11.13) can be rewritten as

$$a_i^j(n+1) = \Gamma_i^j\left(a_i^j(n) + b(n)\xi_1(n)\right),$$

where $\xi_1(n) = o(1)$ as a consequence of (11.12). Thus, along the faster timescale, the iterates (11.13) asymptotically track (11.15).

Now in lieu of (11.15), one can let $a_i^j(t) = a_i^j, t \geq 0$. Similarly, $\triangle_i^j(t) = \triangle_i^j, t \geq 0$. Hence, $\pi^l(t) \equiv \pi^l, t \geq 0, l = 1,2$. Thus, (11.16) can be rewritten as

$$\dot{V}_t^l(i) = R^{\pi^l} + (\alpha P^{\pi^l} - I)V_t^l(i) \tag{11.17}$$

It is easy to see that (11.17) has $V^{l,*}(i) = (I - \alpha P^{\pi^l})^{-1}R^{\pi^l}$ as its globally asymptotically stable equilibrium.

Define sequences $\{J_i^l(n),\ n \geq 1\}$, $l = 1,2,\ i \in S$, according to $J_i^l(n) \stackrel{\triangle}{=} \sum_{k=0}^{n-1} b(k)$ $[(r(i,\pi_i^l(k)) + \alpha V_k^l(Y_k^l(i,\pi_i^l(k)))) - \Sigma_{j\in S}\, p(i,j,\pi_i^l(k))\,(r(i,\pi_i^l(k)) + \alpha V_k^l(j))]$, $i \in S$, $l = 1,2,\ n \geq 1$. It is easily seen that these are martingale sequences with respect to the filtration $\mathscr{F}_n, n \geq 0$. Let $M_i^l(n+1) = (J_i^l(n+1) - J_i^l(n)), n \geq 0$.

Lemma 11.2. *The iterates $V_k^r(i)$, $r = 1,2$, satisfy* $\sup_k \parallel V_k^r(i) \parallel < \infty\ \forall i \in S$. *Further, $V_k^r(i) \to V^{r,*}(i)$ as $k \to \infty$ almost surely.*

Proof. Note that (11.14) can be rewritten as

$$V_{k+1}^l(i) = V_k^l(i) + b(n)\left(r(i,\pi_i^l(k)) + \alpha\sum_{j\in S} p(i,j,\pi_i^l(k))V_k^l(j) - V_k^l(i)\right) + M_i^l(n+1).$$

Note that one can rewrite $M_i^l(n+1) = b(n)N_i^l(n+1)$ where $(N_i^l(n),\mathscr{F}_n))$ is a martingale difference sequence. It is easy to see that

$$E[|N_i^l(n+1)|^2 \mid \mathscr{F}_n] \leq K(1+ \parallel V_n^l \parallel^2),$$

for some constant $K > 0$. Further, the origin is the globally asymptotically stable equilibrium point for the ODE

$$\dot{V}_t^l(i) = (\alpha P^{\pi^l} - I)V_t^l(i), \tag{11.18}$$

since all the eigenvalues of the matrix $(\alpha P^{\pi^l} - I)$ have negative real parts. Assumptions D.1 and D.2 [11, (A1) and (A2)] are now satisfied and the claim now follows from Theorem D.1 [11, Theorems 2.1(i)-Theorem 2.2]. □

Consider now the slower timescale recursion (11.13). The ODE associated with that recursion is

$$\dot{a}_i(t) = \hat{\Gamma}_i\left(-\nabla^i V^{\pi(t)}(i)\right),\ i \in S. \tag{11.19}$$

Let $\mathscr{K}$ denote the set of local minima of $V^{(\cdot)}(i)$. Also, for given $\varepsilon > 0$,

$$\mathscr{K}^\varepsilon = \{\pi \mid \parallel \pi - \pi_0 \parallel < \varepsilon,\ \pi_0 \in \mathscr{K}\}.$$

The following theorem now follows from standard arguments involving consistency of the SPSA estimator as well as the Kushner–Clark theorem (Theorem E.1).

Theorem 11.3. *Given $\varepsilon > 0$, $\exists \delta_0 > 0$ such that $\forall \delta \in (0, \delta_0]$, the algorithm (11.13)-(11.14) converges to M^ε with probability one.*

11.4.2 The Q-Learning Algorithm and a Simultaneous Perturbation Variant for Infinite Horizon Discounted Cost MDPs

The Q-learning algorithm [26] is based on the Q-value iteration technique. We review it first below and subsequently present a multi-timescale variant based on one-simulation (Hadamard matrix based deterministic perturbation) SPSA that is particularly useful when the action space is large.

11.4.2.1 The Q-Learning Algorithm

Let the Q-value function under an admissible policy $\psi = \{\mu_0, \mu_1, \mu_2, \ldots\}$ be defined as follows: $\forall i \in S,\ a \in A(i)$,

$$Q_\psi(i,a) = E\left[\sum_{k=0}^{\infty} \alpha^k r(X_k, \mu_k(X_k)) \mid X_0 = i, Z_0 = a\right] \tag{11.20}$$

$$= r(i,a) + E\left[\sum_{k=1}^{\infty} \alpha^k r(X_k, \mu_k(X_k)) \mid X_0 = i, Z_0 = a\right]. \tag{11.21}$$

Let the optimal Q-values be defined according to

$$Q^*(i,a) = \min_{\psi \in \Psi} Q_\psi(i,a),\ i \in S, a \in A(i). \tag{11.22}$$

Here Ψ denotes the set of all admissible policies. It can be shown (as described previously) that an optimal policy that is an SDP exists. Hence,

$$Q^*(i,a) = \min_{\mu \in \Psi_S} Q_\mu(i,a), \tag{11.23}$$

where Ψ_S is the class of all SDP μ. It follows as in (11.20) that

$$Q_\mu(i,a) = r(i,a) + \alpha E\left[\sum_{k=1}^{\infty} \alpha^{k-1} r(X_k,\mu(X_k)) \mid X_0 = i, Z_0 = a\right] \tag{11.24}$$

$$= r(i,a) + \alpha \sum_{j\in S} p(i,j,a) E\left[\sum_{k=1}^{\infty} \alpha^{k-1} r(X_k,\mu(X_k)) \mid X_1 = j,\mu\right] \tag{11.25}$$

$$= r(i,a) + \alpha \sum_{j\in S} p(i,j,a) V_\mu(j), \tag{11.26}$$

where $V_\mu(j)$, $j \in S$, is the value function under SDP μ. It follows from (11.23) that

$$\begin{aligned} Q^*(i,a) = \min_{\mu\in\Psi_S} Q_\mu(i,a) &= \min_{\mu\in\Psi_S}\left(r(i,a) + \alpha\sum_{j\in S} p(i,j,a)V_\mu(j)\right), \\ &\geq r(i,a) + \alpha \sum_{j\in S} p(i,j,a) \min_{\mu\in\Psi_S} V_\mu(j), \\ &= r(i,a) + \alpha\sum_{j\in S} p(i,j,a)V^*(j). \end{aligned} \tag{11.27}$$

Further,

$$Q^*(i,a) = \min_{\mu\in\Psi_S} Q_\mu(i,a) \leq r(i,a) + \alpha \sum_{j\in S} p(i,j,a) V_{\mu'}(j) \;\forall \mu' \in \Psi_S.$$

Hence,

$$\begin{aligned} Q^*(i,a) &\leq r(i,a) + \alpha\sum_{j\in S} p(i,j,a) \min_{\mu'\in\Psi_S} V_{\mu'}(j) \\ &= r(i,a) + \alpha\sum_{j\in S} p(i,j,a)V^*(j). \end{aligned} \tag{11.28}$$

The Bellman equation (11.4) now corresponds to

$$V^*(i) = \min_{a\in A(i)} Q^*(i,a),\; i \in S. \tag{11.29}$$

It follows from (11.27)-(11.28) and (11.29) that

$$Q^*(i,a) = \left(r(i,a) + \gamma\sum_{j\in S} p(i,j,a) \min_{v\in A(j)} Q^*(j,v)\right). \tag{11.30}$$

(11.30) is also referred to as the Q-Bellman equation.

The Q-learning Update

The Q-learning algorithm [26] aims to solve the Q-Bellman equation (11.30) using stochastic approximation by assuming lack of information on the transition probabilities $p(i,j,a)$ and proceeds in the following manner: $\forall i \in S,\ a \in A(i)$,

$$Q_{n+1}(i,a) = Q_n(i,a) + c(n)\left(r(i,a) + \gamma \min_{v \in A(Y_n(i,a))} Q_n(Y_n(i,a),v) - Q_n(i,a)\right). \tag{11.31}$$

Here, $Y_n(i,a)$ is a simulated next state when the current state is i and action $a \in A(i)$ is chosen. The random variables $Y_n(i,a)$, $n \geq 0$ are assumed independent and have the distribution $p(i,a,\cdot)$, $i \in S$, $a \in A(i)$. Further, $c(n), n \geq 0$ are step-sizes that satisfy $c(n) > 0\ \forall n \geq 0$ and

$$\sum_n c(n) = \infty, \ \sum_n c(n)^2 < \infty. \tag{11.32}$$

The algorithm (11.31) works in the case of full-state representations. The Q-learning algorithm for full-state representations can become computationally cumbersome in cases when the cardinality of the action sets is high because of the requirement of explicit minimization in (11.31). It is also known to suffer from the problem of improper convergence if all actions are not explored sufficiently. In practice, this problem is tackled by selecting actions as suggested by the algorithm with a high probability, however, with a small probability, actions not suggested by the algorithm are explored to learn the best actions.

11.4.2.2 Two-Timescale Q-learning Algorithm

The algorithm below avoids the computational difficulty with large action sets (discussed above) by incorporating two timescales. Further, it updates randomized policies, as a result of which, actions that are not the 'current best' actions also get selected with a certain probability. A recursion similar to (11.31) but without the explicit minimization is run on the slower timescale, while on the faster scale, the minimization operation is conducted through a gradient search procedure. This algorithm has been presented in [8] and incorporates deterministic perturbations SPSA with the perturbation sequences obtained using the Hadamard matrix-based construction.

Let S and $A \stackrel{\triangle}{=} \cup_{i \in S} A(i)$ be finite sets with each set $A(i)$ assumed to contain exactly $(N+1)$ elements (for simplicity). Let $a_i^0, a_i^1, \ldots, a_i^N$ denote the elements of $A(i)$. Let $Q_n(\cdot,\cdot)$ denote the nth update of the Q-value function. Let $\pi_i(n) \stackrel{\triangle}{=} (\pi_{i,a}(n), a \in A(i))^T$, $i \in S$ denote the nth update of the randomized policy, where

$\pi_{i,a}(n)$ is the probability at the nth update of the randomized policy of picking action $a \in A(i)$ in state $i \in S$. Further, let $\hat{\pi}_i(n) \stackrel{\triangle}{=} (\pi_{i,a}(n), a \in A(i)\backslash\{a_i^0\})^T$ denote the vector of probabilities of actions $a \in A(i)$ other than that of action a_i^0. In the scheme below, we shall update $\hat{\pi}_i(n)$ using our algorithm, while the probability $\pi_{i,a_i^0}(n)$ will get automatically specified via $\pi_{i,a_i^0}(n) = 1 - \sum_{a\in A(i)\backslash\{a_i^0\}} \pi_{i,a}(n)$. Let $\pi_i'(n) = \Gamma(\hat{\pi}_i(n) - \delta\Delta_n(i))$ denote the perturbed SRP corresponding to $\hat{\pi}_i(n)$, where $\delta > 0$ is a given small constant. Also, $\Delta_n(i) \stackrel{\triangle}{=} (\Delta_n(i,a_i^1), \ldots, \Delta_n(i,a_i^N))^T$, $n \geq 0$, $i \in S$ denotes the perturbation vector obtained from the Hadamard matrix construction. We use perturbed SRPs in the simulations in order to estimate the gradient of the Q-function and, therefore, to also update the parameters. The algorithm below incorporates one-simulation deterministic SPSA with Hadamard matrix-based perturbations. Let $PS \subset \mathbb{R}^N$ be the simplex

$$PS = \{(y_1, \ldots, y_N)^T | y_i \geq 0, 1 \leq i \leq N \text{ and } \sum_{i=1}^{N} y_i \leq 1\},$$

in which $\hat{\pi}_i, i \in S$, take values. Further, let $\Gamma : \mathbb{R}^N \to PS$ denote the projection map.

Let

$$(\Delta_n(i))^{-1} \stackrel{\triangle}{=} \left(\frac{1}{\Delta_n(i,a_i^1)}, \ldots, \frac{1}{\Delta_n(i,a_i^N)}\right)^T, \ \forall n \geq 0, \ \forall i \in S.$$

Let $\psi_n(j)$ denote the action chosen from the set $A(j)$ according to the distribution given by $\pi_j'(n)$, with probability of picking action $a_0^j \in A(j)$ automatically specified from the latter. Also, as before, let $Y_n(i,a)$ be the 'next' state of the MDP when the current state is i and action a is chosen. We let $b(n)$ and $c(n)$, $n \geq 0$ be two step-size sequences that satisfy (11.12).

Update Rule

For all $i \in S, a \in A(i)$, initialize $Q_0(i,a)$ and $\hat{\pi}_i(0)$. Then $\forall i \in S, a \in A(i)$,

$$Q_{n+1}(i,a) = Q_n(i,a) + c(n)\left(r(i,a) + \alpha Q_n(Y_n(i,a), \psi_n(Y_n(i,a))) - Q_n(i,a)\right), \tag{11.33}$$

$$\hat{\pi}_i(n+1) = \Gamma\left(\hat{\pi}_i(n) + b(n)\frac{Q_n(i,\psi_n(i))}{\delta}(\Delta_n(i))^{-1}\right). \tag{11.34}$$

A Sketch of Convergence

As with other multi-timescale algorithms, recursion (11.33) being the slower recursion is quasi-static when viewed from the timescale of $b(n)$, $n \geq 0$. Hence one can treat $Q_n(i,a) \approx Q(i,a)$ (i.e., independent of n) when analyzing (11.34).

Using standard arguments, one can show that (11.34) asymptotically tracks the trajectories of the following ODE in the limit as $\delta \to 0$:

$$\dot{\hat{\pi}}_i = \hat{\Gamma}(-\nabla Q^{\hat{\pi}_i(t)}(i)), \tag{11.35}$$

The asymptotically stable fixed points of (11.35) lie within the set $M = \{\hat{\pi}_i \mid \hat{\Gamma}(\nabla Q^{\hat{\pi}_i}(i)) = 0\}$. Hence, let $\pi_i^u(n) \to \pi_i^{u,*}$ as $n \to \infty$. Note that if u^* corresponds to a unique optimal action in state i, then the following will be true: $\pi_i^{u,*} = 1$ for $u = u^*$ and $\pi_i^{u,*} = 0$ for all $u \in U(i)$, $u \neq u^*$. Else, if the optimal action is not unique, then one expects policy π^* in state i to assign equal positive mass to all optimal actions (i.e., those whose Q values are equal and uniformly lower compared to those of the other actions). Let $Q^*(i,u)$, $i \in S, u \in U(i)$ correspond to the unique solution of (11.30).

Theorem 11.4. *For all $i \in S, u \in U(i)$, the recursions $Q_n(i,u)$ converge almost surely to $Q^*(i,u)$ in the limit as $\delta \to 0$.*

11.4.3 Actor-Critic Algorithms for Long-Run Average Cost MDPs

We now consider the case of MDPs under the long-run average cost criterion. As with the infinite horizon discounted cost setting, we also consider two cases here — (a) when the action sets are (non-discrete) compact and (b) when the action sets are discrete and finite. The algorithms here are from [1].

Assumption 11.2. The Markov process $\{X(n)\}$ under any SDP is ergodic.

Under Assumption 11.2, the long-run average cost for any given SDP (and hence also SRP) is well defined.

11.4.3.1 An Algorithm for (Non-Discrete) Compact and Convex Action Sets

Here, one can directly perform a gradient search in the space of SDPs. We apply the one-simulation SPSA algorithm with Hadamard matrix-based perturbations. We assume that the single-stage cost $r(i,a)$ and the transition probabilities $p(i,j,a)$ are continuously differentiable functions of a. We let each action set $A(i)$, $i \in S$ to be a compact and convex subset of $\mathbb{R}^N$. Let Γ_i, $i \in S \stackrel{\triangle}{=} \{1,2,\ldots,s\}$ denote the projection map from $\mathbb{R}^N$ to $A(i)$. As for the discounted cost case, let $\pi(n) \stackrel{\triangle}{=} (a_i(n), i \in S)^T$ with each $a_i(n) \stackrel{\triangle}{=} (a_i^1(n),\ldots,a_i^N(n))^T$ denote the nth update of SDP π. Let $\triangle_i(n) \stackrel{\triangle}{=} (\triangle_i^1(n),\ldots,\triangle_i^N(n))^T \in \mathbb{R}^N$ be a vector of ± 1-valued variables $\triangle_i^j(n)$, $j=1,\ldots,N$, $i \in S$ that are obtained from the Hadamard matrix construction. Let $\Gamma_i : \mathbb{R}^N \rightarrow A(i)$ be the projection operator and $\delta > 0$ be a given constant. Let $\pi_i'(n) = \Gamma_i(\pi_i(n) + \delta\Delta_i(n))$ denote the perturbed policy at instant n.

Update Rule

Let $\{b(n)\}$ and $\{c(n)\}$ be two step-size sequences that satisfy (11.12). Also, let $\{Y_n'(i,a)\}$ be a family of i.i.d. random variables, each having the distribution $p(i,\cdot,a)$. The algorithm is as follows: For all $i \in S$, we initialize $h_i'(0) = 0$. Fix a reference state $i_0 \in S$ arbitrarily. The algorithm proceeds as follows: $\forall i \in S$, we have

$$\pi_i(n+1) = \Gamma_i\left(\pi_i(n) - c(n)\frac{(h_i'(n) + h_{i_0}'(n))}{\delta}(\Delta_i(n))^{-1}\right), \tag{11.36}$$

$$h_i'(n+1) = (1-b(n))h_i'(n) + b(n)(r(i,\pi_i'(n)) - h_{i_0}'(n) + h_{Y_n'(i,a_i')}'(n)). \tag{11.37}$$

This algorithm is also of the policy iteration type and performs policy evaluation on a faster timescale as compared to policy improvement.

A Sketch of Convergence

Lemma 11.5. *The iterates $h_i'(n)$ governed according to (11.37) stay uniformly bounded, i.e.,* $\sup_{n \geq 0} |h_i^r(n)| < \infty, \forall i \in S$.

Proof. (*Sketch:*) A detailed proof of this result is given in [15, Section 6.1]. The main idea there is to first show that the iterates (11.37) stay uniformly bounded for a given initial condition and then show that if they stay bounded for one initial condition, they remain bounded under all initial conditions. This also gives an alternative stability criterion, other than the one in Appendix D, and is presented in Chapter 3 of [10]. □

Lemma 11.6. *For a given SDP $\pi_i'(n) \equiv \pi_i'$, the iterates $h_i'(n)$ obtained from (11.37) asymptotically converge to $h_{\pi'}(i)$, where $h_{\pi'}(i)$ is the solution to the Poisson equation*

$$h_{\pi'}(i) + h_{\pi'}(i_0) = r(i, \pi'_i) + \sum_{j \in S} p(i, j, \pi'_i) h_{\pi'}(j), \tag{11.38}$$

$i \in S$, where π'_i is the action prescribed by the SDP π' in state i.

Proof. For given $\pi'_i(n) \equiv \pi'_i$, the ODE associated with (11.37) is

$$\dot{h}_{\pi'}(i) = r(i, \pi'_i) + \sum_{j \in S} p(i, j, \pi'_i) h_{\pi'}(j) - h_{\pi'}(i) - h_{\pi'}(i_0). \tag{11.39}$$

Now (11.39) has $h_{\pi'}(i)$ as its unique globally asymptotically stable equilibrium, where $h_{\pi'}(i)$ is the unique solution of (11.38). The result can now be shown from an application of the Borkar–Meyn theorem (Theorem D.1). □

Consider now the ODE along the slower timescale:

$$\dot{\pi}_i(s) = \hat{\Gamma}_i(-\nabla_i h_{\pi(s)}(i) - \nabla_i h_{\pi(s)}(i_0)) \tag{11.40}$$

for all $i \in S$. Let $M = \{\pi \mid \hat{\Gamma}_i(\nabla_i h_{\pi(s)}(i) + \nabla_i h_{\pi(s)}(i_0)) = 0, \ \forall i \in S\}$ be the set of all fixed points of (11.40). Also, given $\varepsilon > 0$, let M^ε be the ε-neighborhood of M.

The following is the main result that follows from an application of the Kushner–Clark theorem (Theorem E.1).

Theorem 11.7. *Given $\varepsilon > 0$, there exists $\delta_0 > 0$ such that for all $\delta \in (0, \delta_0]$, $\{\pi(n)\}$ converges to M^ε with probability 1.*

11.4.3.2 Algorithms for Finite Action Sets

Here, both the state space S and the action sets $A(i)$, $i \in S$ are discrete-valued and finite. There are two ways in which one can deal with such action sets. One possibility is to construct the closed convex hull of these sets which will result in the sets being compact and convex, and apply the algorithm for compact and convex action spaces using SDP-based updates. This will require suitable extensions to the transition dynamics, i.e., the transition probabilities $p(i, j, a)$ as well as the single-stage costs $r(i, a)$ so that these quantities are well defined over the afore-mentioned closed convex hulls. Thus, while the algorithm will update SDPs over the closed and convex hull, the actual actions that are picked during the process will be obtained by projecting the continuous-valued update to the discrete set, see Chapter 9 for ideas along these lines in the case of simulation-based discrete parameter optimization.

The other alternative (as we do here) is to search for the optimum within the space of SRPs. Let Γ_i, $i \in S \stackrel{\triangle}{=} \{1, 2, \ldots, s\}$ now denote the projection map from $\mathbb{R}^N$ to the probability simplex PS (defined previously). For simplicity, as before, we assume that each action set $A(i)$ comprises exactly N elements. Let $\pi_i(n) \stackrel{\triangle}{=} (\pi_i^1(n), \ldots, \pi_i^N(n))^T$ denote the vector of probabilities of picking individual actions, i.e., the elements of $A(i)$ as per the policy update at time n. We let $\pi(n) = (\pi_i(n)^T, i \in S)^T$ denote the nth update of SRP π. Again let $\triangle_i(n) \stackrel{\triangle}{=} (\triangle_i^1(n), \ldots, \triangle_i^N(n))^T \in \mathbb{R}^N$ be a vector of ± 1-valued variables $\triangle_i^j(n)$, $j = 1, \ldots, N$, $i \in S$ that are obtained from the Hadamard matrix construction and let $\pi_i'(n) \stackrel{\triangle}{=} \Gamma_i(\pi_i(n) + \delta \triangle_i(n))$ denote the perturbed policy at the nth update of the algorithm. Also, let $\{Y_n'(i,a)\}$ be a family of i.i.d. random variables, each having the distribution $p(i,\cdot,a)$. Further, $\xi_n'(i,\pi_i')$, $n \geq 0$ will denote i.i.d random variables having the distribution π_i' over the action set $A(i)$, $i \in S$.

Update Rule

The algorithm is as follows: For all $i \in S, n \geq 0$,

$$\pi_i(n+1) = \Gamma_i\left(\pi_i(n) - c(n)\frac{(h_i'(n) + h_{i_0}'(n))}{\delta}(\Delta_i(n))^{-1}\right), \tag{11.41}$$

$$h_i'(n+1) = (1 - b(n))h_i'(n) + b(n)(r(i, \xi_n'(i, \pi_i'(n))) - h_{i_0}'(n) + h'_{Y_n'(i,\xi_n'(i,\pi_i'(n)))}(n)) \tag{11.42}$$

A Sketch of Convergence

Similar results as those in the previous case are obtained here.

Lemma 11.8. *The iterates $h_i'(n)$ governed according to (11.42) stay uniformly bounded, i.e.,* $\sup_{n\geq 0} |h_i^r(n)| < \infty, \forall i \in S$.

Proof. (*Sketch:*) Follows in a similar manner as Lemma 11.5 (see Section 6.1 of [15]). □

Lemma 11.9. *For a given SRP $\pi_i'(n) \equiv \pi_i'$, the iterates $h_i'(n)$ obtained from (11.42) asymptotically converge to $h_{\pi'}(i)$, where $h_{\pi'}(i)$ is the solution to the Poisson equation*

$$h_{\pi'}(i_0) + h_{\pi'}(i) = \left(\sum_{a\in A(i)} \pi_{i,a}'(r(i,a) + \sum_{j\in S} p(i,j,a)h_{\pi'}(j)\right), \ \forall i \in S. \tag{11.43}$$

where $\pi_{i,a}'$ is the probability of action a being picked in state i under SRP π'.

Proof. Follows in a similar manner as the proof of Lemma 11.6. □

Now let $M = \{\pi \mid \hat{\Gamma}_i(\nabla_i h_{\pi(s)}(i) + \nabla_i h_{\pi(s)}(i_0)) = 0, \ \forall i \in S\}$ be the set of all fixed points of a similar ODE as (11.40) except that $\hat{\Gamma}_i$ is now defined for the simplex PS. The following is the main result that can be shown using the Kushner–Clark theorem (Theorem E.1).

Theorem 11.10. *Given $\varepsilon > 0$, there exists $\delta_0 > 0$ such that for all $\delta \in (0, \delta_0]$, $\{\pi(n)\}$ converges to M^ε with probability 1.*

11.5 Reinforcement Learning Algorithms with Function Approximation

We now consider a class of problems where the number of states and also possibly actions are large so that the algorithms for solving the Bellman equation by simulating state transitions from every state at each instant become highly (computationally) inefficient. In particular, if the state space is very large as happens when the dimension of the state vector increases, even storing a vector of the size of the state space might become impossible. For example, consider a communication network with 10 nodes, where the state is the vector of number of packets at each of these nodes. If each node can accommodate 100 equal-sized packets, then the size of the state space becomes $100^{10} = 10^{20}$. Further, if the number of nodes is increased by 10, i.e., if the total number of nodes is now 20, then the size of the state space becomes $100^{20} = 10^{40}$ which is an exponential increase in the state-space size. Nevertheless, storing vectors of large sizes becomes an infeasible task. In such cases, one often resorts to suitable parameterizations of the value functions and/or policies.

11.5.1 Temporal Difference (TD) Learning with Discounted Cost

The temporal difference learning algorithm is a popular algorithm for the problem of prediction, i.e., estimating the value function corresponding to a given policy, see [22, 25]. The value function $V^\pi(\cdot)$ under an SDP π is approximated here with the parametrized function

$$w(i,v) = v^T f_i, \tag{11.44}$$

with parameter $v \stackrel{\triangle}{=} (v_1, \ldots, v_d)^T \in \mathbb{R}^d$, where $f_i \stackrel{\triangle}{=} (f_i(1), \ldots, f_i(d))^T$ is the feature vector corresponding to state $i \in S$. Let Φ be a $|S| \times d$-matrix whose kth column $(k = 1, \ldots, d)$ is $f(k) \stackrel{\triangle}{=} (f_i(k), i \in S)^T$. Here $|S|$ denotes the cardinality of S. The following are standard requirements [25].

Assumption 11.3. The Markov chain $\{X(n)\}$ under SDP π is ergodic.

Assumption 11.4. The basis functions $\{f(k), k = 1, \ldots, d\}$ are linearly independent. Further, $d \leq |S|$.

The functions $f(1), \ldots, f(d)$ are the basis functions from S to $\mathbb{R}$. The idea here is to tune v suitably so that $w(i,v)$ is 'close' to $V^{\pi}(i)$. The gradient of the parametrized function in (11.44) is

$$\nabla w(i,v) = f_i.$$

Representations of the type (11.44) are called linear representations or architectures. Nonlinear representations such as those based on sigmoidal functions or neural networks have also been widely studied in the literature [5]. However, analytical/convergence results for the TD algorithm are known mainly in the case of linear architectures and so we will be primarily concerned with these (architectures) here.

The TD algorithm works with an infinite Markov sequence of states $i_0, i_1, i_2, \ldots$ obtained by picking actions according to the SDP π. Let v_n be the nth update of the parameter. At instant n, let δ_n be the 'temporal difference' that is defined according to

$$\begin{aligned} \delta_n &= r(i_n, \mu(i_n)) + \alpha w(i_{n+1}, v_n) - w(i_n, v_n) \\ &= r(i_n, \mu(i_n)) + \alpha v_n^T f_{i_{n+1}} - v_n^T f_{i_n}. \end{aligned} \tag{11.45}$$

Let $\{\gamma(n)\}$ be a step-size sequence satisfying the following requirement:

Assumption 11.5. The step-sizes $\gamma(n)$ satisfy $\gamma(n) > 0 \forall n$ and

$$\sum_n \gamma(n) = \infty, \ \sum_n \gamma^2(n) < \infty.$$

The TD(λ) algorithm for the infinite horizon discounted cost case is the following:

$$v_{n+1} = v_n + \gamma(n) \delta_n z_n, \tag{11.46}$$

where $z_n \in \mathbb{R}^d$ is called the eligibility trace vector and is defined as

$$z_n = \sum_{m=0}^{n} (\alpha\lambda)^{n-m} f_{i_m}. \tag{11.47}$$

Here $\lambda \in [0,1]$ is a given parameter.

The vectors z_n, $n \geq 0$ can be obtained recursively according to

$$z_{n+1} = \alpha\lambda z_n + f_{i_{n+1}}. \tag{11.48}$$

Convergence of TD

We analyze here the convergence of the TD recursion using the Borkar–Meyn theorem (Theorem D.1). We consider here specifically the case of $\lambda = 0$ for simplicity. In this case, $z_n = f_{i_n}$ in (11.46) define the operator $T^{\pi} : \mathbb{R}^{|S|} \to \mathbb{R}^{|S|}$ as follows:

$$T^{\pi}(J)(i) = r(i,\pi_i) + \alpha \sum_{j \in S} p(i,j,\pi_i)J(j)),$$

$\forall i \in S$, where π_i is the action selected in state i using the SDP π. Let

$$R = (r(i,\pi_i),\ i \in S)^T,$$

denote the column vector of single-stage costs under SDP π. Also, let $P = [[p(i,j,\pi_i]]_{i,j \in S}$ denote the transition probability matrix under SDP π. Let $d \stackrel{\triangle}{=} (d(i), i \in S)^T$ denote the stationary distribution of the corresponding Markov chain under SDP π and D denote the $(|S| \times |S|)$-diagonal matrix with entries $d(i)$, $i \in S$ along the diagonal. The proof of the following result has been shown in detail in [25]. We show the same below using the stability and convergence result of [11] (Theorem D.1).

Theorem 11.11. *Under Assumptions 11.3 – 11.5, $v_n, n \geq 0$ governed by recursion (11.46) satisfy $v_n \to v^{\pi}$ as $n \to \infty$ with probability one. Also, v^{π} is the unique solution to the following system of equations:*

$$\Phi^T D \Phi v^{\pi} = \Phi^T D^{\pi} T^{\pi}(\Phi v^{\pi}). \tag{11.49}$$

In particular,

$$v^{\pi} = -(\Phi^T D(\alpha P - I)\Phi)^{-1} \Phi^T DR. \tag{11.50}$$

Proof. The ODE associated with (11.46) for $\lambda = 0$ is the following:

$$\dot{v} = \sum_{i \in S} d(i)[r(i,\pi_i) + \alpha v^T \sum_{j \in S} p(i,j,\pi_i) f_j - v^T f_i] f_i. \tag{11.51}$$

In vector-matrix notation, the above ODE is analogous to

$$\dot{v} = \Phi^T D(T^{\pi}(\Phi v) - \Phi v) \stackrel{\triangle}{=} g(v). \tag{11.52}$$

It is easy to see that $g(v)$ is Lipschitz continuous in v. Now define $g_{\infty}(v)$ as

$$g_{\infty}(v) \stackrel{\triangle}{=} \lim_{n \to \infty} \frac{g(nv)}{n} = \Phi^T D(\alpha P - I)\Phi v,$$

where I is the identity matrix. Consider now the ODE

$$\dot{v} = g_\infty(v). \tag{11.53}$$

For $x \in \mathbb{R}^{|S|}$, define the weighted Euclidean norm $\| x \|_D$ according to $\| x \|_D = (x^T Dx)^{1/2}$. Note that

$$\| x \|_D^2 = x^T Dx = \| (D)^{1/2} x \|^2 .$$

Now for any function $V \in \mathbb{R}^{|S|}$, we have

$$\| PV \|_D^2 = V^T P^T DPV = \sum_{i \in S} d(i) E^2[V(X(n+1)) \mid X(n) = i, \pi]$$

$$\leq \sum_{i \in S} d(i) E[V^2(X(n+1)) \mid X(n) = i, \pi] = \sum_{i \in S} d(j) V^2(j) = \| V \|_D^2 .$$

The inequality above follows from the conditional Jensen's inequality while the second last equality on upon evaluating the conditional expectation on its LHS and noting that $d^T = d^T P$. We thus have

$$\| \alpha PV \|_D \leq \alpha \| V \|_D .$$

Now,

$$V^T D\alpha PV = \alpha V^T (D)^{1/2} (D)^{1/2} PV \leq \alpha \| (D)^{1/2} V \| \| (D)^{1/2} PV \|$$

$$= \alpha \| V \|_D \| PV \|_D \leq \alpha \| V \|_D^2 = \alpha V^T DV.$$

Thus,

$$V^T D(\alpha P - I)V \leq (\alpha - 1) \| V \|_D^2 < 0, \ \forall V \neq 0,$$

implying that $D(\alpha P - I)$ is negative definite. Thus, $\Phi^T D(\alpha P - I)\Phi$ is also negative definite since Φ is a full rank matrix by Assumption 11.4. Thus, (11.53) has the origin as its unique globally asymptotically stable equilibrium.

Next, define $N(n)$, $n \geq 0$ according to

$$N(n) = \delta_n f_{X(n)} - E[\delta_n f_{X(n)} \mid \mathscr{F}(n)],$$

where $\mathscr{F}(n) = \sigma(v_r, N(r), r \leq n)$. It is easy to see that

$$E[\| N(n+1) \|^2 \mid \mathscr{F}(n)] \leq \hat{C}(1 + \| v_n \|^2), \ n \geq 0, \tag{11.54}$$

for some $0 < \hat{C} < \infty$.

Now let $\hat{v} = v^\pi$ be a solution to

$$g(\hat{v}) = \Phi^T D(T^\pi(\Phi\hat{v}) - \Phi\hat{v}) = 0. \tag{11.55}$$

Note that (11.55) corresponds to the linear system of equations

$$\Phi^T DR + \Phi^T D(\alpha P - I)\Phi\hat{v} = 0. \tag{11.56}$$

Now since we have already shown that $\Phi^T D(\alpha P - I)\Phi$ is negative definite, it is of full rank and invertible. Hence v^{π} is the unique solution to (11.56) and corresponds to (11.50). The claim now follows from Theorems 2.1–2.2(i) of [11] (cf. Theorem D.1). □

11.5.2 An Actor-Critic Algorithm with a Temporal Difference Critic for Discounted Cost MDPs

We now consider the problem of control in discounted cost MDPs and present here an actor-critic algorithm that incorporates TD in the critic and policy gradients [17, 13, 12, 23] in the actor. The TD algorithm solves the problem of prediction by incorporating temporal differences in its update. We now restrict attention to SRPs π that depend on a parameter $\theta \stackrel{\triangle}{=} (\theta_1, \ldots, \theta_N)^T \in \mathbb{R}^N$ and consider the problem of finding the optimum θ. We let θ take values in a compact and convex subset C of $\mathbb{R}^N$. Let $\pi^\theta \stackrel{\triangle}{=} (\pi_i^\theta, i \in S)$ denote the parametrized SRP. Here, $\pi_i^\theta = (\pi_{i,a}^\theta, i \in S, a \in A(i))$ is the distribution over the set of actions $A(i)$ that are feasible in state i. Here, $\pi_{i,a}^\theta$ is the probability of picking action a in state i under policy π^θ. By an abuse of notation, we let π itself denote the parametrized SRP π^θ.

Assumption 11.6. The Markov chain $\{X(n)\}$ under SRP π^θ for any $\theta \in C$ is ergodic.

Assumption 11.7. For any $a \in A(i)$, $i \in S$, $\pi(i,a)$ is continuously differentiable in θ.

Assumption 11.8. Let $b(n), c(n), n \geq 0$ be two step-size sequences that satisfy $b(n), c(n) > 0 \forall n$ and

$$\sum_n b(n) = \sum_n c(n) = \infty, \ \sum_n (b^2(n) + c^2(n)) < \infty, \ c(n) = o(b(n)).$$

Let $\Gamma : \mathbb{R}^N \to C$ denote the projection operator defined so that for any $x = (x_1, \ldots, x_N)^T \in \mathbb{R}^N$, $\Gamma(x) \stackrel{\triangle}{=} (\Gamma_1(x_1), \ldots, \Gamma_N(x_N))^T$ is the nearest point to x in the set C. Let $\theta(n) \stackrel{\triangle}{=} (\theta_1(n), \ldots, \theta_N(n))^T$ denote the nth update of θ. Further, let $\Delta(n) \stackrel{\triangle}{=} (\Delta_1(n), \ldots, \Delta_N(n))^T$, $n \geq 0$ be a sequence of ± 1-valued variables $\Delta_j(n)$, $j = 1, \ldots, N$ obtained from the Hadamard matrix construction for perturbation sequences (for a one-simulation implementation). A some what similar algorithm incorporating two simulation random perturbation SPSA for MDPs with functional constraints has been presented in [6].

The Algorithm

Let $\{X'_n\}$ denote the simulation governed by the sequence of policy updates $\{\pi'(n)\}$. Here, $\pi'(n) \stackrel{\triangle}{=} (\pi'_i(n), i \in S)$, $n \geq 0$ where $\pi'_i(n)$ is the distribution $(\pi'_{i,a}(n), a \in A(i))$ over the set $A(i)$ of feasible actions in state $i \in S$. The policy updates $\pi'(n)$ are parametrized by $\theta(n) + \delta\Delta(n)$, $n \geq 0$. Let Z'_n denote the action chosen at time n according to the above policy. Let $P(X'_0 = s_0) \stackrel{\triangle}{=} \beta(s_0)$, $s_0 \in S$ denote the initial distribution of the Markov chain under the given policy.

The algorithm is as follows: For $n \geq 0$, $k = 1, \ldots, N$,

$$\delta'_n = r(X'_n, Z'_n) + + \alpha {v'_n}^T f_{X'_{n+1}} - {v'_n}^T f_{X'_n}, \tag{11.57}$$

$$v'_{n+1} = v'_n + b(n)\delta'_n f_{X'_n}, \tag{11.58}$$

$$\theta_k(n+1) = \Gamma_k\left(\theta_k(n) - c(n)\sum_{s_0 \in S}\beta(s_0)\left(\frac{{v'_n}^T f_{s_0}}{\delta\Delta_k(n)}\right)\right). \tag{11.59}$$

The recursions (11.57)–(11.58) correspond to the TD(0) update.

Convergence of the Actor-Critic Algorithm

The analysis of the (faster timescale) TD recursion proceeds in a similar manner as for the TD convergence analyzed previously. Note, however, that the latter was analyzed for the case when the actions in the MDP are selected according to a given SDP and not when they are chosen according to a parametrized SRP. Nevertheless, under Assumption 11.6, the Markov chain under SRP π^θ for any $\theta \in C$ is ergodic. Hence, a similar analysis as before can be carried through in this case.

Since the actor-critic scheme is a multi-timescale stochastic approximation algorithm, one can let $\theta(n) \equiv \theta$ and $\Delta(n) \equiv \Delta$ (where θ, Δ are given constants) when analyzing (11.58). Let π' denote the policy governed by the parameter $\theta + \delta\Delta$. Define the operator $T' : \mathbb{R}^{|S|} \to \mathbb{R}^{|S|}$ as follows:

$$T'(J)(i) = \sum_{a \in A(i)} \pi'_{i,a}(r(i,a) + \alpha \sum_{j \in S} p(i,j,a)J(j)),$$

$\forall i \in S$. Also, let

$$R^{\pi'} = \left(\sum_{a \in A(i)} \pi'(i,a)r(i,a),\ i \in S\right)^T,$$

denote the column vector of single-stage costs under policy π'. Also, define $T(J)$ and R^π in an analogous manner as $T'(J)$ and $R^{\pi'}$, respectively, except with π in place of π' in their definitions, where π is the policy governed by the parameter θ.

Let $d^{\pi'}$ (resp. d^{π}) denote the stationary distribution of the corresponding Markov chain when the underlying parameter is $\theta+\delta\Delta$ (resp. θ). Also, let $D^{\pi'}$ (resp. D^{π}) denote the $(|S|\times|S|)$-diagonal matrix with entries $d^{\pi'}(i)$ (resp. $d^{\pi}(i)$), $i\in S$ along the diagonal.

The proof of the following result can be shown along the same lines as Theorem chap11-rl-prop1-1.

Theorem 11.12. *Under Assumptions 11.4 and 11.6–11.8, with $\theta(n)\equiv\theta$ and $\Delta(n)\equiv\Delta$, (for given θ and Δ), $v'_n, n\geq 0$ governed by recursion (11.58) satisfy $v'_n\rightarrow v^{\pi'}$ as $n\rightarrow\infty$ with probability one. Also, $v^{\pi'}$ is the unique solution to the following system of equations:*

$$\Phi^T D^{\pi'}\Phi v^{\pi'} = \Phi^T D^{\pi'} T'(\Phi v^{\pi'}). \tag{11.60}$$

In particular,

$$v^{\pi'} = -(\Phi^T D^{\pi'}(\alpha P^{\pi'}-I)\Phi)^{-1}\Phi^T D^{\pi'} R^{\pi'}. \tag{11.61}$$

In a similar manner as above, under policy π (i.e., when the governing parameter is $\theta\in C$, one can also obtain v^{π} as the unique solution to the following system of equations:

$$\Phi^T D^{\pi}\Phi v^{\pi} = \Phi^T D^{\pi} T(\Phi v^{\pi}), \tag{11.62}$$

or alternatively,

$$v^{\pi} = -(\Phi^T D^{\pi}(\alpha P^{\pi}-I)\Phi)^{-1}\Phi^T D^{\pi} R^{\pi}. \tag{11.63}$$

Lemma 11.13. *Under Assumptions 11.4 and 11.6–11.7, the solution v^{π} to (11.62) is continuously differentiable in θ.*

Proof. From Assumption 11.7, it is easy to see that R^{π} and P^{π} are continuously differentiable in θ. One can now verify that the stationary distribution d^{π} of a Markov chain $\{X(n)\}$ under the SRP π is also continuously differentiable in θ (see, for instance, Theorem 2, pp. 402–403 of [20]). Hence, D^{π} is also continuously differentiable in θ. Now writing out the inverse of the matrix $\Phi^T D^{\pi}(\alpha P^{\pi}-I)\Phi$ explicitly using Cramer's rule, one can see that v^{π} is continuously differentiable in θ. □

The analysis of the slower recursion now proceeds along expected lines. Consider the following ODE associated with (11.59):

$$\dot{\theta} = \hat{\Gamma}\left(-\sum_{s_0\in S}\beta(s_0)\nabla_\theta v^{\pi T} f_{s_0}\right). \tag{11.64}$$

Let $M\subset\{\theta\in C\mid \hat{\Gamma}\left(\sum_{s_0\in S}\beta(s_0)\nabla_\theta v^{\pi T} f_{s_0}\right)=0\}$ denote the set of asymptotically stable equilibria of (11.64) within the set C, i.e., the local minima of the function

$\sum_{s_0 \in S} \beta(s_0) v^{(\cdot)^T} f_{s_0}$. Let M^ε be the ε-neighborhood of M. The following now follows from the convergence of the one-simulation Hadamard matrix gradient estimates and the Kushner-Clark theorem (Theorem E.1).

Theorem 11.14. *Under Assumptions 11.4 and 11.6–11.8, given $\varepsilon > 0$, $\exists \delta_0 > 0$ such that for all $\delta \in (0, \delta_0)$, $\theta(n)$, $n \geq 0$ obtained according to (11.59) satisfy $\theta(n) \to M^\varepsilon$ as $n \to \infty$, with probability one.*

11.5.3 Function Approximation Based Q-Learning Algorithm and a Simultaneous Perturbation Variant for Infinite Horizon Discounted Cost MDPs

We now describe the Q-learning algorithm with function approximation and its two-timescale variant. Even though it is a popular algorithm, Q-learning with function approximation is known to suffer from the off-policy problem that we describe below. Broadly speaking, the algorithm does not converge in some cases because of the non-linearity in the update equation that comes about because of the minimization operation. Indeed, a convergence analysis of the algorithm under general conditions is not available. We describe in Section 11.5.3.2, a two-timescale variant of Q-learning, where the minimization step is conducted on a faster timescale recursion, while the algorithm without the minimization is run on a slower scale. The latter recursion (without minimization) is then a linear update rule that resembles TD for the joint (state-action) Markov chain. We incorporate a one-simulation, deterministic, Hadamard matrix-based perturbations for the faster recursion and prove its convergence. The two-timescale variant of Q-learning does not suffer from the off-policy problem.

11.5.3.1 The Q-Learning Algorithm with Function Approximation

Recall that the Q-Bellman equation (11.30) holds in the case of full-state representations. The Q-learning algorithm under full state representations tracks the Q-Bellman equation and converges to the optimal Q values. We now discuss the function approximation version of Q-learning.

For $i \in S$, $a \in A(i)$, let $Q^*(i,a) \approx \theta^{*T} \phi_{i,a}$, where $\theta^* \stackrel{\triangle}{=} (\theta^*(1), \ldots, \theta^*(d))^T$ is a d-dimensional parameter and $\phi_{i,a} = (\phi_{i,a}(1), \ldots, \phi_{i,a}(d))^T$ is the associated feature vector. Note that $\phi_{i,a}$ are state–action features and are defined for all tuples $(i,a) \in \mathscr{S}$ where $\mathscr{S} = \{(i,a) \mid i \in S, a \in A(i)\}$ denotes the set of all feasible state-action tuples.

Let Φ now denote a matrix with rows $\phi_{i,a}^T$, $(i,a) \in \mathscr{S}$. Assuming that the total number of states is n and the number of feasible actions in any state i (i.e., the cardinality of the set $A(i)$) is m_i (where $m_i \geq 1$), the number of rows in the matrix

Φ is $\sum_{j=1}^{n} m_j$. The number of columns of this matrix is d. One can also write Φ as $\Phi = (\Phi(k), k = 1, \ldots, d)$, where $\Phi(k)$ is the column vector

$$\Phi(k) = (\phi_{i,a}(k),\ (i,a) \in \mathscr{S})^T,\ k = 1, \ldots, d.$$

Now $Q^* = (Q^*(i,a), (i,a) \in \mathscr{S})^T$ is approximated according to

$$Q^* \approx \sum_{i=1}^{d} \Phi(i)\theta^*(i), \text{ i.e., } Q^* \approx \Phi\theta^*.$$

The estimates $Q_n(i,a)$, $n \geq 0$, of $Q^*(i,a)$, $(i,a) \in \mathscr{S}$ are approximated as $Q_n(i,a) \approx \theta_n^T \phi_{i,a}$, where $\theta_n \stackrel{\triangle}{=} (\theta_n(1), \ldots, \theta_n(d))^T$ is the nth update of the parameter θ.

The Q-Learning Update Rule

The Q-learning algorithm with function approximation updates the parameter θ according to

$$\theta_{n+1} = \theta_n + c(n)\phi_{X(n),Z_n}(r(X(n),Z_n) + \gamma \min_{v \in A(X(n+1))} \theta_n^T \phi_{X(n+1),v} - \theta_n^T \phi_{X(n),Z_n}), \tag{11.65}$$

where θ_0 is set arbitrarily and the step-sizes $c(n)$, $n \geq 0$ satisfy (11.32). It is important to note that like the actor-critic algorithm (11.57)–(11.59), (11.65) is also an on-line scheme as it works with a single trajectory of (feasible) state-action tuples $(X(n), Z_n)$, $n \geq 0$ and updates the parameter θ as new states are observed and actions chosen. Also note that $\nabla_{\theta_n} Q_n(X(n), Z_n) \approx \nabla_{\theta_n} \theta_n^T \phi_{X(n),Z_n} = \phi_{X(n),Z_n}$. The algorithm (11.65), however, is known to suffer from the off-policy problem [2], [22] and may not converge in some cases. This is because the update in (11.65) is nonlinear because of the minimization operation. Note that if actions are picked according to a given policy (and one does not have minimization), then (11.65) is a regular TD(0) scheme for the joint (state–action) Markov chain.

11.5.3.2 Two-Timescale Q-Learning with Function Approximation

We now describe an algorithm based on Q-learning with function approximation that does not suffer from the off-policy problem because it incorporates multiple timescales. Let $\pi_w = (\pi_w(i), i \in S)^T$ represent a class of SRP, parametrized by w, where each $\pi_w(i)$ is the distribution $\pi_w(i) = (\pi_w(i,a), a \in A(i))^T$ over the set of feasible actions $A(i)$ in state i. Here $w \stackrel{\triangle}{=} (w_1, \ldots, w_N)^T \in \mathbb{R}^N$ is a parameter in addition to θ. In what follows, we restrict attention to SRPs that are parametrized by w. We let w take values in a compact and convex set $W \subset \mathbb{R}^N$. We now make the following assumptions.

Assumption 11.9. The Markov chain $\{X(n)\}$ under any SRP π_w is aperiodic and irreducible.

Assumption 11.10. The probabilities $\pi_w(i,a)$, $(i,a) \in \mathscr{S}$ are continuously differentiable in the parameter w with $\nabla_w \pi_w(i,a)$ being Lipschitz continuous. Further, $\pi_w(i,a) > 0 \; \forall i \in S,\, a \in A(i),\, w \in C$.

Assumption 11.11. The basis functions $\{\Phi(k), k = 1,\ldots,d\}$ are linearly independent. Further, $d \leq |\mathscr{S}|$.

Assumptions 11.9 and 11.11 are similar to 11.6 and 11.7, respectively, except for a slight change in the notation being used. Note also that the matrix Φ considered here is a state–action feature matrix unlike the one considered in Sections 11.5.1 and 11.5.2, respectively.. Assumption 11.10 is stronger than Assumption 11.7. However, for the classes of parametrized policies that one normally considers, Assumption 11.7 is seen to hold. A well-studied example of parametrized policies that satisfy Assumption 11.10 or Assumption 11.7 is the parametrized Boltzmann policies given by

$$\pi_w(i,a) = \frac{\exp(w^T \phi_{i,a})}{\sum_{b \in A(i)} \exp(w^T \phi_{i,b})}.$$

Let $\Delta_n = (\Delta_n(1),\ldots,\Delta_n(N))^T$ be certain perturbation vectors obtained from a normalized Hadamard matrix that will be used to perturb the updates w_n of the parameter w. In what follows, we use the one-simulation simultaneous perturbation Hadamard matrix-based updates for the sequence w_n while the parameters θ_n follow a TD(0) update for the state-action Markov chain.

Two-Timescale Q-Learning Update Rule

Let $\Theta \subset \mathbb{R}^d$ be the set in which $\theta_n, n \geq 0$ take values. Also, recall that W is the set in which $w_n, n \geq 0$ take values. We assume that Θ (resp. W) is a compact and convex subset of $\mathbb{R}^d$ (resp. $\mathbb{R}^N$). This requirement on the sets Θ and W essentially ensure that the updates θ_n and w_n below remain uniformly bounded almost surely. Let $\pi_{(w_n+\delta\Delta_n)} \stackrel{\triangle}{=} (\pi_{(w_n+\delta\Delta_n)}(i,a),\, (i,a) \in \mathscr{S}$, where $\delta > 0$ is a given small constant, be the randomized policy followed during the nth update. Note that this randomized policy is governed by the parameter $(w_n + \delta\Delta_n)$.

The algorithm is as follows: $\forall n \geq 0$,

$$\theta_{n+1} = \gamma\left(\theta_n + c(n)\phi_{X(n),Z_n}\left(r(X(n),Z_n) + \gamma\theta_n^T\phi_{X(n+1),Z_{n+1}} - \theta_n^T\phi_{X(n),Z_n}\right)\right), \tag{11.66}$$

$$w_{n+1} = \Gamma\left(w_n - b(n)\left(\frac{\theta_n^T\phi_{X(n),Z_n}}{\delta}\right)(\Delta_n)^{-1}\right), \tag{11.67}$$

where $\gamma : \mathbb{R}^d \to \Theta$ (resp. $\Gamma : \mathbb{R}^N \to W$) is the projection operator that projects any $x \in \mathbb{R}^d$ (resp. $x \in \mathbb{R}^N$) to the set Θ (resp. W).

Convergence Analysis

It is easy to verify that $p_w(i,a;j,b) = p(i,j,a)\pi_w(j,b)$, $(i,a),(j,b) \in \mathscr{S}$ form transition probabilities for the joint process $(X(n),Z_n)$, $n \geq 0$ under the SRP π_w. Under Assumptions 11.9 and 11.10, it is also easy to see that the process $(X(n),Z_n), n \geq 0$ with Z_n, $n \geq 0$ obtained from the SRP π_w, for any $w \in W$, is an ergodic Markov process. Hence, $(X(n),Z_n), n \geq 0$ has a unique stationary distribution $f_w(i,a) = d^{\pi_w}(i)\pi_w(i,a)$, $(i,a) \in \mathscr{S}$. One can also show from an application of Theorem 2 on pp.402–403 of [20] (on smoothness of the stationary distribution for finite state chains) that under Assumptions 11.9 and 11.10, $f_w(i,a)$, $(i,a) \in \mathscr{S}$ are differentiable in $w \in W$ with $\nabla_w f_w(i,a)$ being Lipschitz continuous in w.

As with other multi-scale schemes, the recursion (11.66) is quasi-static when viewed from the faster timescale corresponding to $b(n), n \geq 0$. Hence, let $\theta_n \equiv \theta$ when analyzing (11.67).

Let

$$\bar{Q}(\theta,w) \stackrel{\triangle}{=} \sum_{(i,a)\in\mathscr{S}} f_w(i,a)\theta^T\phi_{i,a}$$

denote the stationary average Q value under the parameters θ and w, respectively.

Lemma 11.15. *The partial derivatives of $\bar{Q}(\theta,w)$ with respect to any $\theta \in \Theta$ and $w \in W$ exist. Further, $\nabla_w\bar{Q}(\theta,w)$ is Lipschitz continuous in $(\theta,w) \in \Theta \times W$.*

Proof. This can be seen from the fact that W and Θ are both compact sets, hence continuous functions on these sets remain uniformly bounded. □

The ODE associated with (11.67) is

$$\dot{w}(t) = \hat{\Gamma}\left(-\nabla_w\bar{Q}(\theta,w(t))\right), \tag{11.68}$$

with θ fixed. Let K_θ denote the set of asymptotically stable equilibria of (11.68) and K_θ^ε be the ε-neighborhood of K_θ. Let

$$K_n^i = \sum_{j=0}^{n-1} a(j) \left(\frac{\theta_j^T \phi_{X_j,Z_j} - E[\theta_j^T \phi_{X_j,Z_j} \mid \mathscr{F}_j]}{\delta \Delta_j^i} \right),$$

with $\mathscr{F}_n = \sigma(X_j, Z_j, j < n; \theta_j, w_j, j \leq n)$, $n \geq 1$, as a sequence of associated sigma fields. The following result now follows from the martingale convergence theorem (Theorem B.2) in a straight forward manner.

Lemma 11.16. *For all $i = 1, \ldots, N$, $(K_n^i, \mathscr{F}_n)$, $n \geq 0$ are almost surely convergent martingale sequences.*

The convergence of the recursion (11.67) now follows as a consequence of the consistency of the Hadamard matrix-based estimator and the Kushner–Clark theorem (Theorem E.1).

Theorem 11.17. *Given $\varepsilon > 0$, there exists $\delta_0 > 0$ such that for all $\delta \in (0, \delta_0]$, $\{w_n\}$ governed by (11.67) converges to K_θ^ε almost surely.*

Let F be a $(|\mathscr{S}| \times |\mathscr{S}|)$-diagonal matrix with entries $f_w(i,a)$, $(i,a) \in \mathscr{S}$ along the diagonal. Let P denote the transition probability matrix of the joint Markov chain $(X(n), Z_n)$, $n \geq 0$ when actions are selected according to the SRP π_w. Also, let R denote the vector of single-stage expected costs $r(i,a)$, $(i,a) \in \mathscr{S}$. The following ODE is associated with (11.66):

$$\dot{\theta}(t) = \hat{\gamma}(\Phi^T \mathscr{F}(T(\Phi\theta(t)) - \Phi\theta(t))), \tag{11.69}$$

where for any bounded and continuous $\zeta : \mathbb{R}^d \to \mathbb{R}^d$,

$$\hat{\gamma}(\zeta(\theta)) = \lim_{\eta \downarrow 0} \left(\frac{\gamma(\theta + \eta\zeta(\theta)) - \theta}{\eta} \right).$$

Note that when $\theta \in \Theta^o$, $\hat{\gamma}(\zeta(\theta)) = \zeta(\theta)$. Also, when $\theta \in \partial\Theta$ such that $\theta + \eta\zeta(\theta) \notin \Theta$ for any $\eta > 0$, $\hat{\gamma}(\zeta(\theta))$ is the projection of $\zeta(\theta)$ to the set Θ.

Let $\bar{M} \stackrel{\triangle}{=} \{\theta \in \Theta \mid \hat{\gamma}(\Phi^T F(T(\Phi\theta) - \Phi\theta)) = 0\}$. Note that if $\theta \in \Theta^o \cap \bar{M}$, $\Phi^T F(T(\Phi\theta) - \Phi\theta) = 0$. We now have the following result on the convergence of the recursion (11.66).

Theorem 11.18. *Under Assumptions 11.9 – 11.11, the quantities $\theta_n, n \geq 0$ governed according to (11.66) satisfy $\theta_n \to \bar{M}$ with probability one.*

Proof. The result can be seen to follow from an application of the Kushner-Clark theorem (Theorem E.1) for projected stochastic approximation. □

Lets denote by $w \equiv w(\theta)$ any point in K_θ and view $w(\cdot)$ as a map from W to $\mathbb{R}^N$.

Lemma 11.19. *The map $w : W \to \mathbb{R}^N$ is Lipschitz continuous.*

Proof. Follows as a consequence of the implicit function theorem (cf. Theorem 1 of [14], also stated as Theorem 1.1 of [21]). □

Let $\mathscr{U} = \{(\theta, w) \mid \theta \in \bar{M}, w \in K_\theta\}$ and given $\varepsilon > 0$, let $\mathscr{U}^\varepsilon = \{(\theta, w) \mid \theta \in \bar{M}, w \in K_\theta^\varepsilon\}$.

Theorem 11.20. *Given $\varepsilon > 0$, there exists $\delta_0 > 0$ such that for all $\delta \in (0, \delta_0]$, the sequence of iterates (θ_n, w_n), $n \geq 0$ satisfy*

$$(\theta_n, w_n) \to \mathscr{U}^\varepsilon,$$

with probability one.

Proof. (*Sketch:*) The result can be shown in a similar manner as Theorem 2, Chapter 6 of [10], with the difference being that since the θ-update does not have a unique fixed point (i.e., a unique globally asymptotically stable equilibrium for the associated ODE), the convergence can only be shown to the set $\mathscr{U}$ using similar techniques. □

11.6 Concluding Remarks

In this chapter, we considered the application of simultaneous perturbation methods for problems of stochastic control under (a) lack of model information and (b) large state-action spaces. We presented various reinforcement learning algorithms based on simultaneous perturbation approaches for this purpose. These algorithms are seen to perform well even over large state-action spaces. For instance, some of these algorithms have been applied in the context of road traffic control in [18, 16] (see Chapter 13), where they have been observed to work well even over very high-dimensional state-action spaces. Finally, in [6], the simultaneous perturbation methodology has been applied to develop an actor-critic algorithm for constrained Markov decision processes that is similar in flavour to the methods developed in Chapter 10.

References

1. Abdulla, M.S., Bhatnagar, S.: Reinforcement learning based algorithms for average cost Markov decision processes. Discrete Event Dynamic Systems 17(1), 23–52 (2007)
2. Baird, L.C.: Residual algorithms: Reinforcement learning with function approximation. In: Proceedings of the Twelfth International Conference on Machine Learning, pp. 30–37 (1995)

3. Bertsekas, D.P.: Dynamic Programming and Optimal Control, 3rd edn., vol. I. Athena Scientific, Belmont (2005)
4. Bertsekas, D.P.: Dynamic Programming and Optimal Control, 3rd edn., vol. II. Athena Scientific, Belmont (2007)
5. Bertsekas, D.P., Tsitsiklis, J.N.: Neuro-Dynamic Programming. Athena Scientific, Belmont (1996)
6. Bhatnagar, S.: An actor-critic algorithm with function approximation for discounted cost constrained markov decision processes. Systems and Control Letters 59, 760–766 (2010)
7. Bhatnagar, S., Abdulla, M.S.: Simulation-based optimization algorithms for finite horizon Markov decision processes. Simulation 84(12), 577–600 (2008)
8. Bhatnagar, S., Babu, K.: New algorithms of the Q-learning type. Automatica 44(4), 1111–1119 (2008)
9. Bhatnagar, S., Kumar, S.: A simultaneous perturbation stochastic approximation based actor-critic algorithm for Markov decision processes. IEEE Transactions on Automatic Control 49(4), 592–598 (2004)
10. Borkar, V.S.: Stochastic Approximation: A Dynamical Systems Viewpoint. Cambridge University Press and Hindustan Book Agency (Jointly Published), Cambridge and New Delhi (2008)
11. Borkar, V.S., Meyn, S.P.: The O.D.E. method for convergence of stochastic approximation and reinforcement learning. SIAM Journal of Control and Optimization 38(2), 447–469 (2000)
12. Cao, X.R.: Stochastic Learning and Optimization: A Sensitivity Based Approach. Springer, New York (2007)
13. Cao, X.R., Chen, H.F.: Perturbation realization, potentials, and sensitivity analysis of Markov processes. IEEE Transactions on Automatic Control 42, 1382–1393 (1997)
14. Clarke, F.H.: On the inverse function theorem. Pacific Journal of Mathematics 64(1), 97–102 (1976)
15. Konda, V.R., Borkar, V.S.: Actor-critic like algorithms for Markov decision processes. SIAM Journal on Control and Optimization 38(1), 94–123 (1999)
16. Prashanth, L.A., Bhatnagar, S.: Threshold tuning using stochastic optimization for graded signal control. Tech. rep., Stochastic Systems Lab, IISc (2012), `http://stochastic.csa.iisc.ernet.in/www/research/files/IISc-CSA-SSL-TR-2012-1.pdf`
17. Marbach, P., Tsitsiklis, J.N.: Simulation-based optimization of Markov reward processes. IEEE Transactions on Automatic Control 46(2), 191–209 (2001)
18. Prashanth, L., Bhatnagar, S.: Reinforcement learning with function approximation for traffic signal control. IEEE Transactions on Intelligent Transportation Systems 12(2), 412–421 (2011)
19. Puterman, M.L.: Markov Decision Processes: Discrete Stochastic Dynamic Programming. John Wiley, New York (1994)
20. Schweitzer, P.J.: Perturbation theory and finite Markov chains. Journal of Applied Probability 5, 401–413 (1968)
21. Sun, D.: A further result on an implicit function theorem for locally Lipschitz functions. Operations Research Letters 28, 193–198 (2001)
22. Sutton, R.S., Barto, A.: Reinforcement Learning: An Introduction. MIT Press, Cambridge (1998)
23. Sutton, R.S., McAllester, D., Singh, S., Mansour, Y.: Policy gradient methods for reinforcement learning with function approximation. In: Advances in Neural Information Processing Systems (NIPS), vol. 12, pp. 1057–1063. MIT Press (2000)

24. Tesauro, G.J.: Temporal difference learning and TD-Gammon. Communications of the ACM 38, 58–68 (1995)
25. Tsitsiklis, J.N., Roy, B.V.: An analysis of temporal difference learning with function approximation. IEEE Transactions on Automatic Control 42, 674–690 (1997)
26. Watkins, C., Dayan, P.: Q-learning. Machine Learning 8, 279–292 (1992)
27. Ng, Y., Coates, A., Diel, M., Ganapathi, V., Schulte, J., Tse, B., Berger, E., Liang, E.: Inverted autonomous helicopter flight via reinforcement learning. In: International Symposium on Experimental Robotics, Singapore (2004)

Part V
Applications

This part deals with engineering applications of simultaneous perturbation methods that have been discussed in previous chapters. Specifically, the engineering applications that we consider are in the domains of (a) service systems, (b) road traffic control and (c) communication networks.

In many service domains such as call centers, one is often interested in dynamically finding the optimal staffing levels based on various service requirements of incoming customers and the desired quality of service (QoS). Prashanth, Prasad, Bhatnagar, Desai and Dasgupta in a few papers presented simultaneous perturbation algorithms based on both SPSA and SF techniques for this problem and observed that these algorithms showed better empirical performance as compared to the current state-of-the-art technique. Chapter 12 discusses the application of the SPSA and SF algorithms to service systems.

To maximize flow of vehicles and minimize congestion near road traffic junctions, it is important to regulate traffic lights in a manner that achieves the desired results. By assuming coarse information the system state (for instance, the level of congestion along a lane as being in the 'high', 'medium' or 'low' regions, Prashanth and Bhatnagar, in a paper in 2011, presented an adaptation of the Q-learning algorithm with function approximation. This algorithm, however, incorporates threshold-type feedback policies where the values of the thresholds are considered fixed. Prashanth and Bhatnagar subsequently presented adaptations of the deterministic SPSA algorithm in order to find optimal thresholds and also presented various other threshold-based schemes for traffic signal control. The combinations of the simultaneous perturbation method for adapting thresholds together with the proposed traffic signal control schemes are seen to result in powerful algorithms for this problem. An advantage here is that the simultaneous perturbation module (on top of the regular algorithms) results in only a minor increase in computational effort. Chapter 13 discusses application of simultaneous perturbation methods to road traffic control.

Simultaneous perturbation approaches have also been found to be highly efficient in the context of communication networks. Chapter 14 discusses some of these applications. We consider, in particular, three applications where simultaneous perturbation methods have been found to be very useful. These applications are on (a) random early detection (RED), (b) multi-access communication and (c) internet pricing. The regular RED scheme prescribes a fixed set of threshold parameters that are not seen to work well over various settings. While there have been several works that aim at designing adaptive algorithms for RED, most of them, like regular RED suffer from the problem of wide oscillations in the instantaneous queue length process. By formulating the problem in the nonlinear optimization setting and by developing deterministic perturbation based Newton SPSA algorithms, Patro and Bhatnagar showed that the resulting scheme is both provably convergent and also results in significantly low variance. Next, in the slotted Aloha multi-access communication protocol, it is observed that the feedback probability parameter is held fixed. It is clearly the case that the same parameter value will not work for all network settings. Bhatnagar, Karmeshu and Mishra designed an SF algorithm for finding the optimal parameter trajectory for controlling a parametrized stochastic differential equation. This algorithm was then applied by them for finding the optimal parameter settings for the slotted Aloha communication protocol. The algorithm is seen to exhibit good performance. Finally, Vemu, Bhatnagar and Hemachandra studied the application of SPSA for finding optimal pricing policies within a given parametrized class of these policies. In particular, threshold-type feedback policies were considered. The resulting algorithm is seen to exhibit good performance.

Chapter 12
Service Systems

12.1 Introduction

A *Service System (SS)* is an organization composed of (i) the resources that support, and (ii) the processes that drive service interactions so that the outcomes meet customer expectations . Here we consider the domain of data-center management, where the customers own data centers and other IT infrastructures supporting their businesses. The size, complexity, and uniqueness of the technology installations drive outsourcing of the management responsibilities to specialized service providers that manage the data-centers from remote locations. These are called *delivery centers* and comprise groups of *service workers (SWs)* skilled in specific technology areas supporting *service requests (SRs)* from customers. Each such group is a SS constituting of the processes, the people, and the customers that drive the operations. A delivery center in general contains multiple SS.

An important problem in the context of service systems is to find the optimal staffing levels subject to Service Level Agreement (SLA) and queue stability constraints and for a given dispatching policy (a map from the service requests to service workers). Given a dispatching policy, there are two fundamental challenges in optimizing the staffing levels, i.e., specifications of numbers of workers across shifts and skill levels. First, given an SS with its operational characteristics, the staffing levels need to be optimized while maintaining steady-state and compliance to aggregate Service Level Agreement (SLA) constraints, e.g., 95% of all urgent SRs in a month from a given customer must be resolved in 4 h. Note that the 4 hour deadline does not apply to all individual SRs, but to 95% of them that arrive in a month. Second, it is also necessary to keep the SR queues stable owing to the fact that SLAs are calculated for completed work and not unresolved SRs. The problem is challenging because analytical modeling of SS operations is difficult due to aggregate SLA constraints and also because the SS characteristics such as work patterns, technologies, and customers supported change frequently.

S. Bhatnagar et al.: Stochastic Recursive Algorithms for Optimization, LNCIS 434, pp. 225–241.
springerlink.com

Our approach is to formulate this problem as a constrained hidden Markov cost process [5] parameterized by the (discrete) worker parameter and develop simultaneous perturbation methods to solve the same. To have a sense of the search space size, an SS consisting of 30 SWs who work in 6 shifts and 3 distinct skill levels corresponds to more than 2 trillion configurations. Apart from the high cardinality of the discrete parameter set, the constrained Markov cost process involves a hidden or unobserved state component. The single-stage cost function for the constrained Markov cost process is designed so as to balance the conflicting objectives of worker under-utilization and SLA under/over-achievement. The performance objective is a long-run average of this single stage cost function and the goal is to find the optimum steady state worker parameter (i.e., the one that minimizes this objective) from a discrete high-dimensional parameter set. Note that the optimum worker parameter is a constrained minimum owing to the queue stability and SLA compliance constraints.

We present algorithms based on the simultaneous perturbation technique for solving the above problem. Simulation is employed for finding the optimum (constrained) worker parameter as the single stage cost function can be estimated only via simulation. Henceforth, we shall refer to these algorithms as *Staff Allocation using Stochastic Optimization with Constraints (SASOC)* algorithms. Both first and second order methods based on the techniques presented in the earlier chapters are described. An important aspect of all SASOC algorithms is that they involve the generalized smooth projection operator, which is essential to project the continuous-valued worker parameter tuned by the SASOC algorithms onto the discrete set. As described in Chapter 9, the generalized projection operator ensures that the underlying transition dynamics of the constrained Markov cost process is itself smooth (as a function of the continuous-valued parameter), which in turn allows one to mimic the continuous constrained optimization techniques, such as those described in Chapter 10, in the context of optimizing staff levels of a SS.

The remaining part of this chapter is organized as follows: We introduce the service system framework in Section 12.2. We formulate the labor cost optimization problem with SLA constraints in Section 12.3. We then discuss several first order simultaneous perturbation algorithms, similar to those in Chapter 10, for solving the afore-mentioned problem in Section 12.5. Thereafter, two second-order methods using SPSA for estimating the gradient and the Hessian are presented in Section 12.6. A discussion on the convergence of all the algorithms is available in Section 12.7 and some representative experimental results on the algorithms described are briefly presented in Section 12.8. The material reported in this chapter is based on [2, 3, 6].

12.2 Service System Framework

Figure 12.1 shows the main components of the operational model of SS. SRs arrive from multiple customers supported by the SS and get classified and queued

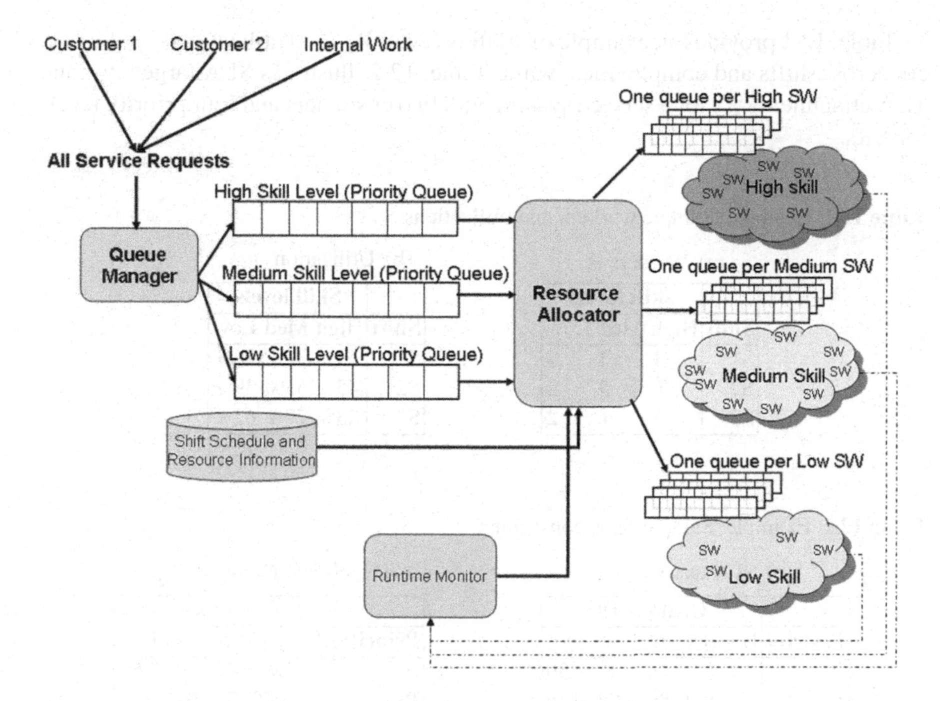

Fig. 12.1 Components of the operational models of SS

into high, medium, or low complexity queues by a queue manager (human or automated). Also, depending on the urgency of the SRs as well as the dispatching policy being used, each individual SR is assigned a priority in each of the complexity queues. SWs are grouped according to their skill levels, viz., high, medium, or low and work according to predefined shift schedules. Depending on the dispatching policy in place, the resource allocator (human or machine based) either pushes the SRs to SWs proactively or the SWs pull the highest priority SR from the complexity queue when it becomes available. In the former case, each of the SWs has an associated priority queue. Generally, SWs work on SRs with complexity matching to their skill levels. However, a *swing* policy may kick-in dynamically and assign higher-skilled workers to lower complexity queues if they grow beyond a threshold. Finally, a *preemption* policy specifies a preemptive action such as an urgent SR preempting all other SRs regardless of their status. A runtime monitor collects statistics on the performance of the SS against the SLAs, monitors the queues for unstable behavior such as unbounded growth, and triggers invocation of the swing policy when swing thresholds are crossed.

Table. 12.1 provides an example of staffing levels $W_{i,j}$ and utilization $u_{i,j}$ of workers across shifts and complexities, while Table. 12.2 illustrates SLA targets $\gamma_{i,j}$ and SLA attainments $\gamma'_{i,j}$ for a service system with two customers and four priority levels for SRs.

Table 12.1 Example: Sample workers and utilizations

(a) Workers θ_i

	Skill levels		
Shift	High	Med	Low
S1	1	3	7
S2	0	5	2
S3	3	1	2

(b) Utilizations $u_{i,j}$

	Skill levels		
Shift	High	Med	Low
S1	67%	34%	26%
S2	45%	55%	39%
S3	23%	77%	62%

Table 12.2 Example: Sample SLA constraints

(a) SLA targets $\gamma_{i,j}$

	Customers	
Priority	Bossy Corp	Cool Inc
P_1	95%4h	89%5h
P_2	95%8h	98%12h
P_3	100%24h	95%48h
P_4	100%18h	95%144h

(b) SLA attainments $\gamma'_{i,j}$

	Customers	
Priority	Bossy Corp	Cool Inc
P_1	98%4h	95%5h
P_2	98%8h	99%12h
P_3	89%24h	90%48h
P_4	92%18h	95%144h

12.3 Problem Formulation

We consider the problem of finding the optimal staffing levels (see Fig. 12.1), while adhering to the SLA constraints and maintaining state-to-state queues. We formulate this as a constrained hidden Markov cost process as follows:

Worker parameter θ: The worker parameter specifies the number of workers for each shift and of each skill level in a SS and is given by

$$\theta = (W_1, \ldots, W_{|A|\times|B|})^T \in \mathscr{D},$$

where W_i indicates the number of service workers whose skill level is $i\%|B|$ and whose shift index is $i/|B|$. The parameter vector θ takes values in the set $\mathscr{D}$, where $\mathscr{D} \triangleq \{0, 1, \ldots, W_{\max}\}^N$. Here, $W_{\max}$ serves as an upper bound for the number of workers in any shift and of any skill level. Note that one can enumerate all the points in $\mathscr{D}$ as $\mathscr{D} = \{D^1, D^2, \ldots, D^p\}$ for some $p > 1$.

State $(X(n),Y(n))$**:** The state consists of the observed part $X(n)$ and the unobserved or hidden part $Y(n)$ and is described by

$$X(n) = (\mathbb{N}(n), u(n), \gamma'(n), q(n)), \quad (12.1)$$
$$Y(n) = (\mathbb{Z}(n)). \quad (12.2)$$

In the above,

- $\mathbb{N}(n) = (\mathbb{N}_1(n), \ldots, \mathbb{N}_{|B|}(n))^T$, where $\mathbb{N}_i(n)$ denotes the number of SRs in the system queue corresponding to skill level $i \in \mathscr{B}$.
- $\mathbb{Z}(n) = (\mathbb{Z}_{1,1,1}(n), \ldots, \mathbb{Z}_{1,1,W_{\max}}(n), \ldots, \mathbb{Z}_{|A|,|B|,W_{\max}}(n))$ is the vector of residual service times. Here, $\mathbb{Z}_{i,j,k}(n)$ denotes the residual service time of the SR currently being processed by the kth worker in shift i and of skill level j. Note that if there is no kth worker corresponding to the shift i and skill level j, then $\mathbb{Z}_{i,j,k} = \kappa$, where κ is a special value used to signify the non-existence of a worker. Considering that the service times follow a truncated log-normal distribution in our setting, the residual service time at any point cannot be precisely estimated and hence, is part of the unobserved or hidden state component $Y(n)$.
- The utilization vector $u(n) = (u_{1,1}(n), \ldots, u_{|A|,|B|}(n))$, where each $u_{i,j}(n) \in [0,1]$ is the average utilization of the workers in shift i and skill level j, at instant n.
- The SLA attainment vector $\gamma'(n) = (\gamma'_{1,1}(n), \ldots, \gamma'_{|C|,|P|}(n))$, where $\gamma'_{i,j}(n) \in [0,1]$ denotes the SLA attainment for customer i and priority j, at instant n.
- $q(n)$ is a single scalar (Boolean) variable that denotes the queue feasibility status of the system at instant n. In other words, $q(n)$ is false if the growth rate of the SR queues (for each complexity) is beyond a threshold and is true otherwise. We need $q(n)$ to ensure system steady-state which is independent of SLA attainments because the latter are computed only on the SRs that were completed and not on those queued up in the system.

Considering that the queue lengths, utilizations and SLA attainments at instant $n+1$ depend only on the state at instant n, i.e., $\{(X(n),Y(n))\}$, we observe that $\{(X(n),Y(n)), n \geq 0\}$ is a constrained hidden Markov cost process for any given (fixed) parameter θ.

Allowing S to denote the state space, we observe that S is compact as the various components of $X(n)$ and $Y(n)$ are closed and bounded. This is because, each element of $u(n)$, $\gamma'(n)$ takes values in $[0,1]$ and $0 \leq q(n) \leq 1$. Further, the system SR queues $\mathbb{N}$ have a finite buffer each and hence, $X(n), n \geq 0$ is closed and bounded. Further, the residual time vector in $Y(n)$ also takes values in a compact set in lieu of the fact that each element of $\mathbb{Z}$ is upper bounded by the total service times at the SR queues and that in turn takes values in $[0, \mathrm{T}]$.

Cost: The single-stage cost function is designed so as to minimize the under-utilization of workers as well as over/under-achievement of SLAs. Here, under-utilization of workers is the complement of utilization and in essence, this is equivalent to maximizing the worker utilizations. The over/under-achievement of SLAs is the distance between attained and the contractual SLAs. Hence, the cost function is designed to balance between two conflicting objectives and has the form:

$$c(X(n)) = r \times \left(1 - \sum_{i=1}^{|A|}\sum_{j=1}^{|B|} \alpha_{i,j} \times u_{i,j}(n)\right) + s \times \left(\sum_{i=1}^{|C|}\sum_{j=1}^{|P|} \left|\gamma'_{i,j}(n) - \gamma_{i,j}\right|\right), \tag{12.3}$$

where $r, s \geq 0$ and $r + s = 1$. Further, $0 \leq \gamma_{i,j} \leq 1$ denotes the contractual SLA for customer i and priority j. Note that the first term in (12.3) uses a weighted sum of utilizations over workers from each shift and across each skill level. The weights $\alpha_{i,j}$ are derived from the workload distribution across shifts and skill levels over a month long period. These weights satisfy $0 \leq \alpha_{i,j} \leq 1$, $\sum_{i=1}^{|A|}\sum_{j=1}^{|B|} \alpha_{i,j} = 1$. Such a prioritization of workers helps in optimizing the worker set based on the workload expected in a particular shift and skill combination.

Constraints: The constraints are on the SLA attainments and queue growth, given by:

$$g_{i,j}(X(n)) = \gamma_{i,j} - \gamma'_{i,j}(n) \leq 0, \forall i = 1, \ldots, |C|, j = 1, \ldots, |P|, \tag{12.4}$$

$$h(X(n)) = 1 - q(n) \leq 0, \tag{12.5}$$

Here (12.4) specifies that the attained SLA levels should be equal to or above the contractual SLA targets for each customer-priority tuple. Further, (12.5) ensures that the SR queues for each complexity in the system stay bounded. In the constrained optimization problem formulated below, we attempt to satisfy these constraints in the long-run average sense (see (12.6)).

System evolution:

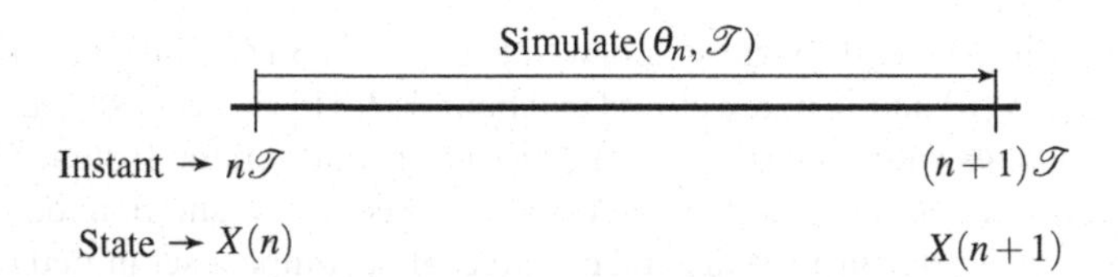

Fig. 12.2 A portion of the time-line illustrating the process

For a typical SS, as described in Section 12.2, the above model translates to a stochastic evolution of the system from one state to another, while incurring a state-dependent single-stage cost as constraint functions described via $c(X(n))$, $g_{i,j}(X(n)), h(X(n)), i = 1,\ldots,|C|, j = 1,\ldots,|P|$. Note that these functions depend explicitly on only the observed part $X(n)$ of the state process $(X(n),Y(n)), n \geq 0$. As illustrated in Fig. 12.2, the nth system transition of this underlying constrained Markov cost process involves a simulation of the service system for a fixed period $\mathscr{T}$ with the current worker parameter θ_n. For instance, in our representative experiments discussed in Section 12.8, $\mathscr{T} = 10$, which translates to a simulation of the service system for a period of 10 months with the staffing levels specified by θ_n. Also, note that this is a continuously running simulation, where at discrete time instants $n\mathscr{T}$ we update the worker parameter θ_n and the simulation output causes a probabilistic transition from the current state $(X(n),Y(n))$ to the next state $(X(n+1),Y(n+1))$, while incurring a single stage cost $c(X(n))$. By an abuse of notation, we refer to the state at instant $n\mathscr{T}$ as $(X(n),Y(n))$.

The Objective: Our aim is to find a θ that minimizes the long-run average cost,

$$
\begin{aligned}
&J(\theta) \triangleq \lim_{n\to\infty} \frac{1}{n} \sum_{m=0}^{n-1} E[c(X(m))] \\
&\text{subject to} \\
&G_{i,j}(\theta) \triangleq \lim_{n\to\infty} \frac{1}{n} \sum_{m=0}^{n-1} E[g_{i,j}(X(m))] \leq 0 \quad \forall i = 1,\ldots,|C|, j = 1,\ldots,|P|, \\
&H(\theta) \triangleq \lim_{n\to\infty} \frac{1}{n} \sum_{m=0}^{n-1} E[h(X(m))] \leq 0.
\end{aligned} \tag{12.6}
$$

Projection Operator: The SASOC algorithms treat the parameter as continuous-valued and tune it accordingly. Let us denote this continuous version of the worker parameter by $\bar{\theta} = (\bar{W}_1,\ldots,\bar{W}_N)$. Note that $\bar{\theta}_i \in [0, W_{\max}], i = 1,2,\ldots,N$. The SASOC algorithms that we present subsequently tune the worker parameter in the convex hull of $\mathscr{D}$, denoted by $\bar{\mathscr{D}}$, a set that can be simply defined as $\bar{\mathscr{D}} = [0, W_{\max}]^N$. The projection operator $\bar{\Gamma}$ projects any $\theta \in \mathbb{R}^N$ onto the set $\bar{\mathscr{D}}$ and is defined as $\bar{\Gamma}(\theta) = (\bar{\Gamma}_1(\theta_1),\ldots,\bar{\Gamma}_N(\theta_N))^T$, where $\bar{\Gamma}_i(\theta_i) = \min(W_{\max}, \max(\theta_i, 0))$, $i = 1,\ldots,N$.

A generalized projection operator $\Gamma(\theta) = (\Gamma_1(W_1),\ldots,\Gamma_N(W_N))^T$ that projects θ on to the discrete set $\mathscr{D}$ is necessary to guide the service system simulation. This projection idea has been described in Chapter 9 for an unconstrained discrete optimization problem. Specifically, $\Gamma_i(W_i)$ is defined in a manner similar to the definition of the generalized projection scheme discussed in Chapter 9.2.3 and we omit the definition here. Recall from Chapter 9 that the Γ-operator ensures that the transition dynamics of the parameter extended Markov process for any $\theta \in \bar{\mathscr{D}}$ is smooth (as desired) and requires a lower computational effort because in a large portion of the parameter space (assuming ζ is small), the Γ-operator is essentially deterministic.

Thus, $\bar{\Gamma}(\cdot)$ keeps the parameter updates within the set $\bar{\mathscr{D}}$ and $\Gamma(\cdot)$ projects them to the discrete set $\mathscr{D}$. The projected updates are then used as the parameter values for conducting the simulation of the service system.

12.4 Solution Methodology

The constrained long-run average cost optimization problem (12.6) can be expressed using the standard Lagrange multiplier theory as an unconstrained optimization problem given below.

$$\max_{\lambda}\min_{\theta} L(\theta,\lambda) \triangleq \lim_{n\to\infty}\frac{1}{n}\sum_{m=0}^{n-1} E\left\{c(X(m)) + \sum_{i=1}^{|C|}\sum_{j=1}^{|P|}\lambda_{i,j}g_{i,j}(X(m)) + \lambda_f h(X(m))\right\}, \tag{12.7}$$

where $\lambda_{i,j} \geq 0, \quad \forall i = 1,\ldots,|C|, j = 1,\ldots,|P|$ represent the Lagrange multipliers corresponding to the constraints $G_{i,j}(\theta) \leq 0$ and λ_f represents the Lagrange multiplier for the constraint $H(\theta) \leq 0$ in the optimization problem (12.6). Also, $\lambda = (\lambda_{i,j}, \lambda_f, i = 1,\ldots,|C|, j = 1,\ldots,|P|)^T$.

As in Chapter 10, we present several simulation optimization methods for obtaining a saddle point of the Lagrangian (12.7). All SASOC algorithms update the worker parameter along a descent direction as follows:

$$\theta(n+1) = \bar{\Gamma}(\theta(n) - b(n)\mathscr{H}_n^{-1}h_n). \tag{12.8}$$

In the above, h_n represents the estimate of the gradient while $\mathscr{H}_n$ is the particular positive definite and symmetric matrix used at update instant n. For the sake of simplicity, we have omitted an additive stochastic noise term in the update (12.8). In other words, all SASOC algorithms can be seen as noisy variants of (12.8) and use either SPSA or SF-based estimates of the gradient and the Hessian of the Lagrangian.

As illustrated in Fig. 12.3, each algorithm involves an iterative procedure, where a proposed candidate solution θ is evaluated using a simulation framework twice - one with unperturbed parameter and another with perturbed parameter. The perturbation $p(n)$ is algorithm-specific and is motivated by the gradient estimate of the given algorithm. The results of the simulation, specifically the attained SLAs $\gamma'_{i,j}$ and the queue stability parameter q are used to tune the parameter in an algorithm-specific descent direction. Algorithm 12.1 gives the structure of all the SASOC algorithms presented in the subsequent sections.

Algorithm 12.1 Skeleton of SASOC algorithms

Input:

- R, a large positive integer;
- $\theta(0)$, initial parameter vector; $p(\cdot)$; Δ
- UpdateRule(), the algorithm-specific update rule for the worker parameter θ and Lagrange multiplier λ.
- Simulate($\theta, \mathscr{T}$) $\rightarrow X$, the simulator of the SS

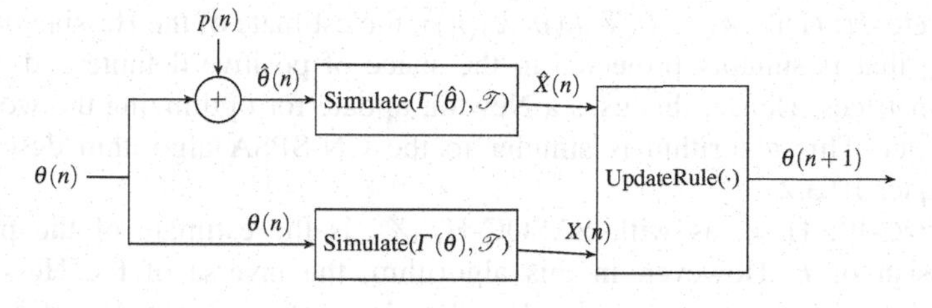

Fig. 12.3 Overall flow of the algorithm 12.1.

Output: $\theta^* \triangleq \Gamma(\theta(R))$.

$\theta \leftarrow \theta(0), n \leftarrow 1$
loop
  $X \leftarrow \text{Simulate}(\Gamma(\theta(n)), \mathscr{T})$.
  $\hat{X} \leftarrow \text{Simulate}(\Gamma(\theta(n) + p(n)), \mathscr{T})$.
  UpdateRule().
  $n \leftarrow n+1$
  if $n = R$ **then**
    Terminate and output $\Gamma(\theta(R))$.
  end if
end loop

The algorithms described next can be categorized as follows:

- Based on whether the algorithms estimate the gradient or the Hessian, they can be categorized as being of the first or the second order.
- Based on the technique employed for estimating the gradient/Hessian, they can be categorized as SPSA or SF-based method. The various choices of $\mathscr{H}_n$ used in our algorithms are described below.

The SASOC algorithms mainly differ in the choice of $\mathscr{H}_n$ in (12.8) and hence the descent direction:

1. *SASOC-G*: Here $\mathscr{H}_n = I$ (identity matrix). This algorithm tunes the worker parameter θ in the negative gradient descent direction, with a one-sided SPSA gradient estimate, as explained in Chapter 10.3.1, being used.
2. *SASOC-SF-N*: Here again $\mathscr{H}_n = I$. However, the gradient estimate incorporates one-sided SF with Gaussian perturbations similar to the estimate used in the CG-SF algorithm described in Chapter 10.3.3.
3. *SASOC-SF-C*: As in SASOC-SF-N, here also we use the SF-based gradient estimate, i.e., $\mathscr{H}_n = I$. However, the perturbations in this case are based on the Cauchy distribution.

4. *SASOC-H*: Here $\mathcal{H}_n = P(\nabla^2 L(\theta,\lambda)(n))$, the estimate of the Hessian of L w.r.t. $\theta(n)$ that is suitably projected to the space of positive definite and symmetric matrices. Hence, this uses a Newton update for optimizing the worker parameter. This algorithm is simular to the CN-SPSA algorithm described in Chapter 10.3.2.
5. *SASOC-W*: Here, as with SASOC-H, $\mathcal{H}_n$ is the estimate of the projected Hessian of L. However, in this algorithm, the inverse of the Hessian matrix is tuned directly using the Woodbury's identity, a procedure described in Chapter 7.4.

Note that while the algorithms of Chapter 10 are for a continuous-valued parameter, the SASOC algorithms are for the discrete-valued worker parameter θ. In essence, the SASOC algorithms use the continuous optimization procedures similar to the ones described in Chapter 10 and the convergence of these algorithms is ensured by employing a generalized projection operator (see Chapter 9.2.3) that makes the underlying transition dynamics smooth. Table. 12.3 summarizes the various features of the SASOC algorithms presented here.

Table 12.3 Summary of SASOC algorithms

Algorithm	**Order**	**Type**	H_n	$p(n)$
SASOC-SPSA	First	SPSA	I	$\delta\Delta(n)$
SASOC-SF-N	First	SF-Gaussian	I	$\beta\eta(n)$
SASOC-SF-C	First	SF-Cauchy	I	$\beta\eta(n)$
SASOC-H	Second	SPSA-Hessian	$P(\nabla^2 L(\theta,\lambda)(n))$	$\delta_1\Delta(n)+\delta_2\widehat{\Delta}(n)$
SASOC-W	Second	SPSA-Woodbury	"	"

We next present the first-order methods that include either SPSA or SF based gradient estimates and in the following section (Section 12.6), we present the second-order methods. A discussion of the convergence of both the first as well as the second order methods is then presented in Section 12.7.

12.5 First Order Methods

12.5.1 SASOC-SPSA

This is a three time-scale stochastic approximation algorithm that does primal descent on the worker parameter while performing dual ascent on the Lagrange multipliers. This algorithm is similar to CG-SPSA described in Chapter 10 and uses a one-sided SPSA gradient estimate. The update rule for this algorithm is given by

$$\left.\begin{aligned}
&W_i(n+1) = \bar{\Gamma}_i\left[W_i(n) + b(n)\left(\tfrac{\bar{L}(n)-\bar{L}'(n)}{\beta\eta_i(n)}\right)\right], \forall i = 1,2,\ldots,N,\\
&\bar{L}(n+1) = \bar{L}(n) + d(n)(l(X(n),\lambda(n)) - \bar{L}(n)),\\
&\bar{L}'(n+1) = \bar{L}'(n) + d(n)(l(\hat{X}(n),\lambda(n)) - \bar{L}'(n)),\\
&\lambda_{i,j}(n+1) = (\lambda_{i,j}(n) + a(n)g_{i,j}(X(n)))^+, \forall i = 1,2,\ldots,|C|, j = 1,2,\ldots,|P|,\\
&\lambda_f(n+1) = (\lambda_f(n) + a(n)h(X(n)))^+,
\end{aligned}\right\} \quad (12.9)$$

where

- $l(X,\lambda) = c(X) + \sum_{i=1}^{|C|}\sum_{j=1}^{|P|} \lambda_{i,j}g_{i,j}(X) + \lambda_f h(X)$ is the single-stage sample of the Lagrangian;
- $X(n)$ represents the state at iteration n from the simulation run with parameter $\Gamma(\theta(n))$ while $\hat{X}(n)$ represents the state at iteration n from the simulation run with the perturbed parameter $\Gamma(\theta(n) + \delta\Delta(n))$. For simplicity, hereafter we use θ to denote $\theta(n)$ and $\theta + \delta\Delta$ to denote $\theta(n) + \delta\Delta(n)$. Also, Γ denotes the generalized projection operator used to project θ onto the discrete set $\mathscr{D}$;
- $\delta > 0$ is a fixed perturbation control parameter while Δ is a vector of perturbation random variables that are independent, zero-mean and have the symmetric Bernoulli distribution;
- The operator $\bar{\Gamma}(\cdot)$ ensures that the updated value for θ stays within the convex hull $\bar{\mathscr{D}}$ and is defined as follows: $\bar{\Gamma}(\theta) = (\bar{\Gamma}_1(\theta_1),\ldots,\bar{\Gamma}_N(\theta_N))^T$, with $\bar{\Gamma}_i(\theta_i) = \min(W_{\max}, \max(\theta_i, 0))$, $i = 1,\ldots,N$.
- $\bar{L}$ and $\bar{L}'$ represent Lagrange estimates corresponding to θ and $\theta + \delta\Delta$ respectively. Thus, for each iteration, two simulations are carried out, one with the normal parameter θ and the other with the perturbed parameter $\theta + \delta\Delta$, the results of which are used to update $\bar{L}$ and $\bar{L}'$.

12.5.2 SASOC-SF-N

This algorithm is also a first-order method like SASOC-SPSA. However, it uses a Gaussian smoothed functional gradient estimate similar to the one in the CG-SF algorithm of 10. The overall update rule for this algorithm is same as that of SASOC-SPSA, except the updates to the parameter θ, which is given by,

$$W_i(n+1) = \bar{\Gamma}_i\left[W_i(n) + b(n)\left(\frac{\eta_i(n)}{\beta}(\bar{L}(n) - \bar{L}'(n))\right)\right], \quad (12.10)$$

for all $i = 1,2,\ldots,|A| \times |B|$. In the above,

- The update equations corresponding to $\bar{L},\bar{L}'$, $\lambda_{i,j}, i = 1,\ldots,|C|, j = 1,\ldots,|P|$ and λ_f are the same as in SASOC-SPSA.
- $\beta > 0$ is a fixed smoothing control parameter.
- $\eta = (\eta_1,\eta_2,\ldots,\eta_{|A|\times|B|})$ is a vector of $|A| \times |B|$ independent $N(0,1)$ random variables;
- The rest of the symbols are the same as in SASOC-SPSA algorithm. Specifically, $X(n)$ represents the state at iteration n from the unperturbed simulation, while $\hat{X}(n)$ represents the state from the perturbed simulation. However, note that the perturbed simulation in this case is run with the parameter $\Gamma(\theta + \beta\eta)$. In other words, the perturbation $p(n)$ in Fig. 12.3 corresponds to $\beta\eta$.

12.5.3 SASOC-SF-C

This algorithm uses the Cauchy instead of the Gaussian density as the smoothing density function. The rest is similar to that in SASOC-SF-N. The Cauchy distribution has a heavier tail as compared to the Gaussian distribution. Hence, it is seen to explore the search space better (see Chapter 6 for a detailed treatment). The update rule of SASOC-SF-C algorithm for the parameter θ, is given by

$$W_i(n+1) = \bar{\Gamma}_i\left[W_i(n) + b(n)\left(\frac{\eta_i(n)(N+1)}{\beta(1+\eta(n)^T\eta(n))}(\bar{L}(n) - \bar{L}'(n))\right)\right], \tag{12.11}$$

for all $i = 1,2,\ldots,|A| \times |B|$. In the above,

- The update equations corresponding to $\bar{L},\bar{L}'$, $\lambda_{i,j}, i = 1,\ldots,|C|, j = 1,\ldots,|P|$ and λ_f are the same as in SASOC-SPSA.
- $\beta > 0$ is a fixed smoothing control parameter while η is an N-dimensional multi-variate Cauchy random vector truncated to some μ;
- The rest of the symbols are the same as in SASOC-SF-N algorithm.

12.6 Second Order Methods

We now present two second-order algorithms — SASOC-H and SASOC-W, which use SPSA based estimates for the gradient and the Hessian. In principle, these algorithms are similar to the Hessian and Woodbury variants of second-order SPSA based algorithms described in Chapter 7.

12.6.1 SASOC-H

This algorithm is similar to CN-SPSA (see Chapter 10.3.2) in terms of the gradient and the Hessian estimates. The update rule of this algorithm is given by

$$W_i(n+1) = \bar{\Gamma}_i\left(W_i(n) + b(n)\sum_{j=1}^{|A|\times|B|} M_{i,j}(n)\left(\frac{\bar{L}(n)-\bar{L}'(n)}{\delta_2\widehat{\triangle}_j(n)}\right)\right), \quad (12.12)$$

$$H_{i,j}(n+1) = H_{i,j}(n) + b(n)\left(\frac{\bar{L}'(n)-\bar{L}(n)}{\delta_1\triangle_j(n)\delta_2\widehat{\triangle}_i(n)} - H_{i,j}(n)\right),$$

for all $i, j = 1, 2, \ldots, |A| \times |B|$. In the above,

* The update equations corresponding to $\bar{L}, \bar{L}'$, $\lambda_{i,j}, i = 1, \ldots, |C|, j = 1, \ldots, |P|$ and λ_f are the same as in SASOC-SPSA.
* $\delta_1, \delta_2 > 0$ are fixed perturbation control parameters while Δ and $\widehat{\Delta}$ are two independent vectors of perturbation random variables that are independent, zero-mean and have the symmetric Bernoulli distribution;
* $\bar{L}$ and $\bar{L}'$ represent the Lagrangian estimates corresponding to θ and $\theta + \delta_1\Delta + \delta_2\widehat{\Delta}$ respectively. Thus, for each iteration, two simulations are carried out, one with the nominal parameter $\Gamma(\theta)$ and the other with the perturbed parameter $\Gamma(\theta + \delta_1\Delta + \delta_2\widehat{\Delta})$, the results of which are used to update $\bar{L}$ and $\bar{L}'$, respectively;
* $H = [H_{i,j}]_{i=1,j=1}^{|A|\times|B|,|A|\times|B|}$ represents the Hessian estimate of the Lagrangian. Here $H(0)$ is set to be a positive definite and symmetric matrix, in particular, $H(0) = cI$, with $c > 0$ and I being the identity matrix; and
* $M(n) = P(H(n))^{-1} = [M(n)_{i,j}]_{i=1,j=1}^{|A|\times|B|,|A|\times|B|}$ represents the inverse of the Hessian estimate H of the Lagrangian, where $P(\cdot)$ is a projection operator ensuring that the Hessian estimates remain symmetric and positive definite, see Chapters 7 and 8.
* The rest of the symbols are the same as in the first-order methods described before.

12.6.2 SASOC-W

The SASOC-H algorithm is more robust than SASOC-G. However, it requires the computation of inverse of the Hessian H at each stage which is a computationally intensive operation. As described in Chapter 7, we develop a second order method that directly tunes the inverse of the Hessian H using the Woodbury's identity. The resulting algorithm has a computational complexity of $O(n^2)$, as compared to SASOC-H, which is $O(n^3)$. The update rule of this algorithm, named SASOC-W, is given by

$$W_i(n+1) = \bar{\Gamma}_i \left(W_i(n) + b(n) \sum_{j=1}^{|A|\times|B|} M_{i,j}(n) \left(\frac{\bar{L}(n) - \bar{L}'(n)}{\delta_1 \widehat{\triangle}_j(n)} \right) \right), \tag{12.13}$$

$$M(n+1) = P\left(\frac{M(n)}{1-b(n)} \left[I - \frac{b(n)\,(\bar{L}'(n) - \bar{L}(n))\,P(n)Q(n)M(n)}{1-b(n)+b(n)\,(\bar{L}'(n)-\bar{L}(n))\,Q(n)M(n)P(n)} \right] \right),$$

where all the symbols are as described in SASOC-H and the update equations corresponding to $\bar{L}, \bar{L}', \lambda_{i,j}, i = 1, \ldots, |C|, j = 1, \ldots, |P|$ and λ_f are the same as in SASOC-SPSA. Note here that M, the inverse of the Hessian H, is directly updated, whereas in SASOC-H, the Hessian was updated first and its inverse was later explicitly computed in order to obtain M.

Remark 12.1. As noted in the previous chapters, an additional averaging over L instants (for a given $L > 1$) of recursions involving data averaging, in between two successive parameter updates, is seen to result in better algorithmic performance, see [2, 3, 6].

12.7 Notes on Convergence

The convergence analysis of all the SASOC algorithms presented above proceed along the lines of their respective counterparts in Chapter 10. However, considering that the problem (12.6) is for a discrete parameter, it is necessary to first smoothen the underlying transition dynamics for any $\theta \in \bar{\mathscr{D}}$ (recall that $\bar{\mathscr{D}}$ is the convex hull of the discrete set $\mathscr{D}$). This can be achieved using a procedure described in Chapter 9. Specifically, a result similar to one in Lemma 9.4 can be shown for the SASOC algorithms. As a consequence, one could mimic a continuous parameter system, allowing the proofs of Chapter 10 to hold.

Some algorithm-specific notes follow:

- The assumptions and the convergence analysis of the SASOC-SPSA algorithm are along the lines of the CG-SPSA algorithm described in Section 10.3.1.
- The convergence analysis of the SF based first order methods - SASOC-SF-N and SASOC-SF-C is similar to the CG-SF algorithm described in Section 10.3.3. In particular, the SF-based algorithms are distinguished by the gradient estimate proceeding on the faster-timescale, the analysis of which can be seen to be along the lines of the Gaussian and Cauchy variants of the SF algorithms described in Chapter 6.
- The convergence analysis of SASOC-H and SASOC-W on the faster timescales (i.e., the recursions corresponding to the worker parameter and the Hessian estimates) proceeds along the lines of the Hessian and the Woodbury variants of the SPSA-based schemes described in Section 7.3.3 and Section 7.4,

respectively. The complete analysis, including the slowest timescale update of Lagrange parameters, can then be seen to be similar to that of the CN-SPSA algorithm of Chapter 10.

12.8 Summary of Experiments

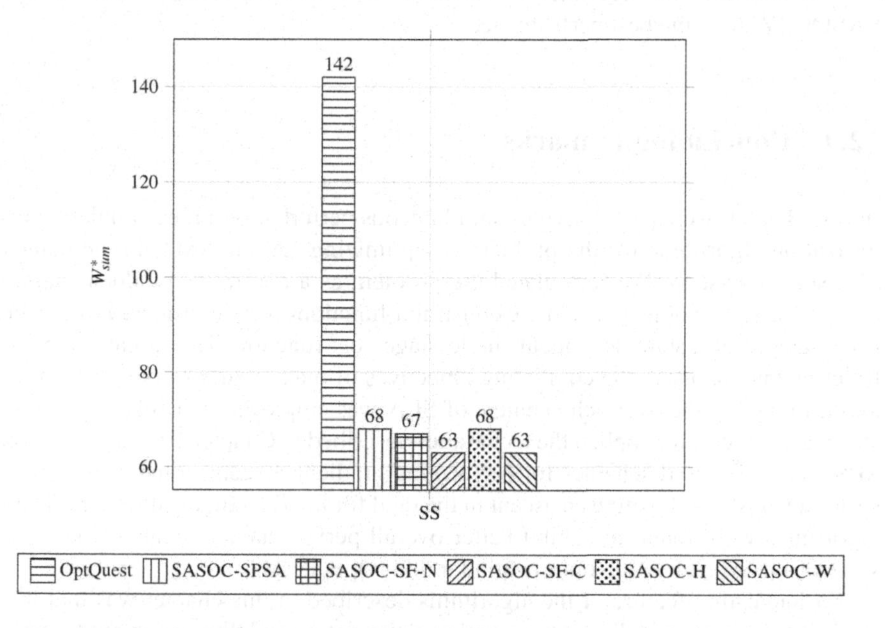

Fig. 12.4 Performance of OptQuest and SASOC algorithms for EDF dispatching policy on a real SS

We present a representative optimum staffing level result obtained using the various SASOC algorithms described previously, on the simulation framework proposed in [1]. The detailed simulation results are available in [3, 6]. The underlying dispatching policy used is the EDF, where the time left to SLA target deadline is used to assign the SRs to the SWs, i.e., the SW works on the SR that has the earliest deadline. As mentioned before, all the SASOC algorithms involve two service system simulations — one with unperturbed parameter and the other with the perturbed parameter. Further, for purposes of comparison, an algorithm for staff allocation using the state-of-the-art optimization tool-kit OptQuest [4] was also implemented. OptQuest is a well-established tool for solving simulation optimization problems and we used a scatter search based algorithm for performance comparisons. The algorithms are compared using W^*_{sum} as the performance metric. Here

$W_{sum} \triangleq \sum_{i=1}^{|A|} \sum_{j=1}^{|B|} W_{i,j}$ is the sum of workers across shifts and skill levels and W_{sum}^* denotes the value obtained upon convergence of W_{sum}.

As evident in Fig. 12.8, the SASOC algorithms in general are seen to exhibit much superior performance compared to OptQuest, as they (a) exhibit more than an order of magnitude faster convergence than OptQuest, (b) consistently obtain solutions of good quality and in most cases better than those found by OptQuest, and (c) show guaranteed convergence even in scenarios where OptQuest does not find a feasible solution even after 5,000 iterations. Amongst the SASOC algorithms, we observe that (the first-order method) SASOC-SF-C and (the second-order method) SASOC-W show the best performance.

12.9 Concluding Remarks

In this chapter, we adapted various simultaneous perturbation-based simulation optimization algorithms for the problem of optimizing staffing levels in the context of a service system. We formulated the problem as a constrained hidden Markov cost process. The objective and the constraint functions were considered to be long run averages of a state dependent single-stage cost function. The single-stage cost function that balanced the conflicting objectives of maximizing worker utilizations and minimizing the over-achievement of SLA was employed. For solving the constrained problem, we applied the techniques described in Chapter 10 to develop both SPSA and SF-based schemes for performing gradient descent in the primal while simultaneously performing an ascent in the dual for the Lagrange multipliers. These algorithms were found to exhibit better overall performance in comparison to the state-of-the-art simulation optimization toolkit OptQuest.

An interesting feature of the algorithms described in this chapter was that they performed constrained discrete parameter optimization and thus are an extension of the algorithms described in Chapter 9 for unconstrained discrete parameter settings as well as the ones described in Chapter 10 for the problem of constrained continuously valued parameters. The developed algorithms are sufficiently general and can be applied for other problems of constrained discrete parameter optimization involving long-run average objective and constraint functions.

References

1. Banerjee, D., Desai, N., Dasgupta, G.: Simulation-based evaluation of dispatching policies in service systems. In: Winter Simulation Conference (2011)
2. Prashanth, L.A., Prasad, H.L., Desai, N., Bhatnagar, S., Dasgupta, G.: Stochastic Optimization for Adaptive Labor Staffing in Service Systems. In: Kappel, G., Maamar, Z., Motahari-Nezhad, H.R. (eds.) Service Oriented Computing. LNCS, vol. 7084, pp. 487–494. Springer, Heidelberg (2011)

3. Prashanth, L.A., Prasad, H., Desai, N., Bhatnagar, S., Dasgupta, G.: Simultaneous perturbation methods for adaptive labor staffing in service systems. Tech. rep., Stochastic Systems Lab, IISc (2012), http://stochastic.csa.iisc.ernet.in/www/research/files/IISc-CSA-SSL-TR-2011-4-rev2.pdf
4. Laguna, M.: Optimization of complex systems with optquest. Opt. Quest for Crystal Ball User Manual, Decisioneering (1998)
5. Marbach, P., Tsitsiklis, J.: Simulation-based optimization of markov reward processes. IEEE Transactions on Automatic Control 46(2), 191–209 (2001)
6. Prasad, H., Prashanth, L.A., Desai, N., Bhatnagar, S.: Adaptive smoothed functional based algorithms for labor cost optimization in service systems. Tech. rep., Stochastic Systems Lab, IISc (2012), http://stochastic.csa.iisc.ernet.in/www/research/files/IISc-CSA-SSL-TR-2012-2.pdf

Chapter 13
Road Traffic Control

13.1 Introduction

In this chapter, we present a few applications of simultaneous perturbation and reinforcement learning techniques developed in the earlier chapters for the problem of maximizing traffic flows through the adaptive control of traffic lights at traffic intersections. We consider two inter-related problems here:

(I) developing suitable traffic light control (TLC) algorithms that use threshold-based coarse information about congestion on the various lanes of the road network as input, and
(II) developing an algorithm to tune the aforementioned thresholds used in any threshold-based TLC algorithm.

Note that any TLC algorithm attempting to maximize traffic flow needs as input - the queue lengths along the individual lanes leading to the intersection. However, precise information about the queue lengths on the individual lanes is hard to obtain in practice, while aggregate information can be obtained using thresholds. For instance, one could use thresholds, say L_1 and L_2, to infer whether or not the traffic congestion on a given lane is in the low (below L_1), medium (between L_1 and L_2) or high (above L_2) range, respectively. The inter-relation between the two problems described above arises from the fact that the thresholds (such as L_1 and L_2) play a crucial role and a problem is to select these thresholds optimally.

Reinforcement learning algorithms are model-free and easy to implement. However, their application to a problem involving high-dimensional state spaces, as is the case with the traffic light control problem here, is nontrivial. Function approximation-based approaches were discussed in Chapter 11 and we specifically make use of Q-learning with a linear function approximation architecture (see Section 11.5.3) to solve the first problem - that of designing a TLC algorithm

S. Bhatnagar et al.: Stochastic Recursive Algorithms for Optimization, LNCIS 434, pp. 243–255.
springerlink.com

that maximizes traffic flow in the long term. The crux of this application is the choice of features used in the Q-learning-based TLC algorithm. We first describe the Q-learning-based TLC algorithm proposed in [4] with its choice of features and then a TLC algorithm that also uses Q-learning with function approximation but with an enhanced feature selection scheme. The enhancement in the feature selection scheme arises from an intelligent combination of the state and action features, as opposed to keeping the state and action features separate, which is a case treated in [4].

The TLC algorithm of [4] is based on certain graded thresholds and the threshold values used in this algorithm are considred fixed and not necessarily optimal. The problem is one of finding an optimal feedback policy within a class of parameterized feedback policies with the underlying parameter, in general, being a vector of the various thresholds. It is thus necessary to design an online algorithm to tune the thresholds on queue lengths and/or elapsed times and thereby tune the parameter of the associated feedback policy. For solving this problem, we consider the one-measurement SPSA algorithm with Hadamard matrix perturbations, described in Chapter 5. This algorithm is easily implementable, converges to the optimal threshold values and most importantly works for any graded threshold-based TLC algorithm. This algorithm is combined with several graded threshold-based TLC algorithms, with each combination resulting in interesting consequences. For instance, when applied together with RL (such as with the Q-learning-based TLC algorithms), our threshold tuning algorithm results in tuning the associated parameterized state-representation features. In the context of RL, developing algorithms for feature adaptation is currently a hot area of research in itself.

The chapter is organized as follows:

- In Section 13.2.1, we formulate the traffic light control problem as a Markov Decision Process (MDP).
- In Section 13.2.2, we describe the Q-learning-based algorithm for solving the above problem.
- In Section 13.3, we formulate the average cost problem for finding the optimal threshold values in any graded threshold-based TLC algorithm and also describe the threshold tuning algorithm based on SPSA for solving the same.
- In Section 13.3.2, we combine the above algorithm with three different graded threshold-based TLC algorithms (including the Q-learning TLC above) and discuss interesting consequences that arise from these combinations.
- In Section 13.3.3, we discuss some of the performance simulation results of the threshold tuning algorithm.

13.2 Q-Learning for Traffic Light Control

13.2.1 Traffic Control Problem as an MDP

We consider a road network with m junctions, $m > 1$. Each junction has multiple cross-roads with each road having j lanes. Our algorithms require a description of states, actions and costs. The state is the vector of queue lengths and the elapsed times since the signal turned red on those lanes that have a traffic signal at the various junctions in the network. Control decisions are made by a centralized controller that receives the state information from the various lanes and makes decision on which traffic lights to switch green during a cycle. This decision is then relayed back to the individual junctions. We assume for simplicity that there are no propagation and feedback delays. The elapsed time counter for a lane with green signal stays at zero till the time the signal turns red. For a network with a total of N signaled lanes, the state at time n is given by

$$s_n = (q_1(n), \ldots, q_N(n), t_1(n), \ldots, t_N(n))^T,$$

where $q_i(n)$ is the queue length on lane i at time n and $t_i(n)$ is the elapsed time for the red signal on lane i at time n.

The actions a_n comprise the sign configuration (which feasible combination of traffic lights to switch) in the m junctions of the road network and have the form: $a_n = (a_1(n), \ldots, a_m(n))^T$, where $a_i(n)$ is the sign configuration at junction i in the time slot n. We consider only sign configurations that are feasible in the action set and not all possible red-green combinations of traffic lights (which would grow exponentially with the number of traffic lights). Thus, the action set $A(s_n) = \{$feasible sign configurations in state $s_n\}$.

The cost function here has two components. The first component is the sum of the queue lengths of the individual lanes and the second component is the sum of the elapsed times since the signal turned red on the lanes on which the signal is red. The elapsed time on lanes for which the signal is green is zero. The idea here is to regulate the flow of traffic so as to minimize the queue lengths, while at the same time ensure fairness so that no lane suffers from being red for a long duration. Further, lanes on the main road are given higher priority over others. We achieve prioritization of main road traffic as follows: Let I_p denote the set of indices of lanes whose traffic should be given higher priority. Then the single-stage cost $k(s_n, a_n)$ has the form

$$\begin{aligned} k(s_n, a_n) = {} & r_1 * \left(\textstyle\sum_{i \in I_p} r_2 * q_i(n) + \sum_{i \notin I_p} s_2 * q_i(n)\right) \\ & + s_1 * \left(\textstyle\sum_{i \in I_p} r_2 * t_i(n) + \sum_{i \notin I_p} s_2 * t_i(n)\right), \end{aligned} \tag{13.1}$$

where $r_i, s_i \geq 0$ and $r_i + s_i = 1, i = 1, 2$. Further, $r_2 > s_2$. Thus, lanes in I_p are assigned a higher cost and hence a cost optimizing strategy must assign a higher priority to these lanes in order to minimize the overall cost.

13.2.2 The TLC Algorithm

Recall the Q-learning algorithm from Chapter 11 with the following update rule:

$$Q_{n+1}(i,a) = Q_n(i,a) + a(n)\left(r(i,a) + \gamma \min_{v \in A(Y_n(i,a))} Q_n(Y_n(i,a),v) - Q_n(i,a)\right). \tag{13.2}$$

Here, $r(i,a)$ is the single-stage cost when state is i and a feasible action a is chosen. Further, $Y_n(i,a)$ is a simulated next state when the current state is i and action $a \in A(i)$ is chosen. One could use the above recursion to find the optimal Q-values and hence, the optimal sign configuration policy for the traffic control MDP described in the previous section.

However, the Q-learning algorithm (13.2) requires a look-up table to store the Q-values for every possible (s,a)-tuple. While this is useful in small state and action spaces, it becomes computationally expensive for larger road networks involving multiple junctions. For instance, in the case of a small road network (e.g. a two-junction corridor) say with 10 signalled lanes, with each lane accommodating 20 vehicles, the number of state-action tuples (and hence the size of the $Q(s,a)$ lookup table) is of the order of 10^{14}. This leads to an extraordinary computation time and space as lookup table representation requires a lot of memory and moreover, the lookup and update operation of $Q(s,a)$ for any (s,a) tuple is expensive because of the number of (s,a)-tuples. For instance, in the case of the ten-lane example above, (13.2) would correspond to a system of 10^{14} equations needed to update $Q_n(i,a)$ for each feasible (i,a)-tuple once. The situation is aggravated when we consider larger road networks such as a grid or a corridor with several junctions, as the sizes of the state and action spaces blow up exponentially. To alleviate this problem of curse of dimensionality, we incorporate feature-based methods. These methods handle the above problem by making computational complexity manageable.

Feature-based methods were introduced in Section 11.5 of Chapter 11. Specifically, the linear function approximation architecture for Q-learning algorithm was described in Section 11.5.3. Recall that the idea there is to approximate the Q-value function $Q(s,a)$ as

$$Q(s,a) \approx \theta^T \sigma_{s,a}, \tag{13.3}$$

where $\sigma_{s,a}$ is a d-dimensional feature (column) vector, with d significantly less in comparison to the cardinality of the set of feasible state-action tuples (s,a). Also, in (13.3) θ is a tunable parameter whose dimension is the same as that of $\sigma_{s,a}$. This approximation thus results in significant complexity gains both in terms of space as well as time. The algorithm QTLC-FA is the function approximation variant of

Q-learning used in the context of traffic control MDP. QTLC-FA thus updates the parameter θ, a d-dimensional quantity using the following update rule (similar to (11.65):

$$\theta(n+1) = \theta(n) + \alpha(n)\sigma_{s_n,a_n}(k(s_n,a_n) + \gamma \min_{v \in A(s_{n+1})} \theta(n)^T \sigma_{s_{n+1},v} - \theta(n)^T \sigma_{s_n,a_n}), \tag{13.4}$$

where θ_0 is set arbitrarily. In (13.4), the action a_n is chosen in state s_n according to $a_n = arg\min_{v \in A(s_n)} \theta_n^T \sigma_{s_n,v}$. Thus, instead of solving a system in $|S \times A(S)|$ variables, we solve here a system in only d variables. Here $S \times A(S) \stackrel{\triangle}{=} \{(i,a) \mid i \in S, a \in A(i)\}$. For instance, in the case of a (3x3)-grid road network, it can be seen that while $|S \times A(S)| \sim 10^{101}$, d is only about 200. This results in significant speed up in the computation time when feature-based representations are used.

A Basic Feature Selection Procedure

Note that σ_{s_n,a_n} are state-action features. The features are chosen based on the queue lengths and elapsed times of each signalled lane of the road network. A basic method for selecting features is to set the features in the following manner: Let

$$\sigma_{s_n,a_n} = (\sigma_{q_1(n)}, \ldots, \sigma_{q_N(n)}, \sigma_{t_1(n)}, \ldots, \sigma_{t_N(n)}, \sigma_{a_1(n)}, \ldots, \sigma_{a_m(n)})^T$$

where

$$\sigma_{q_i(n)} = \begin{cases} 0 & \text{if } q_i(n) < L_1 \\ 0.5 & \text{if } L_1 \le q_i(n) \le L_2 \\ 1 & \text{if } q_i(n) > L_2 \end{cases} \tag{13.5}$$

$$\sigma_{t_i(n)} = \begin{cases} 0 \text{ if } t_i(n) \le T_1 \\ 1 \text{ if } t_i(n) > T_1. \end{cases}$$

Further $\sigma_{a_1(n)}, \ldots, \sigma_{a_m(n)}$ correspond to the actions or sign configurations chosen at each of the m junctions. As before, N is the total number of lanes (inclusive of all junctions) in the network. L_1 and L_2 are thresholds on the queue lengths and T_1 is a threshold on the elapsed time. Note that the parameter θ_n has dimension the same as that of σ_{s_n,a_n}. Again the advantage here is that instead of updating the Q-values for each feasible (s,a)-tuple as before, one estimates these according to the parametrization (13.3).

An advantage in using the above features is that one does not require full information on the queue lengths or the elapsed times. Thresholds L_1 and L_2 can be marked on the lanes and used to estimate low (below L_1), medium (between L_1 and L_2) or high (above L_2) traffic. Likewise the elapsed time can be categorized as being below the threshold (T_1) or above it. While precise queue length information is often hard to obtain, a characterization of traffic at any time as low, medium or high is easier.

13.2.3 Summary of Experimental Results

Results of the simulation experiments of the various TLC algorithms described above, using the Green Light District traffic simulation software [5] are presented in [4]. In particular, the performance of the Q-learning-based TLC algorithm QTLC-FA was compared against various existing TLC algorithms - Fixed timing, Longest queue and SOTL from [2] as well as the Q-learning-based TLC algorithm from [1] that uses full state representation. Four different road networks - a two-junction corridor, a 2x2-grid network, a 3x3-grid network and an eight-junction corridor, were considered for comparing the above TLC algorithms. Using the average junction waiting times (AJWT), i.e., the average time that a user waits at a junction and total arrived road users (TAR), i.e., the number of road users who have completed their trips, as the performance metrics, it was seen there that QTLC-FA consistently shows the best results in all the four road networks studied. QTLC-FA was seen to be easily implementable on larger road network scenarios, and requires much less computation, whereas the algorithm from [1] was implementable only on a two junction corridor and did not scale to larger networks because of the exponential increase in computational complexity with more lanes and junctions. Further, it was also observed that the transient period, i.e., the initial period when QTLC-FA is tuning its parameters before stabilizing on a policy, is only a few cycles and hence, QTLC-FA converges rapidly to a good sign configuration policy.

13.3 Threshold Tuning Using SPSA

The Q-learning-based TLC algorithm as well as the ones that we describe subsequently is based on queue-length thresholds L_1 and L_2 and the elapsed time threshold T_1. More such thresholds may be chosen in practice. However, an increase in the number of thresholds also results in an increase in the computational complexity of the scheme. It is generally observed that this choice of two thresholds (L_1 and L_2) for the queue lengths and one threshold (T_1) for the elapsed time works well. However, a question that remains is how should these threshold parameters be chosen. Ideally, one would want to select them optimally, in a way that optimizes a certain objective criterion. This the goal we consider now.

Thus, our aim here is to find an optimal value for the parameter vector $\theta = (L_1, L_2, T_1)^T$ that minimizes the long run average cost objective. In other words, the aim of the tuning algorithm is to find a θ that minimizes

$$J(\theta) = \lim_{l \to \infty} \frac{1}{l} \sum_{j=0}^{l-1} k(s_j, a_j), \tag{13.6}$$

where $k(s_j, a_j)$ denotes the single stage cost (13.1).

The actions a_j are assumed to be governed by one of the policies that we present below, that in turn will be parameterized by the threshold parameter θ. While it is desirable to find a $\theta^* \in C$ that minimizes $J(\cdot)$, it is, in general, very difficult to achieve a global minimum. We use therefore a local optimization method for which one needs to evaluate $\nabla J(\theta) \equiv (\nabla_1 J(\theta), \nabla_2 J(\theta), \nabla_3 J(\theta))^T$, for all algorithms.

Because of the long-run average nature of the objective, we use a multi-timescale stochastic approximation procedure (cf. Chapter 3).

13.3.1 The Threshold Tuning Algorithm

The threshold tuning algorithm estimates the gradient of the objective function $\nabla_\theta J(\theta)$ using a one-sided SPSA-based estimate with Hadamard matrix perturbations. Let $\triangle(n) = (\triangle_1(n), \triangle_2(n), \triangle_3(n))^T, n \geq 1$ be the perturbation vectors obtained using the Hadamard matrix construction described in Section 5.5.2.2. The recursive update equation for θ is then given by

$$L_1(n+1) = \pi_1\left(L_1(n) - a(n)\left(\frac{\tilde{Z}(nL)}{\delta\triangle_1(n)}\right)\right), \tag{13.7}$$

$$L_2(n+1) = \pi_1\left(L_2(n) - a(n)\left(\frac{\tilde{Z}(nL)}{\delta\triangle_2(n)}\right)\right), \tag{13.8}$$

$$T_1(n+1) = \pi_2\left(T_1(n) - a(n)\left(\frac{\tilde{Z}(nL)}{\delta\triangle_3(n)}\right)\right). \tag{13.9}$$

In the above,

- $L_1(n), L_2(n), T_1(n)$ denote the n-th updates of the thresholds L_1, L_2 and T_1, respectively.
- $\tilde{Z}(nL)$ represents the cost function averaging term obtained by accumulating the single stage cost over L cycles and is specific to the TLC algorithm being used to obtain the sign configuration policy on the faster timescale. These updates will be explained in the TLC algorithms in the next section.
- $L \geq 1$ is a fixed parameter which controls the rate of update of θ in relation to that of $\tilde{Z}$. This parameter allows for accumulation of updates to $\tilde{Z}$ for L iterations in between two successive θ updates. It is usually observed that allowing L to be greater than 1 improves the algorithm's performance.
- $\delta > 0$ is a given small constant.

- The projection operators $\pi_i : \mathbb{R} \to \mathbb{R}, i = 1,2$ are defined as follows: For any $x \in \mathbb{R}$, $\pi_1(x) \triangleq \min(\max(L_{\min},x),L_{\max})$ and $\pi_2(x) \triangleq \min(\max(T_{\min},x),T_{\max})$, respectively. Here, $L_{\min},L_{\max}$ are the bounds on the thresholds L_1 and L_2. Similarly, $T_{\min},T_{\max}$ are the bounds on the threshold T_1.

The complete algorithm is described as under.

Algorithm 13.1. The threshold tuning algorithm

Input:

- R, a large positive integer; θ_0, initial parameter vector; $\delta > 0$; $\triangle$;
- UpdateTheta(), the stochastic update rule discussed in (13.9)
- Simulate(θ) $\to X$: the function that performs one time-step of the road traffic simulation and output the single-stage cost value $k(\hat{s}_n,\cdot)$ (cf. (13.1))
- UpdateAverageCost(): the function that updates the average cost estimate $\tilde{Z}(\cdot)$ used in (13.9) and is specific to the TLC-algorithm.
- UpdateTheta(): the function that updates the threshold parameter θ according to (13.9).

Output: $\theta^* \triangleq \theta_R$.

$\theta \leftarrow \theta_0, n \leftarrow 1$
loop
  $\hat{X} \leftarrow$ Simulate($\theta + \delta\Delta$)
  UpdateAverageCost()
  if $n \,\%\, L = 0$ **then**
    UpdateTheta()
  end if
  $n \leftarrow n+1$
  if $n = R$ **then**
    Terminate with θ.
  end if
end loop

13.3.2 *Traffic Light Control with Threshold Tuning*

Here we describe two TLC algorithms, each based on graded thresholds L_1,L_2 and T_1. While the first algorithm is based on Q-learning and incorporates an enhanced feature selection scheme as compared to [4], the second is a simple priority-based TLC algorithm. The threshold tuning algorithm described in the previous section is combined with each of these TLC algorithms using multi-timescale stochastic approximation. The threshold parameter $\theta = (L_1,L_2,T_1)^T$ is tuned on the slower timescale while the policy is obtained on the faster timescale using one of the TLC algorithms outlined below.

13.3.2.1 Q-Learning TLC with an Enhanced Feature Selection Scheme

It turns out that with QTLC-FS, the curse of dimensionality cannot be fully controlled for larger networks with high-dimensional states. To alleviate this problem, the QTLC-FA algorithm described above made use of feature-based representations and function approximation. We further improve the performance of the QTLC-FA algorithm by incorporating a novel feature selection scheme that uses priorities to intelligently combine the state and action features. In the QTLC-FA algorithm, the feature vector contained a bit each for the congestion estimate, elapsed time estimate and the sign configuration portion, respectively, for each lane of the road network. While each of these attributes is important, the approximation architecture used in the QTLC-FA algorithm did not take into account the dependence between features. We incorporate dependence between the state and the action features while using graded thresholds and also reduce the dimension of the feature vector by more than half as compared to that for the QTLC-FA algorithm. We denote the Q-learning-based TLC with the enhanced feature selection scheme by QTLC-FA-NFS.

The various aspects of the QTLC-FA-NFS algorithm, for instance the function approximation architecture, the Q-learning update rule remain the same as that described for the QTLC-FA algorithm. The key difference is in the choice of features which is explained below.

The features in the QTLC-FA-NFS algorithm are chosen as described below: Let

$$\sigma_{s_n,a_n} = (\sigma_1(n), \ldots, \sigma_K(n))^T, \tag{13.10}$$

where the procedure for selection of feature value $\sigma_i(n)$ corresponding to lane i is explained in Table. 13.1.

Table 13.1 Feature selection ($\sigma_i(n)$) table for lane i

State	Action	Feature
$q_i(n) < L_1$ and $t_i(n) < T_1$	RED	0
	GREEN	1
$q_i(n) < L_1$ and $t_i(n) \geq T_1$	RED	0.2
	GREEN	0.8
$L_1 \leq q_i(n) < L_2$ and $t_i(n) < T_1$	RED	0.4
	GREEN	0.6
$L_1 \leq q_i(n) < L_2$ and $t_i(n) \geq T_1$	RED	0.6
	GREEN	0.4
$q_i(n) \geq L_2$ and $t_i(n) < T_1$	RED	0.8
	GREEN	0.2
$q_i(n) \geq L_2$ and $t_i(n) \geq T_1$	RED	1
	GREEN	0

The feature selection scheme is graded and assigns a value for each lane based on whether the queue length on the lane is below L_1, is between L_1 and L_2, or is above

L_2, on whether the elapsed time is below T_1 or above it and also on whether the sign configuration indicates a RED or GREEN light for the lane. For instance, if both queue length and elapsed time are above the "highest" threshold levels for the lane, then an action of GREEN would result in a feature value of 0 and an action of RED would result in the value 1. In essence, this choice indicates that the TLC algorithm should attempt to switch this lane to green. On the other hand, if both queue length and elapsed time are below the "lowest" threshold level for the lane, then the feature value chosen is just the opposite, i.e., a 0 for RED and 1 for GREEN, implying that it is better to keep this lane red. The feature values corresponding to other decision choices are appropriately graded.

The threshold tuning algorithm (13.9) is combined with the sign configuration policy from QTLC-FA-NFS through multiple time-scale recursions. The recursions on the faster timescale in the case of QTLC-FA-NFS-TT algorithm are as follows: Let $\{\tilde{s}_n, n \geq 0\}$ denote a state-valued process that depends on both the tunable policy as well as the tunable parameter $\tilde{\theta}_l, l \geq 0$, where $\tilde{\theta}_l = \theta_n + \delta\Delta(n)$ for $n = \left[\frac{l}{L}\right]$, and updates of $\theta_n \equiv (L_1(n), L_2(n), T_1(n))^T$ are governed according to (13.9). For $m = nL, \ldots, (n+1)L-1$,

$$\tilde{Z}(m+1) = \tilde{Z}(m) + b(n)\left(k(\tilde{s}_m, \hat{a}_m) - \tilde{Z}(m)\right), \tag{13.11a}$$

$$\omega(m+1) = \omega(m) + b(n)\sigma_{\tilde{s}_m,\hat{a}_m}(k(\tilde{s}_m, \hat{a}_m) + \gamma \min_{v \in \mathscr{A}(\tilde{s}_{m+1})} \omega(m)^T \sigma_{\tilde{s}_{m+1},v} - \omega(m)^T \sigma_{\tilde{s}_m,\hat{a}_m}). \tag{13.11b}$$

The step-size sequences $a(n)$ and $b(n), n \geq 0$ satisfy the standard assumptions for multi-timescale algorithms, i.e.,

$$\sum_n a(n) = \sum_n b(n) = \infty, \sum_n (a(n)^2 + b(n)^2) < \infty, a(n) = o(b(n)).$$

The action $\hat{a}_m$ in (13.11a)-(13.11b) is chosen to be the one that minimizes $\omega_m^T \sigma_{\tilde{s}_m,v}$ over all $v \in \mathscr{A}(\tilde{s}_m)$.

Note that one could combine the QTLC-FA algorithm with the threshold tuning algorithm in a similar manner and we denote the resulting multi-timescale algorithm by QTLC-FA-TT. The difference between QTLC-FA-TT and QTLC-FA-NFS-TT is that the underlying sign configuration policy is derived from QTLC-FA for the former and QTLC-FA-NFS for the latter. The rest of the algorithm, including the update rule for the faster recursion (13.11a)-(13.11b), hold for QTLC-FA-TT as well.

13.3.2.2 Priority-Based TLC

The sign configuration policy is a graded threshold-based policy that assigns different priorities to different policy levels. The thresholds here are on the queue lengths (say L_1 and L_2) and elapsed times since the last switch over of lights to red (say T_1) on individual lanes. The cost assigned to each lane is decided based on whether the queue length on that lane is below L_1, is between L_1 and L_2, or is above L_2 at any instant and also on whether the elapsed time is below T_1 or above it. For instance, if both queue length and elapsed time are above the "highest" threshold levels (L_2 and T_1, respectively) on a given lane, then the policy assigns the highest priority to that lane. The priority assignment for any lane i of the road network based on the queue length q_i and elapsed time t_i is shown in Table. 13.2. The policy then selects the sign configuration with the maximum (over all feasible sign configurations) sum of lane priority values. In essence, the TLC algorithm flushes the traffic on lanes with long waiting queues, while also giving higher priority to lanes that have been waiting on a red signal for a long time. This helps to combine efficiency with fairness.

Table 13.2 Priority assignment for each lane in the TLC policy

Condition	Priority value
$q_i < L_1$ and $t_i < T_1$	1
$q_i < L_1$ and $t_i \geq T_1$	2
$q_i \geq L_1$ and $q_i < L_2$ and $t_i < T_1$	3
$q_i \geq L_1$ and $q_i < L_2$ and $t_i \geq T_1$	4
$q_i \geq L_2$ and $t_i < T_1$	5
$q_i \geq L_2$ and $t_i \geq T_1$	6

As with the previous TLC algorithms, we combine the threshold tuning algorithm (13.9) with PTLC to obtain the PTLC-TT algorithm. The state-valued process $\{\hat{s}_n, n \geq 0\}$ in this case under the priority-based policy described above depends on the tunable parameter sequence $\hat{\theta}_l = \theta_n + \delta\Delta(n), n \geq 0$, where $\theta_n \equiv (L_1(n), L_2(n), T_1(n))^T, n \geq 0$ are updated according to (13.9). The faster timescale recursions here are given as follows: For $m = nL, \ldots, (n+1)L - 1$,

$$\tilde{Z}(m+1) = \tilde{Z}(m) + b(n)(k(\hat{s}_m, \hat{a}_m) - \tilde{Z}(m)). \tag{13.12}$$

The action $\hat{a}_m$ above is selected in state $\hat{s}_m$ based on the priority assignment policy (described above), i.e., select the sign configuration that has the maximum sum of priority values (where the maximum is over all feasible sign configurations) and switch the lanes in the chosen sign configuration to green.

The convergence analysis of the threshold tuning algorithm under standard assumptions proceeds along the lines of the one-measurement SPSA algorithm with Hadamard matrix-based perturbations, discussed in Section 5.5.5. The reader is referred to Theorem 5.11, which provides a complete proof of convergence of the one-measurement SPSA algorithm with Hadamard Matrix Perturbations.

13.3.3 Summary of Experimental Results

Performance of the threshold tuning algorithm, described in Section 13.3, was studied in conjunction with three TLC algorithms that incorporate graded thresholds. These include the Q-learning based algorithms - QTLC-FA and QTLC-FA-NFS, and the priority-based scheme PTLC, respectively. Comparisons drawn were between the tuned variants of the TLC algorithms against their counterparts that involved fixed thresholds (no tuning). We show here the results of some representative experiments on a ten-junction corridor network. The ten-junction corridor consists of 22 edge nodes (where traffic is generated), 10 junctions with traffic lights, 31 roads, with each being 4 lanes wide and when full can house upto 1500 vehicles. The cardinality of the state-action space in this case is of the order of 10^{90}.

Figures 13.2(a) – 13.2(b) show plots comparing PTLC, QTLC-FA and QTLC-FA-NFS algorithms with their tuning counterparts on a ten-junction corridor. It can be observed from the results that incorporating the threshold tuning algorithm results in significant gains for all the TLC algorithms, with the QTLC-FA-NFS algorithm showing the best overall performance. Further, the parameter θ was also seen to converge to the optimal threshold value for all the TLC algorithms.

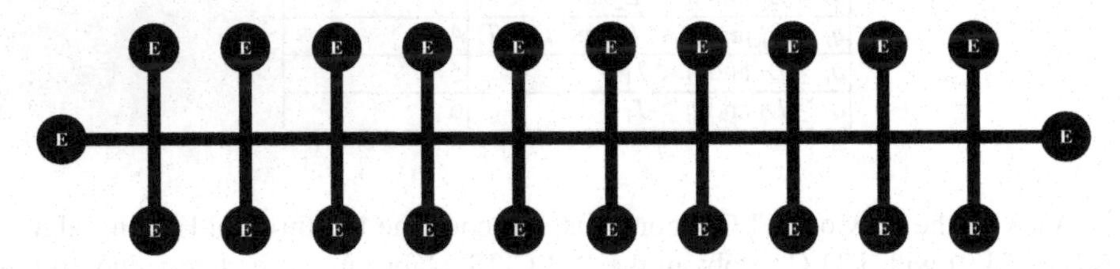

Fig. 13.1 A Ten-Junction Corridor Network - used for our experiments

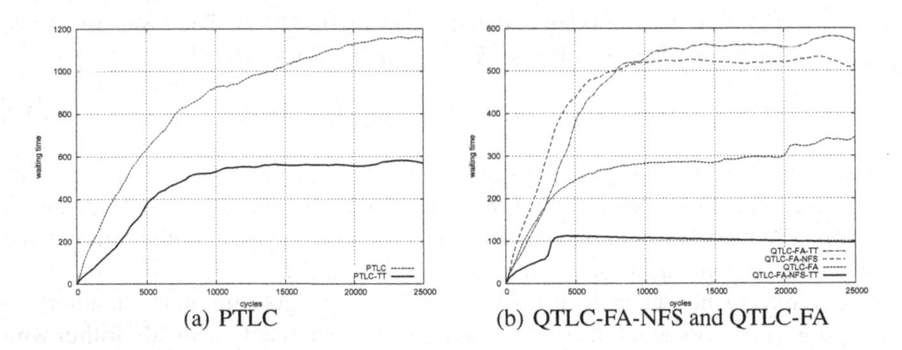

(a) PTLC (b) QTLC-FA-NFS and QTLC-FA

Fig. 13.2 Performance Comparison of TLC Algorithms with their tuning counterparts on a Ten-Junction Corridor Network

13.4 Concluding Remarks

In this chapter, we applied reinforcement learning and simultaneous perturbation methods to the problem of traffic signal control. We described Q-learning with function approximation for traffic light control. Later, we also studied an application of one-measurement SPSA with Hadamard matrix perturbations for finding the optimal threshold values in any graded threshold-based TLC algorithm. The combination of the threshold tuning algorithm with the Q-learning-based TLC algorithms as well as a simple priority-based scheme was found to result in significant performance improvements, in comparison to the TLC algorithms without tuning.

References

1. Abdulhai, B., Pringle, R., Karakoulas, G.: Reinforcement learning for true adaptive traffic signal control. Journal of Transportation Engineering 129, 278 (2003)
2. Cools, S., Gershenson, C., D'Hooghe, B.: Self-organizing traffic lights: A realistic simulation. In: Advances in Applied Self-organizing Systems, pp. 41–50 (2008)
3. Prashanth, L.A., Bhatnagar, S.: Threshold tuning using stochastic optimization for graded signal control. Tech. rep., Stochastic Systems Lab, IISc (2012), `http://stochastic.csa.iisc.ernet.in/www/research/files/IISc-CSA-SSL-TR-2012-1.pdf`
4. Prashanth, L., Bhatnagar, S.: Reinforcement learning with function approximation for traffic signal control. IEEE Transactions on Intelligent Transportation Systems 12(2), 412–421 (2011)
5. Wiering, M., Vreeken, J., van Veenen, J., Koopman, A.: Simulation and optimization of traffic in a city. In: IEEE Intelligent Vehicles Symposium, pp. 453–458 (2004)

Chapter 14
Communication Networks

14.1 Introduction

Simultaneous perturbation methods that we have discussed in the earlier chapters have been found to be useful in the area of communication networks where often the performance metrics depend on certain parameters and one is interested in finding the optimal parameters. Many times, one is interested in optimizing steady-state performance in these settings. Hence, the simultaneous perturbation approaches for the long-run average cost objective play a significant role. Apart from being efficient and scalable, a distinct advantage with these algorithms is that they are independent of the technology and protocols used and hence are widely applicable over a large range of settings.

We consider mainly three different problems in the area of communication networks here. The first problem deals with the random early detection (RED) adaptive queue management scheme for the Internet. The original paper on RED by Floyd [10] proposed a fixed set of parameters for the scheme regardless of the network and traffic conditions. A problem with traditional RED that has been reported in several papers is that of massive queue oscillations. Using our techniques, it is observed that these oscillations dramatically diminish to almost zero variances in the average queue sizes. This problem is dealt with in detail in Section 14.2, and is based on [16, 6].

The second problem that we consider here deals with the problem of finding the optimal retransmission probabilities for the slotted Aloha multi-access communication system [2]. The slotted Aloha protocol prescribes a fixed probability of retransmission for packets that are involved in a collision in a previous slot. In particular, colliding packets attempt retransmission in all subsequent slots after collision has taken place with a certain 'retransmission' probability until successful transmission. The standards, however, specify that these probabilities be held fixed regardless of the network and traffic conditions (including the number of transmitting nodes). We study applications of our techniques on this problem

S. Bhatnagar et al.: Stochastic Recursive Algorithms for Optimization, LNCIS 434, pp. 257–280.
springerlink.com

in Section 14.3. The approach and methodology here follow [15] that is in turn based on [4].

Finally, the third problem deals with the issue of dynamic pricing in the Internet. Pricing is regarded as an effective tool to control congestion and achieve quality of service (QoS) provisioning for multiple differentiated levels of service. We consider the problem of pricing for congestion control in the case of a network of nodes with multiple queues and multiple grades of service, and develop certain graded feedback control policies over which using the simultaneous perturbation algorithms, one obtains the optimal such policies. This part is mainly based on [26]. An important feature when using the approaches developed in the earlier chapters is that they almost always result in significant performance gains over other approaches that do not use these methods. Section 14.5 then provides the concluding remarks.

14.2 The Random Early Detection (RED) Scheme for the Internet

This section deals with the application of the simultaneous perturbation approaches to the RED flow control scheme and is based on [16, 6].

14.2.1 Introduction to RED Flow Control

Enhancing network performance has emerged as a major challenge for today's internet applications given the large volumes of traffic that flows. Various adaptive queue management techniques have been proposed to tackle growing congestion. An important scheme in this direction is the random early detection (RED) [10]. RED uses a weighted average queue length metric as a measure of congestion. Further, it uses two threshold levels, a maximum and a minimum threshold ($\max_{th}$ and $\min_{th}$), to segregate buffers into three regions of low, medium and high congestion intervals. The weighting parameter (w_q) used to compute the weighted average is typically chosen to be much less than 1. Further, a certain "max-probability" parameter ($\max_p$) is chosen. The basic idea in this scheme is as follows: The weighted average queue length is computed each time a packet arrives. This is then compared with the two threshold levels to determine whether the level of congestion is in the aforementioned low, medium or high ranges. If the congestion level is inferred as low (i.e., the weighted average queue length is below the minimum threshold), the arriving packet is not dropped. On the other hand, if the congestion level is high (i.e., the weighted average queue length is above the maximum threshold), the arriving packet is dropped. Finally, if the congestion level is found to be in the medium range, packets are dropped with a certain probability parameter p that increases linearly with the number of packets dropped (until it reaches $\max_p$).

The idea behind using a weighted average queue length as opposed to instantaneous queue length for detecting congestion is to reduce oscillations that would otherwise result in the system. On the other hand, the regular average queue length (i.e., the one computed by assigning the same weight to each observation, and which is the inverse of the total number of observations till that instant) is far less susceptible to large variations in instantaneous queue length than the weighted average queue length (when the number of packets over which the average has been taken is fairly large). Nevertheless, a problem that has been consistently observed with RED is of large oscillations in the weighted average queue behaviour. As such, the parameters w_q, $\min_{th}$, $\max_{th}$ and $\max_p$ have been considered fixed in the original scheme regardless of the network and traffic conditions. This results in poor performance of the scheme. Hence, one needs to tune the parameters in a way as to achieve optimal performance. Various techniques have been proposed for tuning the parameters and many of them are heuristic in nature [9]. Simultaneous perturbation-based approaches have been proposed in [24] and [16]. In [24], a robust SPSA update is used, where the idea is that in order to reduce oscillations in the scheme, one uses the sign of the increment in the update rather than the increment itself. While this results in an interesting alternative, it still does not fully remove the oscillations in the scheme. The algorithms proposed in [16], on the other hand, are geared towards solving a stochastic constrained optimization problem by using the barrier and penalty function methods and incorporate Newton-based updates. The latter schemes are seen to dramatically reduce the queue oscillations in RED. Our treatment here is based entirely on [16]. A proof of convergence of these algorithms has been provided in [6].

14.2.2 The Framework

The aim here is to tune the parameters in a way as to stabilize with a high probability the weighted average queue length (q_{av}) near a given target threshold denoted Q^*. The weighted average queue length as such evolves in the following manner: Let $q_{av}(n)$ denote the weighted average queue length as seen by the nth arriving packet. Then,

$$q_{av}(n) = (1 - w_q) q_{av}(n-1) + w_q q_{inst}(n),$$

where $0 < w_q < 1$ is the weighting factor and $q_{inst}(n)$ is the instantaneous queue length seen by the nth arriving packet. Let θ denote the 4-dimensional parameter $\theta = (w_q, max_p, min_{th}, max_{th})^T$. The optimization problem can now be cast as follows:

$$\min_{\theta} \ f(\theta) = (Eq_{av} - Q^*)^2$$
$$\text{s.t} \quad P[q_{av} \leq Q^*] \geq \alpha. \tag{14.1}$$

Here, q_{av} denotes the steady state average queue length and Eq_{av} denotes its expected value. The objective function aims at bringing the mean steady-state average queue length (Eq_{av}) near the target Q^*. The constraint specifies that the probability of q_{av} being less than or equal to Q^* should not drop below a given $\alpha \in (0,1)$ that in turn can be chosen in a way as to comply with any specific demands on the traffic. Through a proper choice of Q^*, one can maximize throughput while at the same time minimize delays and reduce packet losses, thereby controlling congestion and providing a good quality of service (QoS) to the various flows in the network. We make the following assumptions:

Assumption 14.1. Both the mean $E(q_{av})$ and the variance $Var(q_{av})$ of the average queue length are twice differentiable and have bounded third derivatives w.r.t. the parameter θ.

Assumption 14.2. $E[q_{av}]$ has a *nonlinear* dependency on the RED parameters. Hence, $E[q_{av}] = f_1(\theta)$, where f_1 is some nonlinear function.

Assumption 14.3. The average queue length, q_{av} for given θ has a normal distribution with mean $E(q_{av})$ and variance $Var(q_{av})$ i.e., $q_{av} \sim N(Eq_{av}, Var(q_{av}))$.

Assumption 14.1 is a technical requirement to utilize second-order parameter updation techniques directly. As a consequence of Assumption 14.2, standard nonlinear programming techniques can be applied to the optimization problem. Assumption 14.3 is also a technical requirement that aids in converting the probabilistic constraint into a deterministic one [22]. Note here that since the weighted average queue length process $q_{av}(n), n \geq 0$ evolves using a fixed weighting parameter w_q, one can expect that under certain conditions, the steady-state average queue length q_{av} will have most of its probability mass concentrated in a narrow range around Eq_{av}. Assumption 14.3 has been made only to help formulate the problem in a standard nonlinear programming framework. The requirement on the distribution of q_{av} being normally distributed may, however, be replaced by the more realistic requirement of the same being distributed as per the truncated normal distribution.

Now note that one can rewrite $f(\theta) = (Eq_{av} - Q^*)^2$ as $f(\theta) = \hat{q}^2 + Q^{*2} - 2Q^*\hat{q}$, where $\hat{q} = Eq_{av}$. Further, the constraint $P(q_{av} \leq Q^*) \geq \alpha$ can also be reduced using Assumption 14.3 as follows:

$$P\left(\frac{q_{av} - Eq_{av}}{\sqrt{Var(q_{av})}} \leq \frac{Q^* - \hat{q}}{\hat{\sigma}}\right) \geq \alpha, \tag{14.2}$$

where $\hat{\sigma} = \sqrt{Var(q_{av})} = \sqrt{Eq_{av}^2 - (Eq_{av})^2}$. From Assumption 14.3,

$$\frac{q_{av} - Eq_{av}}{\sqrt{Var(q_{av})}} \sim \text{Normal}(0, 1).$$

Hence, from (14.2),

$$Q^* - \hat{q} \geq \hat{\sigma}\Phi^{-1}(\alpha),$$

where $\Phi^{-1}(\alpha)$ is the inverse of the standard Gaussian c.d.f. evaluated at α. Thus, (14.2) is analogous to

$$\mathscr{C}_3\,\hat{\sigma} + \hat{q} \leq \mathscr{C}_4\ , \text{ or that}$$
$$\mathscr{C}_4 - \mathscr{C}_3\,\hat{\sigma} - \hat{q} \geq 0, \tag{14.3}$$

where $\mathscr{C}_3 = \Phi^{-1}(\alpha)$ and $\mathscr{C}_4 = Q^*$, respectively. Thus, the problem in revised form is the following:

$$\min_{\theta}\,(\hat{q}^2 + \mathscr{C}_1 - \mathscr{C}_2\hat{q})$$
$$\text{s.t } \mathscr{C}_4 - \mathscr{C}_3\hat{\sigma} - \hat{q} \geq 0, \tag{14.4}$$

where $\mathscr{C}_1 = {Q^*}^2$ and $\mathscr{C}_2 = 2Q^*$. Two approaches, the barrier and the penalty function methods, are now incorporated for the solution of this problem. These approaches are first explained below.

14.2.2.1 The Barrier Function Method

This method [12] adds the logarithm of the constraints as the penalty term to the objective function. The minimization problem with the relaxed objective is given as under.

$$\text{Find } \min_{\theta} B(\theta;b) = \hat{q}^2 + \mathscr{C}_1 - \mathscr{C}_2\hat{q} - b\log[\mathscr{C}_4 - \mathscr{C}_3\hat{\sigma} - \hat{q}]. \tag{14.5}$$

One solves the problems (14.5) for a sequence of values for $b = b_k$, $b_k \downarrow 0$ (cf. [19]). A second-order parameter update technique is used for solving (14.5). The gradient and Hessian of $B(\theta;b)$ can be evaluated as below. Representing the 4-dimensional parameter θ as $\theta = (\theta_1, \theta_2, \theta_3, \theta_4)^T$, the gradient $\nabla_\theta B$ of $B(\theta; b)$ is obtained as follows : For $i = 1,2,3,4$ and $b = b_k$,

$$\nabla_{\theta_i} B = 2\hat{q}\hat{q}'_{\theta_i} - \mathscr{C}_2 \hat{q}'_{\theta_i} + b_k \frac{(\mathscr{C}_3 \hat{\sigma}'_{\theta_i} + \hat{q}'_{\theta_i})}{(\mathscr{C}_4 - \mathscr{C}_3 \hat{\sigma} - \hat{q})}, \tag{14.6}$$

Further, the Hessian $\nabla^2_\theta B$ of $B(\theta; b)$ is obtained as follows : For $i, j \in \{1,2,3,4\}$, we have

$$\nabla^2_{\theta_i,\theta_j} B = 2\hat{q}\hat{q}''_{\theta_i,\theta_j} + 2(\hat{q}'_{\theta_i})^2 - \mathscr{C}_2 \hat{q}''_{\theta_i,\theta_j}$$
$$+ b_k \left[\frac{(\mathscr{C}_3 \hat{\sigma}''_{\theta_i,\theta_j} + \hat{q}''_{\theta_i,\theta_j})(\mathscr{C}_4 - \mathscr{C}_3 \hat{\sigma} - \hat{q}) + (\mathscr{C}_3 \hat{\sigma}'_{\theta_i} + \hat{q}'_{\theta_i})^2}{(\mathscr{C}_4 - \mathscr{C}_3 \hat{\sigma} - \hat{q})^2} \right], \tag{14.7}$$

14.2.2.2 The Penalty Function Method

The second approach employed is the penalty method to solve the optimization problem (14.4). Here, the penalty term is taken to be a quadratic function of the constraints [23]. The optimization problem with the relaxed objective obtained afer absorbing the constraints is given by:

$$\min_\theta P(\theta; r) = \hat{q}^2 + \mathscr{C}_1 - \mathscr{C}_2 \hat{q} + \frac{1}{2r}(\mathscr{C}_4 - \mathscr{C}_3 \hat{\sigma} - \hat{q})^2. \tag{14.8}$$

One solves here a sequence of unconstrained minimization problems (14.8) corresponding to values of $r = r_k$, $r_k \uparrow \infty$ (cf. [19]). The gradient $\nabla_\theta P$ is given as follows: For $i = 1,2,3,4$ and $r = r_k$, we have

$$\nabla_{\theta_i} P = 2\hat{q}\hat{q}'_{\theta_i} - \mathscr{C}_2 \hat{q}'_{\theta_i} - \frac{1}{r_k} \left[(\mathscr{C}_4 - \mathscr{C}_3 \hat{\sigma} - \hat{q})(\mathscr{C}_3 \hat{\sigma}'_{\theta_i} + \hat{q}'_{\theta_i}) \right]. \tag{14.9}$$

Also, the Hessian $\nabla^2_\theta P$ of $P(\theta; b)$ is obtained as follows: For $i, j \in \{1,2,3,4\}$, we have

$$\nabla^2_{\theta_i,\theta_j} P = 2\hat{q}\hat{q}''_{\theta_i,\theta_j} + 2(\hat{q}'_{\theta_i})^2 - \mathscr{C}_2 \hat{q}''_{\theta_i,\theta_j}$$
$$- \frac{1}{r_k} \left[(\mathscr{C}_3 \hat{\sigma}''_{\theta_i,\theta_j} + \hat{q}''_{\theta_i,\theta_j})(\mathscr{C}_4 - \mathscr{C}_3 \hat{\sigma} - \hat{q}) - (\mathscr{C}_3 \hat{\sigma}'_{\theta_i} + \hat{q}'_{\theta_i})^2 \right], \tag{14.10}$$

14.2.3 The B-RED and P-RED Stochastic Approximation Algorithms

In order to apply Newton-based algorithms using the barrier and penalty function methods, one requires estimates of the derivatives of quantities such as $\hat{q}$ and $\hat{\sigma}$ that involve steady-state expectations (see Chapter 10). Thus, the relationships of these quantities with θ are not analytically known. A way out is to use multi-timescale stochastic approximation. Let B_i and P_i denote the ith sample observations of the barrier and penalty objective functions obtained either through real network observations or simulation of quantities $\hat{q}$, $\hat{q}^2$ and $\hat{\sigma}$. For given θ and b (resp. r), let B_i (resp. P_i) be i.i.d. We thus need to perform the optimization in (14.5)-(14.8) under only the available sample observations and without any model information being known. The objective functions $B(\theta;b)$ and $P(\theta;r)$ are estimated from the sample observations B_i, P_i, $i = 1,2,\ldots$ as

$$B(\theta;b) = \lim_{n\to\infty} \frac{1}{n}\sum_{i=1}^{n} B_i,\ P(\theta;r) = \lim_{n\to\infty} \frac{1}{n}\sum_{i=1}^{n} P_i.$$

Let $[\theta_{i,\min}, \theta_{i,\max}]$, $\theta_{i,\min} < \theta_{i,\max}$, correspond to the constraint interval for parameter θ_i, $i = 1,\ldots,4$. Thus θ takes values in the set $C \stackrel{\triangle}{=} \prod_{i=1}^{4}[\theta_{i,\min}, \theta_{i,\max}]$. Let $\Gamma_i : \mathscr{R} \to [\theta_{i,\min}, \theta_{i,\max}]$ defined by $\Gamma_i(x) = \max(\min(x, \theta_{i,\max}), \theta_{i,\min})$ denote the projection operator. Let $\{\Delta(n)\}$ and $\{\hat{\Delta}(n)\}$ be two $\{\pm 1\}^4$-valued perturbation sequences with $\Delta(n) = (\Delta_1(n),\ldots,\Delta_4(n))^T$ and $\hat{\Delta}(n) = (\hat{\Delta}_1(n),\ldots,\hat{\Delta}_4(n))^T$, respectively, that are generated using the Hadamard matrix-based construction described in Chapter 5.5.2.1.

Let δ_1, $\delta_2 > 0$ be given small constants. Consider four parallel simulations that are, respectively, governed by parameters $\theta(n) - \delta_1\Delta(n)$, $\theta(n) + \delta_1\Delta(n)$, $\theta(n) - \delta_1\Delta(n) + \delta_2\hat{\Delta}(n)$, and $\theta(n) + \delta_1\Delta(n) + \delta_2\hat{\Delta}(n)$. Let $\{q^-(n)\}$, $\{q^+(n)\}$, $\{q^{-+}(n)\}$ and $\{q^{++}(n)\}$, respectively, denote the instantaneous queue length processes associated with these simulations. Let $\{a(n)\}$, $\{b(n)\}$, $\{c(n)\}$ and $\{d(n)\}$ correspond to four step-size sequences that satisfy

$$\sum_n d(n) = \sum_n b(n) = \sum_n c(n) = \sum_n a(n) = \infty, \tag{14.11}$$

$$\sum_n \left(d(n)^2 + b(n)^2 + c(n)^2 + a(n)^2\right) < \infty, \tag{14.12}$$

$$a(n) = o(c(n)),\ c(n) = o(b(n)) \text{ and } b(n) = o(d(n)). \tag{14.13}$$

14.2.3.1 The B-RED Algorithm

We now describe the B-RED algorithm that incorporates the estimates of the barrier function. For all $w \in \{-,+,-+,++\}$, $i,j = 1,\ldots,4$,

$$Z_q^w(n+1) = (1-d(n))Z_q^w(n) + d(n)q_{av}^w(n) \tag{14.14}$$

$$Z_{q^2}^w(n+1) = (1-d(n))Z_{q^2}^w(n) + d(n)(q_{av}^w(n))^2 \tag{14.15}$$

$$\hat{\sigma}^w(n) = (Z_{q^2}^w(n) - (Z_q^w(n))^2)^{1/2} \tag{14.16}$$

$$\hat{q}_i'(n+1) = (1-b(n))\hat{q}_i'(n) + b(n)G_i(Z_q(n)) \tag{14.17}$$

$$\hat{\sigma}_i'(n+1) = (1-b(n))\hat{\sigma}_i'(n) + b(n)G_i(\hat{\sigma}(n)) \tag{14.18}$$

$$\hat{q}_{j,i}''(n+1) = (1-c(n))\hat{q}_{j,i}''(n) + c(n)H_{j,i}(Z_q(n)) \tag{14.19}$$

$$\hat{\sigma}_{j,i}''(n+1) = (1-c(n))\hat{\sigma}_{j,i}''(n) + c(n)H_{j,i}(\hat{\sigma}(n)) \tag{14.20}$$

$$\theta_i(n+1) = \Gamma_i\left(\theta_i(n) - a(n)(\hat{\lambda}_n)^{-1}\hat{\nabla}_i B\right). \tag{14.21}$$

We now explain the various expressions used above:

- In (14.21), $\hat{\nabla}_i B = 2\hat{q}(n)\hat{q}_i'(n) - \mathscr{C}_2\hat{q}_i'(n) + b(\mathscr{C}_3\hat{\sigma}_i'(n) + \hat{q}_i'(n)) / (\mathscr{C}_4 - \mathscr{C}_3\hat{\sigma}(n) - \hat{q}(n))$ estimated from (14.5) where $\hat{q}(n) = \frac{1}{2}(Z_q^+(n) + Z_q^-(n))$ and $\hat{\sigma}(n) = \frac{1}{2}(\hat{\sigma}^+(n) + \hat{\sigma}^-(n))$.
- In (14.21), $\hat{\lambda}_n$ is obtained as per the procedure of [28] as explained below: Let $\Pi_n \equiv \text{diag}\left[\lambda_{1,n}, \cdots, \lambda_{q,n}, \lambda_{q+1,n}, \cdots, \lambda_{4,n}\right]$, where $\lambda_{i,n}, i = 1, \cdots, 4$ are eigen-values of $\hat{\nabla}^2_{\theta(n)} B$, such that $\lambda_{i,n} > \lambda_{i+1,n}$, $\forall i = 1, \ldots, 4$. Further, $\lambda_{q,n} > 0$, and $\lambda_{q+1,n} \leq 0$, for $q \in \{1, \cdots, 4\}$. Now set $\hat{\lambda}_{q,n} = \eta\lambda_{q-1,n}, \hat{\lambda}_{q+1,n} = \eta\hat{\lambda}_{q,n}, \cdots$, $\hat{\lambda}_{4,n} = \eta\hat{\lambda}_{3,n}$, where $\eta = (\frac{\lambda_{q-1,n}}{\lambda_{1,n}})^{q-2}$. If all $\lambda_{j,n} > 0$, $j \in \{1, \ldots, 4\}$, let $\hat{\lambda}_{j,n} = \lambda_{j,n}$. Now $\hat{\lambda}_n$ denotes the geometric mean $\hat{\lambda}_n = [\lambda_{1,n}\lambda_{2,n} \cdots \lambda_{q-1,n}\hat{\lambda}_{q,n}\hat{\lambda}_{q+1,n} \cdots \hat{\lambda}_{4,n}]^{\frac{1}{4}}$.
- In (14.14)-(14.15), $Z_q^w(n)$ and $Z_{q^2}^w(n)$ are the estimates of Eq_{av}^w and $E(q_{av}^w)^2$, respectively.
- In (14.17), $G_i(Z_q(n)) = (Z_q^+(n) - Z_q^-(n))/(2\delta\Delta_i(n))$ is an estimate of the ith component of the gradient of Eq_{av}.
- In (14.19), $H_{j,i}(Z_q(n)) = (4\delta_1\delta_2)^{-1}[(\triangle_j(n)\hat{\triangle}_i(n))^{-1} + (\triangle_i(n)\hat{\triangle}_j(n))^{-1}]$ $[Z_q^{++}(n) - Z_q^{+} - Z_q^{-+} + Z_q^{-}]$ is an estimate of the (i, j)th component of the Hessian of Eq_{av}.
- In (14.18), (14.20), $G_i(\hat{\sigma}(n))$, $H_{j,i}(\hat{\sigma}(n))$ are similarly the gradient and Hessian estimates for $(Var(q_{av}))^{1/2}$.

Remark 14.1. The algorithm presented above is a second-order method where neither the Hessian update is projected to the space of positive definite and symmetric matrices nor is the inverse of the latter (projected Hessian) computed. Instead, one computes the eigen-values of the Hessian update and projects them in a way as to make them positive (in case they are not). Next the inverse of the geometric mean of these (projected) eigen-values is used in place of the Hessian inverse in the algorithm. This method has been proposed in [28] as an efficient alternative to regular Newton methods.

Remark 14.2. As with the multi-timescale algorithms described in earlier chapters, it is observed in [6] that an additional averaging over a certain number $L > 1$ of epochs for the recursions (14.14)–(14.18) in between two successive updates of the other (slower) recursions is seen to improve the empirical performance of the scheme.

Remark 14.3. Note that while $G_i(Z_q(n))$ and $G_i(\hat{\sigma}(n))$ are true estimates of the gradients of Eq_{av}, $(Var(q_{av}))^{1/2}$ (see Chapter 7), however, $H_{j,i}(Z_q(n))$, $H_{j,i}(\hat{\sigma}(n))$ turn out to be biased estimates of the Hessians of the associated quantities because the bias terms in the Hessians do not cancel when Hadamard matrix perturbations are used. Note that in Chapter 7 [21] a similar four-simulation estimate of the Hessian has been presented that incorporates randomized perturbations. Such an estimate is seen to be asymptotically unbiased unlike the estimate above where the biases in the Hessian estimate may not become asymptotically negligible. This, however, does not affect the analysis as the overall scheme still converges to a local minimum. A Hadamard matrix-based construction for both the gradient and the Hessian has been used in these algorithms as it is seen to exhibit significant improvements in empirical performance.

14.2.3.2 The P-RED Algorithm

The P-RED algorithm is obtained in a similar manner with the only change being the use of $\hat{\nabla}_i P$ in place of $\hat{\nabla}_i B$ in (14.21) that can be estimated from (14.8).

14.2.3.3 A Sketch of Convergence

Let for any bounded and continuous function $v : \mathscr{R} \to \mathscr{R}$,

$$\hat{\Gamma}_i(v(y)) = \lim_{0<\eta\to 0} \left(\frac{\Gamma_i(y+\eta v(y)) - \Gamma_i(y)}{\eta} \right), \; i = 1,\ldots,4.$$

Corresponding to B-RED, consider the system of ODEs: For $i = 1,\ldots,4$,

$$\dot{\theta}_i = \hat{\Gamma}_i(-(\hat{\lambda}_\theta)^{-1}\nabla_{\theta_i} B). \tag{14.22}$$

The stable fixed points of (14.22) lie in the set $K_B = \{\theta \in C \mid \hat{\Gamma}_i\,((\hat{\lambda}_\theta)^{-1}\,\nabla_{\theta_i} B) = 0,$ $i = 1,\ldots,4\}$. For $\eta > 0$, let K_B^η denote the η-neighborhood of K_B.

Theorem 14.1. *Given $\eta > 0$, there exists a $\hat{\delta} > 0$, such that for all δ_1, δ_2 $\in (0,\hat{\delta}]$, the parameters $\theta(n)$, $n \geq 0$ given by the algorithm B-RED converge to K_B^η with probability one.*

Proof. (*Sketch*) The proof of convergence of (14.14)-(14.20) proceeds as with other multi-timescale schemes, see Chapter 7. Hence consider (14.21). Let $P(\hat{\nabla}^2_{\theta(n)}B)$ denote the spectral radius of $\hat{\nabla}^2_{\theta(n)}B$, i.e., the maximum of the magnitudes of the eigenvalues of $\hat{\nabla}^2_{\theta(n)}B$. Then by Proposition A.15 of [1], we have that $P(\hat{\nabla}^2_{\theta(n)}B) \leq \|\hat{\nabla}^2_{\theta(n)}B\|$. Note also that $\sup_n \|\hat{\nabla}^2_{\theta(n)}B\| < \infty$. This follows since all the updates (14.14)-(14.20) are uniformly bounded as they are convex combinations of uniformly bounded quantities in these recursions. Hence $\sup_n \hat{\lambda}_n < \infty$ w.p. 1. Further, from construction, $\sup_n \hat{\lambda}_n > 0$. Also, since the eigenvalues of $\hat{\nabla}^2_{\theta(n)}B$ are uniformly continuous functions of the elements of this matrix, these converge as $\theta(n) \to \theta$ for some $\theta \in C$ (since then $\hat{\nabla}^2_{\theta(n)}B \to \hat{\nabla}^2_{\theta}B$). Let $\hat{\lambda}_\theta$ denote the geometric mean of the projected eigenvalues of $\hat{\nabla}^2_{\theta}B$. Then $\infty > \sup_{\theta\in C}\hat{\lambda}_\theta > 0$. Now the first ODE in (14.22) corresponds to the recursion (14.21). Because of the projection to a compact set, (14.21) is uniformly bounded w.p. 1. The rest now follows as in Chapter 7. □

Remark 14.4. The convergence of the P-RED algorithm follows along exactly the same lines as for the B-RED algorithm and similar conclusions as those of Theorem 14.1 continue to hold.

14.2.4 Summary of Experimental Results

Results of experiments over different networks with multiple nodes have been presented in [16]. In particular, the experiments were conducted using the ns2.26 network simulator [13] by changing the router code. In fact, the standard RED code implemented over the router in ns2.26 was replaced by the code for the B-RED and P-RED algorithms. The four-simulation algorithm was implemented using a single simulation run on the simulation platform as described below: First, data averaging is performed for the perturbed parameters corresponding to the parameter $(\theta(0) - \delta_1\Delta(0))$ for the first L packet arrivals (cf. Remark 14.2). For the next L arrivals, data averaging with parameter $(\theta(0) + \delta_1\Delta(0))$ is conducted. Subsequently, the same is done for $(\theta(0) - \delta_1\Delta(0) + \delta_2\hat{\Delta}(0))$ and finally for $(\theta(0) + \delta_1\Delta(0) + \delta_2\hat{\Delta}(0))$. At the end of $4L$ packet arrivals, the parameter θ is updated and then the next cycle of data averaging over $4L$ packet arrivals is performed. Thus, the algorithms spend majority of the time in on-line data averaging which is a simple operation. Because of the sequential implementation procedure described above, the B-RED and P-RED algorithms are amenable to online implementation in a real network scenario, involving only real data and no simulated outcomes. The value of L was selected to be 64 in the experiments. At the start of the simulations, the algorithms are set in the active state.

Detailed experimental comparisons have been shown in [16] between B-RED and P-RED algorithms with various other well-studied algorithms in the literature, over different network topologies and settings, as well as traffic parameters. The performance was also studied under both (a) given load conditions as well as (b) dynamically increasing loads. The B-RED and P-RED algorithms not only result in good performance by yielding low delays and high throughput but also rapidly stabilize the oscillations in the average queue lengths in all settings (even under dynamically increasing loads). This was much unlike other algorithms in the literature, many of which seem to yield large oscillations. The simultaneous perturbation methods are seen to be highly useful in such settings.

14.3 Optimal Policies for the Retransmission Probabilities in Slotted Aloha

In this section, we study the application of the smoothed functional algorithm for the problem of optimizing the retransmission probabilities for the slotted Aloha multi-access communication protocol. We formulate the problem as a parameterized stochastic differential equation (SDE) and then find the optimal parameter trajectory using a smoothed functional algorithm. The material in this section is based on [15, 4].

14.3.1 Introduction to the Slotted Aloha Multiaccess Communication Protocol

The slotted Aloha multiple access communication scheme [2] is an efficient algorithm for bursty traffic. It divides time into slots of fixed size and each node can send at most one packet at the beginning of each timeslot. We consider a network with N transmitting nodes sending packets on a common broadcast channel. Packets arrive at each node independently with probability p. All packets are assumed to be of equal size and which is the same as the slot length. We assume that there is no buffer available at any of the nodes, i.e., at most one packet can be sent on the channel in any slot by a given node. A new packet received in the current slot at a node is transmitted in the immediate next slot. A packet arriving at a node when a transmission from that node is in progress is dropped, i.e., it immediately leaves the system. A transmission is successful if only one packet is transmitted in a slot. Collision occurs if two or more packets are transmitted in the same slot. Colliding packets are considered backlogged and each such packet is retransmitted with probability q at the beginning of each subsequent slot (by the corresponding nodes that are also referred to as 'backlogged nodes') until such packets are retransmitted successfully. Note that a collision results in at least two nodes (those that transmitted the colliding packets) becoming backlogged at the end of the slot in which the collision took place. New packets can thus only be admitted at unbacklogged nodes. At the end of each slot, the channel broadcasts to each node whether zero, one, or more than one packets were transmitted during the previous slot. The channel as such is assumed to be error free. For instance, if only one packet is transmitted during a slot, it is successfully received.

14.3.2 The SDE Framework

Let $K(n)$ denote the number of backlogged nodes ($0 \le K(n) \le N$) at the beginning of the nth slot. Each of these (backlogged) nodes transmits a packet in the nth slot with probability q independent of the other nodes. The remaining $u(n) = N - K(n)$ nodes are unbacklogged in the nth slot and transmit a packet in the slot if one (or more) packets arrived during the previous slot at these nodes. Note that $K(n), n \ge 0$ satisfies the update rule

$$K(n+1) = K(n) + A(n) - I_n, \tag{14.23}$$

where $A(n)$ is the number of new arrivals admitted to the system (aggregated over all nodes) in the nth slot and I_n is the indicator random variable

$$\begin{aligned} I_n &= 1 \text{ if transmission in the } n\text{th slot is successful.} \\ &\quad\; 0 \text{ otherwise} \end{aligned}$$

It is clear from the above that $K(n), n \ge 0$ is a discrete time Markov chain.

Under the identification $X^N(t) \equiv K([Nt])/N$ where $[Nt]$ denotes the integer part of Nt, it is argued in [14] that for a large but finite number N of users, the behaviour of the system can be approximated by the following SDE:

$$dX^N(t) = \mu(X^N(t))dt + \frac{1}{\sqrt{N}}\sigma(X^N(t))dW(t), \tag{14.24}$$

where the drift and the diffusion terms $\mu(\cdot)$ and $\sigma(\cdot)$, respectively, are given by

$$\begin{aligned} \mu(X^N(t)) &= Np(1 - X^N(t)) \\ &- (Np(1-X^N(t)) + NqX^N(t))\exp(-Np(1-X^N(t)) - NqX^N(t)), \qquad (14.25) \\ \sigma^2(X^N(t)) &= (Np)^2(1-X^N(t))^2 + Np(1-X^N(t)) \\ &- (Np(1-X^N(t)) - NqX^N(t))\exp(-Np(1-X^N(t)) - NqX^N(t)). \qquad (14.26) \end{aligned}$$

In (14.24), $W(\cdot)$ denotes the one-dimensional Brownian motion. It is important to note that both the drift and the diffusion terms, i.e., $\mu(X^N(t))$ and $\sigma(X^N(t))$, respectively, depend on the parameter q. Thus, we explicitly consider the parameterization of these terms and the resulting parameterized SDE takes the following form:

$$dX^N(t) = \mu(X^N(t), q)dt + \frac{1}{\sqrt{N}}\sigma(X^N(t), q)dW(t), \tag{14.27}$$

with $\mu(X^m(t),q)$ and $\sigma(X^m(t),q)$ defined according to (14.25) and (14.26), respectively. In [15], two different cost formulations, the expected finite horizon cost as well as the long-run average cost, have been considered. We focus here on the expected finite horizon cost structure.

14.3.2.1 The Expected Finite Horizon Cost

Let $[0,T]$ for some $0 < T < \infty$ be the interval over which we consider the evolution of the SDE (14.27). Let $\bar{g} : [0,T] \times \mathbb{R} \to \mathbb{R}$ represent the associated cost function. Let

$$\bar{J}_{X_0^m}(q(\cdot)) \triangleq E\left[\int_0^T \bar{g}(t,X(t))dt \mid X^m(0) = X_0^m\right]. \tag{14.28}$$

Here, $q(t) \in [0,1]$ is the retransmission probability prescribed by the trajectory $q(\cdot)$ at time $t \in [0,T]$. The objective is to find a function $q^* : [0,T] \to \mathscr{R}$ with $q^*(t) \in [0,1]$, $\forall t \in [0,T]$ that minimizes (14.28) over all functions $q : [0,T] \to \mathbb{R}$, given the initial state X_0^m.

For computational purposes, we shall consider a suitable discretization of the SDE (14.27) and recast the problem in the discrete time framework.

14.3.2.2 The Discretized Problem

Let $T = Mh$ for some $M > 0$, where h is a small time element. The Euler-Milstein discretization of the SDE (14.27) [11, pp.340-343] corresponds to:

$$\begin{aligned} X_{j+1}^N &= X_j^N + \mu(X_j^N, q_j)h + \frac{1}{\sqrt{N}}\sigma(X_j^N, q_j)\sqrt{h}Z_{j+1} \\ &\quad + \frac{1}{2N}\sigma_X'(X_j^N, q_j)\sigma(X_j^N, q_j)h(Z_{j+1}^2 - 1), \end{aligned} \tag{14.29}$$

where $\sigma_X'(\cdot,\cdot)$ is the partial derivative of $\sigma(\cdot,\cdot)$ with respect to the first argument (X). Also, $q_j \equiv q(jh)$ is the retransmission probability parameter at instant jh and Z_{j+1}, $j \geq 0$ are independent $N(0,1)$-distributed random variables.

Let $g_j(X_j^N) \equiv \bar{g}(jh, X^N(jh))$, $j = 1,\ldots,M$ be Lipschitz continuous functions. Given the initial state X_0^N of the SDE, the aim in the discretized setting is to find parameters $q_0, q_1, \ldots, q_{M-1} \in [0,1]$ that minimize the finite horizon cost

$$J_{X_0^N}(q_0,\ldots,q_{M-1}) \triangleq E_{X_0^N}\left[\sum_{j=1}^M g_j(X_j^N)\right] h. \tag{14.30}$$

The quantity h (the slot length) is a constant and does not play a role and hence can be dropped from (14.30). Thus, the final form of the discretized objective is

$$J_{X_0^N}(q_0,\ldots,q_{M-1}) \stackrel{\triangle}{=} E_{X_0^N}\left[\sum_{j=1}^{M} g_j(X_j^N)\right]. \tag{14.31}$$

14.3.3 The Algorithm

The algorithm incorporates two step-size schedules $a(n), b(n), n \geq 0$ such that $a(n) = o(b(n))$. More generally, these schedules satisfy Assumption 3.6. Let $q(n) \stackrel{\triangle}{=} (q_0(n),\ldots,q_{M-1}(n))^T$ denote the parameter trajectory at instant n. Let $\beta > 0$ be a small constant. Also, let $\eta_i(n), n \geq 0$, $i = 0,1,\ldots,M-1$ be independent $N(0,1)$-distributed random variables and let $\eta(n) \stackrel{\triangle}{=} (\eta_0(n),\ldots,\eta_{M-1}(n))^T$, $n \geq 0$. Generate two independent SDE trajectories $X^+(n) \stackrel{\triangle}{=} \{X_0^+(n), X_1^+(n),\ldots,X_{M-1}^+(n)\}$ and $X^-(n) \stackrel{\triangle}{=} \{X_0^-(n), X_1^-(n),\ldots,X_{M-1}^-(n)\}$ that are, respectively, governed by the parameter trajectories or vectors $q^+(n) = q(n) + \beta\eta(n)$ and $q^-(n) = q(n) - \beta\eta(n)$. The algorithm is as follows: $\forall j = 0,\ldots,M-1$,

$$Y_j(n+1) = (1-b(n))Y_j(n) + b(n)\frac{\eta_j(n)}{2\beta}\sum_{i=j}^{M-1}(g_i(X_i^+(n)) - g_i(X_i^-(n))), \tag{14.32}$$

$$q_j(n+1) = \Gamma(q_j(n) - a(n)Y_j(n)). \tag{14.33}$$

The algorithm prescribes a retransmission probability $q_j(n)$ at the nth iteration in the jth stage. The value of this parameter affects the evolution of the system from the jth stage onwards. Consider the ODE

$$\dot{q}(t) = (\tilde{\Gamma}(-\nabla_1 J_{X_0^N}(q(t))),\ldots,\tilde{\Gamma}(-\nabla_M J_{X_0^N}(q(t))))^T, \tag{14.34}$$

where for any $y \in \mathbb{R}$ and a bounded, continuous function $v : \mathbb{R} \to \mathbb{R}$,

$$\tilde{\Gamma}(v(y)) = \lim_{\eta \to 0}(\Gamma(y + \eta v(y)) - \Gamma(y))/\eta.$$

The stable fixed points of this ODE lie within the set

$$\mathscr{K} \stackrel{\triangle}{=} \{\theta \in [0,1]^M \mid (\tilde{\Gamma}(-\nabla_1 J_{X_0^N}(\theta)),\ldots,\tilde{\Gamma}(-\nabla_M J_{X_0^N}(\theta)))^T = (0,0,\ldots,0)^T\}.$$

Given $\varepsilon > 0$, let $\mathscr{K}^\varepsilon$ denote the ε-neighborhood of $\mathscr{K}$.

Theorem 14.2. *Given $\varepsilon > 0$, there exists a $\hat{\beta} > 0$, such that for all $\beta \in (0, \hat{\beta}]$, $q(n) = (q_0(n), q_1(n), \ldots, q_{M-1}(n))^T \to K^\varepsilon$ as $n \to \infty$ with probability one.*

Proof. Using a standard two-timescale argument, one can let $q_j(n) \equiv q_j$, $n = 0, 1, \ldots, M-1$ when analyzing the faster recursion (14.32). Now $\{M_j(p), p \geq 1\}$, $j \in \{0, 1, \ldots, M-1\}$ defined according to

$$M_j(p) = \sum_{n=1}^{p} b(n)(\frac{\eta_j(n)}{2\beta} \sum_{i=j}^{N} (g_i(X_i^+(n)) - g_i(X_i^-(n)))$$

$$-E[\frac{\eta_j(n)}{2\beta} \sum_{i=j}^{N} (g_i(X_i^+(n)) - g_i(X_i^-(n))) \mid \mathscr{F}(n-1)]),$$

can be seen to be martingale sequences, where $\mathscr{F}(k) = \sigma(q_j(n),\ \eta_j(n),\ X_j^+(n),\ X_j^-(n), n \leq k,\ j = 0, 1, \ldots, M-1)$, $k \geq 1$, is the associated filtration. Since $g_i(\cdot)$ are Lipschitz continuous functions,

$$\| g_i(x) \| - \| g_i(0) \| \leq \| g_i(x) - g_i(0) \| \leq L \| x \|,$$

where $L > 0$ is the Lipschitz coefficient for the function $g_i(\cdot)$. Thus,

$$\| g_i(x) \| \leq \bar{K}(1 + \| x \|),$$

where $\bar{K} = \max(\| g_i(0) \|, L) > 0$. This together with the square summability of the sequence $b(n), n \geq 0$ and the fact that the fourth moment of the $N(0,1)$ random variable is finite implies that the quadratic variation processes of the above martingales are almost surely convergent and by the martingale convergence theorem (Theorem B.2), $M_j(p), p \geq 0$, $j = 0, 1, \ldots, M-1$ are almost surely convergent as well. Consider now the following system of ODEs along the faster timescale: For $j = 0, 1, \ldots, M-1$,

$$\dot{Y}_j(t) = D_\beta^j J_{X_0}(q) - Y_j(t), \tag{14.35}$$

where $D_\beta^j J_{X_0}(q) = E[\frac{\eta_j}{2\beta}(J_{X_0}^j(q + \beta\eta) - J_{X_0}^j(q - \beta\eta))]$, with $J_{X_0}^j(\hat{q}) = E_{X_0^N}[\sum_{i=j}^{M} g_i(X_i^N)]$, for $\hat{q} = q + \beta\eta$ or $\hat{q} = q - \beta\eta$, respectively. Here, $q = (q_0, q_1, \ldots, q_{M-1})^T$ is the trajectory (or vector) of re-transmission probabilities over the M stages and $\eta = (\eta_0, \eta_1, \ldots, \eta_{M-1})^T$ is the M-dimensional vector of independent $N(0,1)$-random variables. One can now show that almost surely $\| Y_j(n) - D_\beta^j J_{X_0}(q(n)) \| \to 0$ as $n \to \infty$. From Taylor's expansions of $J_{X_0}^j(q(n) + \beta\eta(n))$ and $J_{X_0}^j(q(n) - \beta\eta(n))$ around $q(n)$, it is easy to see that $\| D_\beta^j J_{X_0}(q(n)) - \nabla_j J_{X_0}^j(q(n)) \| \to 0$ as $\beta \to 0$ almost surely. We thus obtain

$$\| Y_j(n) - \nabla_j J_{X_0}^j(q(n)) \| \to 0 \text{ as } n \to \infty \text{ and } \beta \to 0, \tag{14.36}$$

almost surely.

Finally, consider the slower timescale recursion (14.33). In lieu of the above, one can rewrite (14.33) as follows: $\forall j = 0, 1, \ldots, M-1$,

$$q_j(n+1) = \Gamma(q_j(n) - a(n)\nabla_j J^j_{X_0}(q(n)) + a(n)\xi_2(n)), \tag{14.37}$$

where (because of (14.36)), $\xi_2(n) = (\nabla_j J^j_{X_0}(q(n)) - Y_j(n)) \to 0$ as $n \to \infty$ and $\beta \to 0$. The ODEs associated with (14.37) correspond to $\dot{q}_j(t) = \tilde{\Gamma}(-\nabla_j J^j_{X_0}(q(t)))$, $j = 1, \ldots, M$, for which $\mathscr{K}$ is the set of asymptotically stable attractors with $V(q) = J_{X_0}(q)$ as the associated strict Lyapunov function. The claim follows from the Kushner-Clark theorem (Theorem E.1). □

14.3.4 Summary of Experimental Results

In [15], results of experiments with varying number of nodes and the net arrival rate ($\lambda = Np$) are shown. The discretization constant h is chosen as 0.01 and the total number of epochs is $M = 40$. The performance is measured in terms of the fraction of backlogged nodes and the average throughput. Two results of two sets of experiments have been shown in [15]. In the first of these, λ is varied for a given number ($N = 200$) of nodes. It is observed here that as λ is increased, the fraction of backlogged nodes increases as well while the throughput decreases. This happens because of an increase in collisions that result from a higher value of λ.

In the second set of experiments, for a fixed value of λ ($\lambda = 0.4$), the number of nodes N is varied from 100 to 500. It is interesting to observe that in this case (as the number of nodes is increased while keeping the net arrival rate constant), the throughput increases while the fraction of backlogged nodes decrease. This is possibly because for fixed λ, a higher value of N results in a lower value of p.

These results point to the need to dynamically adapt the retransmission probability parameter in a slotted Aloha multi-access communication system for improved performance. Similar enhancements in the case of other multi-access communication protocols such as CSMA, CSMA/CD, etc. can be made along similar lines.

14.4 Dynamic Multi-layered Pricing Schemes for the Internet

In this section, we describe the problem of finding a dynamic optimal pricing scheme in the presence of multiple queues and multiple grades of service. We consider a class of multi-layered price feedback policies and apply the SPSA algorithm with a Hadamard matrix construction for finding an optimal policy within the prescribed class of policies. This portion is based on [26].

14.4.1 Introduction to Dynamic Pricing Schemes

Network pricing has been recognized as an effective means for managing congestion and providing better quality of service to the internet users. There is a large body of work on internet pricing where the emphasis is on the network adjusting the prices of its resources based on the demand and the users adapting their transmission rates suitably to optimize a certain utility function. The idea in the differentiated services architecture is to suitably divide the available bandwidth amongst the various competing flows.

The work conserving Tirupati Pricing (TP) scheme is proposed in [17]. A stochastic approximation-based pricing scheme is used in [27] for a single-node system. This scheme increases the price if congestion is above a certain threshold and lowers it otherwise. In [20], the TP pricing is observed to perform better than another pricing scheme that goes by the name of Paris Metro Pricing (PMP) over a single-node model. A stochastic approximation-based adaptive pricing methodology is considered in [8] to bring the congestion along any route to a certain prescribed level. Unlike [27], the objective function there depends on the price and not the actual congestion levels. However, prices for the entire routes and not of individual queues along the route are considered. The latter scenario incorporating prices for individual queues along a route is considered in [25] and is seen to result in performance improvements as it allows for greater flexibility since packets from one service grade at one link can shift to another service grade on another link.

In [26], the TP pricing scheme is adopted and a state-dependent multi-layered pricing scheme is considered that clusters together states in each queue into various levels with prices assigned to each such level. Thus, the queue manager charges a price to an incoming packet joining that queue on the basis of the level of congestion within the queue. The material in this section is based on [26].

14.4.2 The Pricing Framework

Consider a network having N links with the ith link providing J_i possible grades of service to the packets. The transmission capacity on the ith link is assumed to be μ_i. There is a separate queue for each service grade. Thus, there are a total of J_i queues on the ith link and packets desiring a particular grade of service join the corresponding queue. Let $b_{i,j}$ denote the buffer size in the jth queue on the ith link. A route r is denoted by a sequence of tuples $r := [(i_1, j_1), (i_2, j_2), \cdots, (i_{n_r}, j_{n_r})]$ comprising n_r links $(i_1, \ldots, i_{n_r})$ and the corresponding service grades $(j_1, \ldots, j_{n_r})$ used on each of these. The service grades could be different for different links on a route. Let K be the total number of routes. Each of the J_i queues at link i can be serviced according to any policy that provides the required QoS, e.g., round robin, weighted fair queuing, etc.

Let $Z_{ij}(t)$ denote the queue length or buffer occupancy of the jth queue on the ith link at time t. Let $Z^r(t) \in \Re$ denote the total congestion along route r at time t, i.e., the sum of $Z_{ij}(t)$ over all queues on the route r. Thus,

$$Z^r(t) = \sum_{k=1}^{n_r} Z_{i_k j_k}(t).$$

Let $Z_i(t) \triangleq [Z_{i1}(t), \ldots, Z_{iJ_i}(t)]$ denote the vector of queue lengths of all queues on the ith link. The state of the network at time t shall be denoted by $Z(t) \triangleq [Z_1(t), \ldots, Z_N(t)]$. In order to meet the desired QoS, the network service provider selects an operating point Z^* for the vector of congestion levels, where $Z^* = [Z_1^*, \ldots, Z_K^*]$ with each $Z_i^* = [Z_{i1}^*, \ldots, Z_{iJ_i}^*]$.

Let $p^i(t) = [p_{i1}(t), \ldots, p_{iJ_i}(t)]$ be the vector of prices for unit traffic on link i at time t, where $p_{ij}(t)$ denotes the price (for unit traffic) on link i for service class j at time t. The vector of prices $p^i(t)$ is posted by the service provider for each link $i = 1, 2, \ldots, N$. The price vector $p^i(t)$ is updated periodically every T time instants (for given $T > 0$) using an SPSA-based algorithm that is described below.

Each user sends packets along the route with the least cost. The cost function is assumed to be an increasing function of both price and congestion. Further, the users strictly follow the routes prescribed by their associated cost functions. Let $C_s(x, Z_{ij}(t), p_{ij}(t))$ denote the cost to user s for sending x units of traffic on link i using service grade j. It is assumed that instantaneous values of the quantities $Z_{ij}(t)$ and $p_{ij}(t)$ for all tuples (i, j) along a route are known to the users. Let

$$j^i = \arg \min_{j \in \{1, \ldots, J_i\}} C_s(x, Z_{ij}(t), p_{ij}(t))$$

denote the least cost service grade on the ith link for user s at time t. The user s would then select its least cost route corresponding to

$$\arg \min_{r = [(1, j^1), \ldots, (n_r, j^{n_r})]} \sum_{i=1}^{n_r} C_s(x, Z_{ij^i}(t), p_{ij^i}(t)),$$

assuming 1 denotes the source node and n_r the destination node. The minimum above is taken over all feasible routes from the source to destination for the sth user. An example of a cost function $C_s(\cdot, \cdot, \cdot)$ is

$$C_s(x, Z_{ij}(t), p_{ij}(t)) = x(p_{ij}(t) - U_s(x, Z_{ij}(t))),$$

where $U_s(x, Z_{ij}(t))$ is the utility of user s in sending x units of traffic using the jth service grade on the ith link when the congestion level there is $Z_{ij}(t)$. In general both the cost function and the utility are different for different users. Thus, the optimal routes for two users sending packets from the same source to the same destination node under identical conditions of congestion could, in general, be different. For instance, a user transmitting real-time video might be more interested in getting a low-delay path to transmit packets even if it means that he needs to pay more for it. On the other hand, another user transmitting data packets might be more interested in getting a low price path to send packets even if it takes a longer time for packets to reach their destination.

14.4.3 The Price Feed-Back Policies and the Algorithms

We first describe two different price feed-back policies here.

14.4.3.1 SPSA-Based Link Route Pricing (SPSA-LRP)

Let Z_{ij}^{1*} and Z_{ij}^{2*} be two queue length thresholds with $Z_{ij}^{1*} > Z_{ij}^{2*}$. Let

$$p_{ij}(n+1) = \begin{cases} \theta_{ij}^{1}(n) \text{ if } B \geq Z_{ij}(n) \geq Z_{ij}^{1*} \\ \theta_{ij}^{2}(n) \text{ if } Z_{ij}^{2*} < Z_{ij}(n) \leq Z_{ij}^{1*} \\ \theta_{ij}^{3}(n) \text{ if } 0 \leq Z_{ij}(n) < Z_{ij}^{2*}. \end{cases} \tag{14.38}$$

14.4.3.2 SPSA-Based Weighted Average Link Route Pricing (SPSA-WA-LRP)

$$p_{ij}(n+1) = \begin{cases} \theta_{ij}^{1}(n) \text{ if } B \geq \widehat{Z}_{ij}(n) \geq Z_{ij}^{1*} \\ \theta_{ij}^{2}(n) \text{ if } Z_{ij}^{2*} < \widehat{Z}_{ij}(n) \leq Z_{ij}^{2*} \\ \theta_{ij}^{3}(n) \text{ if } 0 \leq \widehat{Z}_{ij}(n) < Z_{ij}^{2*}, \end{cases} \tag{14.39}$$

where $\widehat{Z}_{ij}(n)$ denotes the 'weighted average' congestion in the jth queue at the ith link at instant nT, $n \geq 0$, and is obtained recursively as

$$\widehat{Z}_{ij}(n+1) = (1-w_z)\widehat{Z}_{ij}(n) + w_z Z_{ij}(n), \tag{14.40}$$

with $0 < w_z < 1$ being a given small constant. The idea behind using a 'weighted average' queue length in the pricing policy SPSA-WA-LRP as opposed to the regular queue length in SPSA-LRP is to reduce oscillations that would otherwise result in frequent adjustments to the price levels. This is in the spirit of the RED algorithm (cf. Section 14.2).

Remark 14.5. Using an appropriate choice of the threshold levels Z_{ij}^{1*} and Z_{ij}^{2*} for the jth queue on the ith link, one can effectively classify congestion in the queue at any instant as being in the 'low', 'medium' or 'high' ranges. The policies described above would then assign a different price depending on the aforementioned congestion levels. In general, more layers for the pricing policies (14.38) and (14.39) may be used.

14.4.3.3 The Cost Formulation

Let $\theta_{ij} \stackrel{\Delta}{=} (\theta^1_{ij}, \theta^2_{ij}, \theta^3_{ij})$ denote the 'price' parameter vector associated with the jth queue on the ith link, and $\theta = (\theta_{ij}, j = 1,\ldots,J_i, i = 1,\ldots,N)$. The components of these parameters correspond to the price levels obtained through one of the policies SPSA-LRP or SPSA-WA-LRP. Let θ take values in the set $C \subset \mathbb{R}^d$, where $d = 3\gamma$, with γ being the dimension of $Z(t)$. The set C has the form $C = [\underline{A}, \overline{A}]^d$, where $0 < \underline{A} < \overline{A} < \infty$.

Let $\{Z(t), t \geq 0\}$ be an ergodic Markov process for any given $\theta \in C$. In many interesting scenarios, $Z(t)$, $t \geq 0$, may not be Markov but the appended process $\{(Z(t), W(t)), t \geq 0\}$ is Markov, where $\{W(t), t \geq 0\}$ is a suitable additional process. The methodology and analysis here easily extend to the latter case as well as long as $\{(Z(t), W(t)), t \geq 0\}$ is ergodic for any given θ.

No specific model for the demand is assumed except that at any instant, the demand may depend on the current congestion (i.e., state) and price (i.e., parameter) levels. Further, given the current congestion and price levels, the demand at the current instant is independent of the previous values of the demand.

Controllers at individual queues update the prices associated with their queues based on local congestion information pertaining to their queues. Let $h_{ij}(\cdot)$ denote the single-stage cost associated with link-service grade tuple (i, j) that depends at any instant t only on the state $Z_{ij}(t)$ of the queue. Let queue lengths be observed every T instants of time, for some fixed $T > 0$, and based on this information, prices at individual queues are instantly updated. Let $Z_{ij}(k) \stackrel{\Delta}{=} Z_{ij}(kT)$ denote the queue length at the jth queue on the ith link at instant kT.

For any given $\theta \in C$, let

$$J(\theta) = \sum_{\substack{j \in \{1,\ldots,J_i\},\\ i \in \{1,\ldots,N\}}} J_{ij}(\theta), \tag{14.41}$$

$$\text{where} \quad J_{ij}(\theta) = \lim_{n \to \infty} \frac{1}{n} \sum_{k=1}^{n} h_{ij}(Z_{ij}(k)).$$

The aim is to find a parameter $\theta^* \in C$ that minimizes $J(\cdot)$.

14.4.3.4 The Algorithm

Let $a(n), b(n), n \geq 0$ be two-step size sequences that satisfy Assumption 3.6. Let $\delta > 0$ be a given (small) constant. Let $\Delta_{ij}(n) \stackrel{\Delta}{=} (\Delta^1_{ij}(n), \Delta^2_{ij}(n), \Delta^3_{ij}(n))$ be a vector of $\{\pm 1\}^3$-valued perturbations obtained via the one-simulation Hadamard matrix-based construction described in Chapter 5.5.2.2.

Let the queue length process $Z_{ij}(n), n \geq 0$ of the jth queue on the ith link be governed by the parameter sequence $(\theta_{ij}(n) + \delta\Delta_{ij}(n)), n \geq 0$. Then, for all $j = 1,\ldots,J_i; i = 1,\ldots,N; k = 1,2,3$, we have

$$Y_{ij}(n+1) = Y_{ij}(n) + b(n)\left(h_{ij}(Z_{ij}(n)) - Y_{ij}(n)\right) \tag{14.42}$$

$$\theta_{ij}^k(n+1) = \Gamma\left(\theta_{ij}^k(n) - a(n)\frac{Y_{ij}(n)}{\delta\Delta_{ij}^k(n)}\right). \tag{14.43}$$

In the above, $Y_{ij}(n)$, $n \geq 0$ are quantities used to average the single-stage cost $h_{ij}(\cdot)$ in order to estimate $J_{ij}(\cdot)$. Also, $\Gamma(\cdot)$ is a projection operator that projects each price update to the interval $[\underline{A}, \overline{A}]$.

Consider now the following system of ODEs: For $j = 1,\ldots,J_i$; $i = 1,\ldots,N$,

$$\dot{\theta}_{ij}(t) = \hat{\Gamma}(-\nabla J_{ij}(\theta_{ij}(t))), \tag{14.44}$$

where $\hat{\Gamma}(\cdot)$ is defined according to

$$\hat{\Gamma}(v(y)) = \lim_{\gamma \downarrow 0}\left(\frac{\Gamma(y + \gamma v(y)) - \Gamma(y)}{\gamma}\right),$$

for any bounded and continuous function $v(\cdot)$. The stable fixed points of (14.44) lie within the set $M = \{\theta_{ij} \mid \hat{\Gamma}(\nabla J_{ij}(\theta_{ij})) = 0\}$. Let for given $\varepsilon > 0$, M^ε denote the ε-neighborhood of M. We have the following main convergence result:

Theorem 14.3. *Given any $\varepsilon > 0$, there exists a $\delta_0 > 0$ such that for all $\delta \in (0, \delta_0]$, $\theta_{ij}(n)$ converges as $n \to \infty$ to a point in M^ε.*

Proof. Follows from a standard two-timescale argument as in Chapter 5. □

14.4.4 Summary of Experimental Results

The results of several experiments over a setting involving a four-node network have been presented in [26]. All simulations were carried out using the network simulator with each simulation run for one thousand seconds. Detailed performance comparisons were drawn between SPSA-LRP, SPSA-WA-LRP and an algorithm from [25] that was, in turn, seen to be significantly better in comparison to [8]. Performance comparisons between these algorithms were drawn in terms of both the throughput and the delay metrics. Amongst the three algorithms, SPSA-LRP shows the best results followed by SPSA-WA-LRP. It is observed that SPSA-LRP exhibits a throughput improvement in the range of 67-82 percent for all routes over the algorithm of

[25], while SPSA-WA-LRP shows a similar improvement in the range of 34 to 69 percent. This happens because both SPSA-LRP and SPSA-WA-LRP use a combination of congestion-based feed-back control policies that are tuned using SPSA and hence utilize network resources in a better manner as compared to the algorithm of [25] that does not use any of the simultaneous perturbation approaches.

14.5 Concluding Remarks

We considered the applications of simultaneous perturbation approaches on problems of control and optimization in communication networks. Specifically, we studied the applications of these approaches on the following three problems: (a) finding optimal parameters in the case of the RED scheme for the internet, (b) finding optimal retransmission probabilities in the case of the slotted Aloha multi-access communication protocol, and (c) finding optimal strategies for network pricing in the internet.

The problem of RED flow control was formulated using a constrained nonlinear programming framework. The barrier and penalty function objectives were used and two multi-timescale Newton-based stochastic approximation algorithms that incorporated the Newton SPSA technique but with Hadamard matrix perturbations were presented. These algorithms are seen to show significantly better performance when compared with many other algorithms in the literature as they are seen to considerably bring down the queue oscillations – a problem consistently reported in many other studies. A different formulation of the RED problem has also been studied in [24], where a 'robust' version of gradient SPSA has been developed. The idea there is to replace the increment in the SPSA update with the sign of the same (i.e., $+1$ if the increment is positive, -1 if it is negative, and 0 otherwise). This helps in bringing down the queue oscillations over regular RED but is not as effective as the B-RED and P-RED schemes.

Next, the problem of finding the optimal retransmission probabilities in slotted Aloha was formulated in the setting of parameterized SDEs over a finite horizon and a gradient SF algorithm was used to find the optimal parameter trajectory. In [15], the same problem for the long-run average cost objective has also been addressed. The resulting algorithm in such a case results in a scalar (retransmission probability) parameter. The slotted Aloha problem in a different setting (without an SDE formulation) has also been studied in [7]. The regular gradient SPSA algorithm has been incorporated there. However, traffic from sources is individually considered, unlike [15] where the aggregate behaviour under a large number of sources is taken into account using an SDE framework. Thus, the number of sources considered in [7] is only of the order of a few tens, unlike [15] where the same is in the order of a few hundreds.

Finally, the problem of internet pricing was studied using two classes of closed-loop feedback policies that assigned price levels depending on the levels of congestion. Whereas in SPSA-LRP, instantaneous congestion levels were considered, in SPSA-WA-LRP, weighted average queue lengths were considered for the pricing levels. The latter policies are reminiscent of the RED flow control mechanism and lead to less oscillations that would otherwise result from rapid price changes. The one-simulation SPSA algorithm with Hadamard matrix perturbations was employed here and is seen to result in significantly better performance over the other algorithms.

Simultaneous perturbation methods have been studied in various other applications in communication networks. For instance, in [3, 5], applications of SPSA to available bit rate (ABR) flow control in asynchronous transfer mode (ATM) networks have been studied. Also, in [18], SPSA has been applied over the problem of finding optimal slot assignment to slaves in bluetooth networks for both piconets as well as scatternets. From these applications, it is clear that simultaneous perturbation methods play a significant role in problems of performance optimization in communication networks. An important characteristic of these methods is that they are independent of the technology and protocols used, are scalable (as they can be applied in high-dimensional settings), and hence are widely applicable.

References

1. Bertsekas, D.P.: Nonlinear Programming. Athena Scientific, Belmont (1999)
2. Bertsekas, D.P., Gallager, R.G.: Data Networks. Prentice-Hall, New York (1991)
3. Bhatnagar, S., Fu, M.C., Marcus, S.I., Fard, P.J.: Optimal structured feedback policies for ABR flow control using two-timescale SPSA. IEEE/ACM Transactions on Networking 9(4), 479–491 (2001)
4. Bhatnagar, S., Karmeshu, Mishra, V.: Optimal parameter trajectory estimation in parameterized sdes: an algorithmic procedure. ACM Transactions on Modeling and Computer Simulation (TOMACS) 19(2), 8 (2009)
5. Bhatnagar, S., Kumar, S.: A simultaneous perturbation stochastic approximation based actor-critic algorithm for Markov decision processes. IEEE Transactions on Automatic Control 49(4), 592–598 (2004)
6. Bhatnagar, S., Patro, R.K.: A proof of convergence of the B-RED and P-RED algorithms in random early detection. IEEE Communication Letters 13(10), 809–811 (2009)
7. Chakraborty, A., Bhatnagar, S.: Optimized policies for the retransmission probabilities in slotted aloha. Simulation 86(4), 247–261 (2010)
8. Garg, D., Borkar, V.S., Manjunath, D.: Network pricing for QoS: A 'regulation' approach. In: Abed, E.H. (ed.) Advances in Control, Communication Networks and Transportation Systems (in honour of Pravin Varaiya), pp. 137–157 (2005)
9. Floyd, S., Gummadi, R., Shenker, S., et al.: Adaptive red: An algorithm for increasing the robustness of reds active queue management (2001), `http://www.icir.org/floyd/papers.html`
10. Floyd, S., Jacobson, V.: Random early detection gateways for congestion avoidance. IEEE/ACM Transactions on Networking 1(4), 397–413 (1993)
11. Glasserman, P.: Monte Carlo methods in financial engineering, vol. 53. Springer (2004)
12. Gockenbach, M.: Analysis of the logarithmic barrier method (preprint), `http://www.math.mtu.edu`
13. Issariyakul, T., Hossain, E.: Introduction to network simulator NS2. Springer (2008)

14. Kaj, I.: Stochastic modeling in broadband communications systems. Society for Industrial Mathematics, vol. 8 (2002)
15. Karmeshu, Bhatnagar, S., Mishra, V.: An optimized sde model for slotted aloha. IEEE Transactions on Communications 59(6), 1502–1508 (2011)
16. Patro, R.K., Bhatnagar, S.: A probabilistic constrained nonlinear optimization framework to optimize RED parameters. Performance Evaluation 66(2), 81–104 (2009)
17. Dube, P., Borkar, V.S., Manjunath, D.: Differential join prices for parallel queues: Social optimality, dynamic pricing algorithms and application to internet pricing. In: Proceedings of IEEE Infocom, N.Y., vol. 1, pp. 276–283 (2002)
18. Ramana Reddy, G., Bhatnagar, S., Rakesh, V., Chaturvedi, V.: An efficient algorithm for scheduling in bluetooth piconets and scatternets. Wireless Networks 16(7), 1799–1816 (2010)
19. Rao, S., Rao, S.: Engineering optimization: theory and practice. Wiley (2009)
20. Tandra, R., Hemachandra, N., Manjunath, D.: Diffserv node with join minimum cost queue policy and multiclass traffic. Perform. Eval. 55(1-2), 69–91 (2004), `http://dx.doi.org/10.1016/S0166-5316(03)00105-6`
21. Spall, J.C.: Adaptive stochastic approximation by the simultaneous perturbation method. IEEE Trans. Autom. Contr. 45, 1839–1853 (2000)
22. Taha, H., Taha, H.: Operations research: an introduction, vol. 8. Prentice Hall, Upper Saddle River (1997)
23. Uri, A.: Penalty, barrier and augmented lagrangian methods (preprint), `www.cs.ubc.ca/spider/ascher/542/chap10.pdf`
24. Vaidya, R., Bhatnagar, S.: Robust optimization of random early detection. Telecommunication Systems 33(4), 291–316 (2006)
25. Vemu, K., Bhatnagar, S., Hemachandra, N.: Link-Route Pricing for Enhanced QoS. Technical Report 2007-8, Department of Computer Science and Automation. Indian Institute of Science, Bangalore (2007); Shorter version in Proceedings of IEEE Conference on Decision and Control, New Orleans, USA, December 12-14, pp. 504–1509 (2007), `http://archive.csa.iisc.ernet.in/TR/2007/8/`
26. Vemu, K.R., Bhatnagar, S., Hemachandra, N.: Optimal multi-layered congestion based pricing schemes for enhanced qos. Computer Networks (2011), `http://dx.doi.org/10.1016/j.comnet.2011.12.004`
27. Borkar, V.S., Manjunath, D.: Charge-based control of diffserv-like queues. Automatica 40(12), 2043–2057 (2004)
28. Zhu, X., Spall, J.C.: A modified second-order SPSA optimization algorithm for finite samples. Int. J. Adapt. Control Signal Process. 16, 397–409 (2002)

Part VI
Appendix

This part puts together five appendices on (a) convergence notions for a sequence of random vectors, (b) results on martingales and their convergence, (c) ordinary differential equations, (d) the Borkar and Meyn stability result, and (e) a result on convergence of projected stochastic approximation due to Kushner and Clark.Some of the background material as well as the main results used in other chapters have been summarized here.

Appendix A
Convergence Notions for a Sequence of Random Vectors

We briefly discuss here the various notions of convergence for random vectors. Let $(\Omega, \mathscr{F}, P)$ denote the underlying probability space, where Ω is the sample set, $\mathscr{F}$ the sigma field and P the probability measure, see for instance, [1] for a good account of probability theory. Let $X_n, n \geq 0$ denote a sequence of $\mathbb{R}^N$–valued random vectors on $(\Omega, \mathscr{F}, P)$. Suppose X is another $\mathbb{R}^N$–valued random vector on $(\Omega, \mathscr{F}, P)$. Further, let $x \in \mathbb{R}^N$ be an N-dimensional vector. Let $F_{X_n}(\cdot)$, $F_X(\cdot)$, $n \geq 0$ denote the corresponding distribution functions associated with the random vectors X_n, X, $n \geq 0$. Suppose $X_n = (X_n^1, \ldots, X_n^N)^T, X = (X^1, \ldots, X^N)^T$ and $x = (x^1, \ldots, x^N)^T$, respectively, where X_n^i, X^i, x^i, $i = 1, \ldots, N$ are $\mathbb{R}$-valued. Then $F_{X_n}(x) = P(X_n^i \leq x^i, i = 1, \ldots, N)$ and $F_X(x) = P(X^i \leq x^i, i = 1, \ldots, N)$, respectively. The following are standard notions of convergence:

1. **Deterministic Convergence:** We say that $X_k \to X$ as $k \to \infty$ deterministically if $X_k(w) \to X(w)$ as $k \to \infty$ for all $w \in \Omega$.
2. **Uniformly:** $X_k \to X$ uniformly as $k \to \infty$ if for all $\varepsilon > 0$ there exists an $N > 1$ such that $\forall n \geq N, \forall w \in \Omega$, $\|X_k(w) - X(w)\| < \varepsilon$.
3. **Almost Sure (a.s.) or With Probability One (w.p.1) Convergence:** We say that $X_k \to X$ as $k \to \infty$ almost surely (a.s.) or with probability one (w.p.1) if

$$P\left(w \in \Omega \mid \lim_{k \to \infty} \| X_k(w) - X(w) \| = 0 \right) = 1.$$

4. **Probabilistic or In Probability Convergence:** We say that $X_k \to X$ as $k \to \infty$ probabilistically or in probability if

$$\lim_{k \to \infty} P(w \in \Omega \mid \| X_k(w) - X(w) \| \geq \varepsilon) = 0 \; \forall \varepsilon > 0.$$

5. **Convergence in L^p:** Let for $p \geq 1$,

$$L^p(\Omega, \mathscr{F}, P) = \{X | E|X|^p < \infty\},$$

denote a set of all $\mathbb{R}^N$-valued random variables on $(\Omega, \mathscr{F}, P)$ which have finite p^{th} moment. We say that $X_k \to X$ as $k \to \infty$ in L^p for $p \geq 1$, if

$$\lim_{k \to \infty} E\left(\| X_k(w) - X(w) \|^p\right) = 0.$$

When $p = 2$, the L^p convergence is also referred to as *mean-square convergence*.

6. **In Distribution Convergence:** We say that $X_k \to X$ as $k \to \infty$ in distribution if

$$\lim_{k \to \infty} F_{X_k}(x) = F_X(x) \text{ at all points } x \text{ of continuity of } F_X(x).$$

7. **Nearly uniformly:** $X_k \to X$ nearly uniformly as $k \to \infty$ if $\forall \varepsilon > 0, \exists A \in F$ such that $P(A) < \varepsilon$ and on $A^c, X_k \to X$ uniformly.

Theorem A.1 (Egorov). *If $X_k \xrightarrow{a.s.} X$, then, $X_k \xrightarrow{n.u.} X$. The result is true for any measure μ with $\mu(\Omega) < \infty$.*

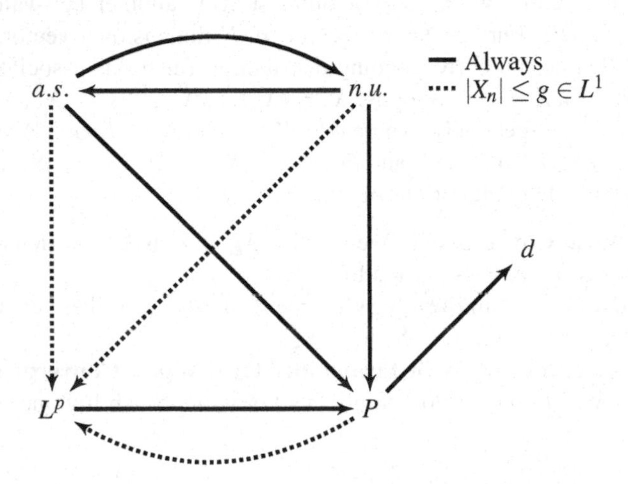

Fig. A.1 Relationship between the various notions of convergence. We use the following abbreviations - a.s. to denote “almost surely”, n.u. for “nearly uniformly”, P for “convergence in probability”, d for “convergence in distribution”, L^p for “convergence in L^p”.

A general relationship between the various notions of convergence is shown in Figure A.1. In the figure, a directed arrow from “A” to “B” i.e., $A \to B$ indicates that “A is stronger than B”. Further, we assume that a transitivity property holds in that $A \to B$ and $B \to C$ implies that $A \to C$, even when an arrow from “A” to “C” is not explicitly shown. Note that a.s. or w.p.1 convergence implies that there exists a set of zero probability on which the said convergence does not hold. Deterministic convergence can be viewed as a special case of a.s. convergence as here the above zero-probability set is in fact empty. Also, while in Figure A.1, there is no arrow between a.s. convergence and m.s. convergence, the former implies the latter under certain conditions on the random vectors X_n, X, $n \geq 1$. As an example, if the said random vectors are uniformly bounded by a L^1 function, then L^p convergence follows from a.s. convergence.

Reference

1. Chow, Y.S., Teicher, H.: Probability Theory: Independence, Interchangeability and Martingales, 3rd edn. Springer, New York (1997)

Appendix B
Martingales

As before, let $(\Omega, \mathscr{F}, P)$ be a given probability space. Let $\{\mathscr{F}_n\}$ be a family of increasing sub-$\sigma-$fields of $\mathscr{F}$ (also called a filtration), i.e.,

$$\mathscr{F}_0 \subset \mathscr{F}_1 \subset \mathscr{F}2_{\subset} \cdots \subset \mathscr{F}_n \subset \mathscr{F}_{n+1} \subset \cdots \subset \mathscr{F}.$$

Definition B.1. 1. A sequence of $\mathscr{R}$-valued random variables $X_n, n \geq 0$ defined on $(\Omega, \mathscr{F}, P)$ is said to be a martingale w.r.t. the filtration $\{\mathscr{F}_n\}$ if each X_n is integrable and measurable with respect to $\mathscr{F}_n$.
2. Further,

$$E[X_{n+1} \mid \mathscr{F}_n] = X_n \text{ w.p.1 } \forall n \geq 0. \tag{B.1}$$

Definition B.2. A sequence of random variables $X_n, n \geq 0$ as in Definition 1 is said to be a submartingale w.r.t. the filtration $\{\mathscr{F}_n\}$ if the first part in Definition 1 holds. In addition, the equality in (B.1) is replaced with "$\geq$".

Definition B.3. A sequence of random variables $X_n, n \geq 0$ as in Definition 1 is said to be a submartingale w.r.t. the filtration $\{\mathscr{F}_n\}$ if the first part in Definition 1 holds. In addition, the equality in (B.1) is replaced with "$\leq$".

Many times, one identifies the martingale (alternatively, sub- or super-martingale) with the sequence of tuples $(X_n, \mathscr{F}_n)$, $n \geq 0$ instead of just $\{X_n\}$ itself.

Definition B.4. For a martingale sequence $X_n, n \geq 0$, the sequence M_{n+1}, $n \geq 0$ obtained as $M_{n+1} = (X_{n+1} - X_n)$, $n \geq 0$ with $M_0 = X_0$, is called a martingale difference sequence.

Note that

$$\begin{aligned} E[M_{n+1} \mid \mathscr{F}_n] &= E[(X_{n+1} - X_n) \mid \mathscr{F}_n] \\ &= (E[X_{n+1} \mid \mathscr{F}_n] - X_n) = 0 \text{ w.p.1}, \end{aligned}$$

from (B.1).

Definition B.5. A vector martingale (also many times referred to as a martingale) is a sequence of $\mathscr{R}^N$−valued random vectors $X_n = (X_n^1, \ldots, X_n^N)$ such that each of its component processes $X_n^i, n \geq 0$ $(i = 1, \ldots, N)$ is a martingale.

We recall the following important result due to Doob, see, for instance, [1, Theorem 3.2.2 on pp. 49].

Theorem B.1 (Doob decomposition). *A submartingale $(X_n, \mathscr{F}_n), n \geq 0$, can be decomposed as $X_n = Y_n + A_n$, $n \geq 0$, where $(Y_n, \mathscr{F}_n)$, $n \geq 0$, is a zero-mean martingale and $A_n, n \geq 0$ is a non-decreasing process, i.e., $A_n \leq A_{n+1}$ almost surely for all $n \geq 0$. Further, A_n is $\mathscr{F}_{n-1}$-measurable for all $n \geq 0$, where $\mathscr{F}_{-1} = \{\phi, \Omega\}$. This decomposition is almost surely unique.*

There are various convergence results for martingales but the one that we often use in this book is based on the convergence of the quadratic variation process associated with the martingale $X_n, n \geq 0$ (see below). Let X_n, $n \geq 0$ be a square integrable (scalar) martingale, i.e., it is a martingale for which $E[X_n^2] < \infty$ for all $n \geq 0$. It is easy to see that in this case, $(X_n^2, \mathscr{F}_n)$, $n \geq 0$ forms a submartingale. Hence from the Doob decomposition theorem (cf. Theorem B.1), it follows that $X_n^2 = Y_n + A_n$, $n \geq 0$, where $\{Y_n\}$ and $\{A_n\}$ satisfying the properties in Theorem B.1. It is easy to see that

$$\begin{aligned} A_n &= \sum_{m=1}^{n} \left(E\left[X_m^2 \mid \mathscr{F}_{m-1}\right] - X_{m-1}^2\right) + E\left[X_0^2\right] \\ &= \sum_{m=0}^{n-1} E\left[(X_{m+1} - X_m)^2 \mid \mathscr{F}_m\right] + E\left[X_0^2\right], \end{aligned} \tag{B.2}$$

$\forall n \geq 0$. As mentioned above, $A_n, n \geq 0$ is called the quadratic variation process associated with the martingale $X_n, n \geq 0$.

Theorem B.2 (Martingale Convergence Theorem). *Let $(X_n, \mathscr{F}_n)$, $n \geq 0$ be a square-integrable martingale with $A_n, n \geq 0$ as its quadratic variation process. Let $A_\infty = \lim_{n\to\infty} A_n$. Then $\{X_n\}$ converges with probability one on the set $\{A_\infty < \infty\}$ and $X_n = o(f(A_n))$ on $\{A_\infty = \infty\}$ for every increasing $f : [0,\infty) \to [0,\infty)$ satisfying $\int_0^\infty (1 + f(t))^{-2} dt < \infty$.*

The proof of this result is available for instance on pp. 53-54 of [1] (cf. Theorem 3.3.4). Detailed treatments of martingales can be found, for instance, in the texts of Breiman [2], Neveu [3] and Borkar [1].

References

1. Borkar, V.S.: Probability Theory: An Advanced Course. Springer, New York (1995)
2. Breiman, L.: Probability. Addison-Wesley, Reading (1968)
3. Neveu, J.: Discrete Parameter Martingales. North Holland, Amsterdam (1975)

Appendix C
Ordinary Differential Equations

We begin with a definition of the O and o notation as this has been used at various places the text.

Definition C.1. Let $\{a_n\}$ and $\{b_n\}$ be two sequences of real numbers such that $b_n \geq 0, \forall n$.

1. We say $a_n = O(b_n)$ if there exists a constant $L > 0$ such that $|a_n| \leq Lb_n$ for all n.
2. We say $a_n = o(b_n)$ if $\lim_{n\to\infty} \frac{a_n}{b_n} = 0$.

Definition C.2. A function $h : \mathscr{R}^d \to \mathscr{R}^d$ is said to be Lipschitz continuous if $\exists M > 0$ such that

$$\| h(x) - h(y) \| \leq M \| x - y \|, \ \forall x, y \in \mathscr{R}^d.$$

The Gronwall inequality plays an important role in the proof of convergence of stochastic approximation algorithms. We give the result below, whose proof can be found in several texts, see for instance, Appendix B of [1].

Lemma C.1 (Gronwall inequality). *For continuous functions $f(\cdot), g(\cdot) \geq 0$ and scalars $K_1, K_2, T \geq 0$,*

$$f(t) \leq K_1 + K_2 \int_0^t f(s)g(s)ds \ \ \forall t \in [0,T], \tag{C.1}$$

implies

$$f(t) \leq K_1 e^{K_2 \int_0^T g(s)ds}, \ t \in [0,T].$$

Consider the ODE given by

$$\dot{\theta}(t) = L(\theta(t)), \ \theta(0) = \theta_0. \tag{C.2}$$

Definition C.3. The ODE (C.2) is said to be well-posed if starting from any $\theta(0) = \theta_0$, the trajectory $\theta(\cdot) = \{\theta(t), t \geq 0\}$ of (C.2) is unique. Further, the map $\theta_0 \to \theta(\cdot)$ is continuous.

Theorem C.2. *A sufficient condition for (C.2) to be well-posed is if the function $L : \mathbb{R}^N \to \mathbb{R}^N$ is Lipschitz continuous.*

Proof. See Theorem 5 on pp.143 of [1]. □

Definition C.4. A closed set $H \subset \mathbb{R}^N$ is called an invariant set for the ODE (C.2) if whenever the initial point $\theta(0) \in H$, then $\theta(t) \in H$ for all $t \geq 0$, i.e., if the ODE trajectory is initiated in H, it stays in H for all time.

Definition C.5. A closed set $H \subset \mathbb{R}^N$ is called an attractor for the ODE (C.2) if

(i) H is an invariant set, and
(ii) there is an open set M containing H (i.e., M is an open neighborhood of H) such that if the ODE trajectory is initiated in M, it stays in M and converges to H.

Definition C.6. The largest possible open set M that is an open neighborhood of H such that any ODE trajectory initiated in M stays in M and converges to H is called the Domain of Attraction of H.

Given $\eta > 0$, let

$$H^{\eta} = \{\theta \in \mathbb{R}^N \mid \parallel \theta - \bar{\theta} \parallel < \eta,$$

denote the η-neighborhood of H, i.e., the set of all points within a distance η from the set H.

Definition C.7. A closed invariant set H is Lyapunov stable if for any $\varepsilon > 0$, there exists $\delta > 0$ such that every trajectory initiated in H^{δ} stays in H^{ε} for all time (i.e., if $\theta(0) \in H^{\delta}$, then $\theta(t) \in H^{\varepsilon}$ for all t).

Definition C.8. A closed invariant set H is asymptotically stable if it is both Lyapunov stable and an attractor.

Definition C.9. A closed invariant set H is globally asymptotically stable if H is asymptotically stable and an attractor. All trajectories of the ODE in this case converge to H. Thus, the domain of attraction of H when it is globally asymptotically stable is $\mathbb{R}^N$.

The following theorem gives a criterion to verify asymptotic stability of the set H.

Theorem C.3. *The set H is asymptotically stable for the ODE (C.2) if one can find a function $V : \mathbb{R}^N \to \mathbb{R}$ such that the following hold:*

(i) $V(\theta) \geq 0 \, \forall \theta \in \mathbb{R}^N$,

(ii) There exists an open neighborhood M of H such that $V(\theta) \to \infty$ *as* $\theta \to \partial O$ *(i.e., the boundary of M),*

(iii) $\dfrac{dV(\theta(t))}{dt} = \nabla V(\theta(t))^T \dot{\theta}(t) = \nabla V(\theta(t))^T L(\theta(t)) \leq 0, \, \forall \theta(\cdot) \in M.$

In particular, $\dfrac{dV(\theta(t))}{dt} = 0$ *if and only if* $\theta(t) \in H$.

The following result on convergence of an ODE trajectory is due to Lasalle [3].

Theorem C.4 (Lasalle Invariance Theorem). *Let H be the globally asymptotically stable attractor set for the ODE (C.2). Let* $V : \mathbb{R}^N \to \mathbb{R}$ *be a function such that* $V(\theta) \geq 0 \; \forall \theta \in \mathbb{R}^N$. *Further,* $V(\theta) \to \infty$ *as* $\| \theta \| \to \infty$ *and* $\nabla V(\theta)^T L(\theta) \leq 0 \; \forall \theta$. *Then any trajectory* $\theta(\cdot)$ *must converge to the largest invariant set contained in*

$$\{\theta \mid \nabla V(\theta)^T L(\theta) = 0\}.$$

Definition C.10 ((T,Δ)-perturbation). Given T, $\Delta > 0$, we call a bounded, measurable $y(\cdot) : \mathbb{R}^+ \cup \{0\} \to \mathbb{R}^N$, a (T,Δ)-perturbation of (C.2) if there exist $0 = T_0 < T_1 < T_2 < \cdots < T_r \uparrow \infty$ with $T_{r+1} - T_r \geq T \; \forall r$ and solutions $\theta^r(t)$, $t \in [T_r, T_{r+1}]$ of (C.2) for $r \geq 0$, such that

$$\sup_{t \in [T_r, T_{r+1}]} \| \theta^r(t) - y(t) \| < \Delta.$$

Again let H be the globally asymptotically stable attractor set for (C.2) and H^ε be the ε-neighborhood of H. The following result due to Hirsch [2] (Theorem 1, pp.339) describes convergence to H^ε of a function that closely approximates the ODE trajectory.

Lemma C.5 (Hirsch Lemma). *Given* ε, $T > 0$, $\exists \bar{\Delta} > 0$ *such that for all* $\Delta \in (0, \bar{\Delta})$, *every* (T,Δ)*-perturbation of (C.2) converges to* H^ε.

References

1. Borkar, V.S.: Stochastic Approximation: A Dynamical Systems Viewpoint. Cambridge University Press and Hindustan Book Agency (Jointly Published), Cambridge and New Delhi (2008)
2. Hirsch, M.W.: Convergent activation dynamics in continuous time networks. Neural Networks 2, 331–349 (1989)
3. Lasalle, J.P., Lefschetz, S.: Stability by Liapunov's Direct Method with Applications. Academic Press, New York (1961)

Appendix D
The Borkar-Meyn Theorem for Stability and Convergence of Stochastic Approximation

While there are various techniques to show stability of stochastic iterates, we review below the one by Borkar and Meyn [2] (see also [1], Chapter 3) as it is seen to be widely applicable in a large number of settings. They analyze the N-dimensional stochastic recursion

$$X_{n+1} = X_n + a(n)(h(X_n) + M_{n+1}),$$

under the following assumptions:

Assumption D.1.

(i) The function $h : \mathscr{R}^N \to \mathscr{R}^N$ is Lipschitz continuous and there exists a function $h_\infty : \mathscr{R}^N \to \mathscr{R}^N$ such that

$$\lim_{r \to \infty} \frac{h(rx)}{r} = h_\infty(x), x \in \mathscr{R}^N.$$

(ii) The origin in $\mathscr{R}^N$ is an asymptotically stable equilibrium for the ODE

$$\dot{x}(t) = h_\infty(x(t)). \tag{D.1}$$

(iii) There is a unique globally asymptotically stable equilibrium $x^* \in \mathscr{R}^N$ for the ODE D.1.

Assumption D.2. The sequence $\{M_n, \mathscr{G}_n, n \geq 1\}$ with $\mathscr{G}_n = \sigma(X_i, M_i, i \leq n)$ is a martingale difference sequence. Further for some constant $C_0 < \infty$ and any initial condition $X_0 \in \mathscr{R}^N$,

$$E[\| M_{n+1} \|^2 | \mathscr{G}_n] \leq C_0(1 + \| X_n \|^2), n \geq 0.$$

Further, the step-sizes $a(n), n \geq 0$ satisfy

$$a(n) > 0 \forall n, \ \sum_n a(n) = \infty, \ \sum_n a(n)^2 < \infty.$$

The main result of [2] (see Theorems 2.1(i)-2.2 of [2]) is the following:

Theorem D.1 (Borkar and Meyn Theorem). *Suppose Assumptions D.1 and D.2 hold. For any initial condition $X_0 \in \mathscr{R}^N$, $\sup_n \| X_n \| < \infty$ almost surely (a.s.). Further, $X_n \to x^*$ a.s. as $n \to \infty$.*

[2] also contains a result for bounded step-size sequences (not tapering to zero). However, for our purposes, we only require the result for diminishing step-sizes. Assumptions D.1 and D.2 are seen to be satisfied in many cases, for instance, in reinforcement learning algorithms.

References

1. Borkar, V.S.: Stochastic Approximation: A Dynamical Systems Viewpoint. Cambridge University Press and Hindustan Book Agency (Jointly Published), Cambridge and New Delhi (2008)
2. Borkar, V.S., Meyn, S.P.: The O.D.E. method for convergence of stochastic approximation and reinforcement learning. Journal of Control and Optimization 38(2), 447–469 (2000)

Appendix E
The Kushner-Clark Theorem for Convergence of Projected Stochastic Approximation

We review here an important result due to Kushner and Clark [3] (cf. Theorem 5.3.1 on pp. 191-196 of [3]) that shows the convergence of projected stochastic approximations. While the result, as stated in [3], is more generally applicable, we present its adaptation here that is relevant to the setting that we consider.

Let $C \subset \mathscr{R}^N$ be a compact and convex set and $\Gamma : \mathscr{R}^N \to C$ denote a projection operator that projects any $x = (x_1, \ldots, x_N)^T \in \mathscr{R}^N$ to its nearest point in C. Thus, if $x \in C$, then $\Gamma(x) \in C$ as well. For instance, if C is an N-dimensional rectangle having the form $C = \prod_{i=1}^{N} [a_{i,\min}, a_{i,\max}]$, where $-\infty < a_{i,\min} < a_{i,\max} < \infty$, $\forall i = 1, \ldots, N$, then a convenient way to identify $\Gamma(x)$ is according to $\Gamma(x) = (\Gamma_1(x_1), \ldots, \Gamma_N(x_N))^T$, where the individual operators $\Gamma_i : \mathscr{R} \to \mathscr{R}$ are defined by $\Gamma_i(x_i) = \min(a_{i,\max}, \max(a_{i,\min}, x)), i = 1, \ldots, N$.

Consider the following the N-dimensional stochastic recursion

$$X_{n+1} = \Gamma(X_n + a(n)(h(X_n) + \xi_n + \beta_n)), \tag{E.1}$$

under the assumptions listed below. Also, consider the following ODE associated with (E.1):

$$\dot{X}(t) = \bar{\Gamma}(h(X(t))). \tag{E.2}$$

Let $\mathscr{C}(C)$ denote the space of all continuous functions from C to $\mathscr{R}^N$. The operator $\bar{\Gamma} : \mathscr{C}(C) \to \mathscr{C}(\mathscr{R}^N)$ is defined according to

$$\bar{\Gamma}(v(x)) = \lim_{\eta \to 0} \left(\frac{\Gamma(x + \eta v(x)) - x}{\eta} \right), \tag{E.3}$$

for any continuous $v : C \to \mathscr{R}^N$. The limit in (E.3) exists and is unique since C is a convex set. In case this limit is not unique, one may consider the set of all limit points of (E.3). Note also that from its definition, $\bar{\Gamma}(v(x)) = v(x)$ if $x \in C^o$ (the interior of C). This is because for such an x, one can find $\eta > 0$ sufficiently small so that $x + \eta v(x) \in C^o$ as well and hence $\Gamma(x + \eta v(x)) = x + \eta v(x)$. On the other hand, if $x \in \partial C$ (the boundary of C) is such that $x + \eta v(x) \notin C$, for any small $\eta > 0$, then $\bar{\Gamma}(v(x))$ is the projection of $v(x)$ to the tangent space of ∂C at x.

Consider now the assumptions listed below.

Assumption E.1. The function $h : \mathscr{R}^N \to \mathscr{R}^N$ is continuous.

Assumption E.2. The step-sizes $a(n), n \geq 0$ satisfy

$$a(n) > 0 \forall n, \ \sum_n a(n) = \infty, \ a(n) \to 0 \text{ as } n \to \infty.$$

Assumption E.3. The sequence $\beta_n, n \geq 0$ is a bounded random sequence with $\beta_n \to 0$ almost surely as $n \to \infty$.

Let $t(n), n \geq 0$ be a sequence of positive real numbers defined according to $t(0) = 0$ and for $n \geq 1$, $t(n) = \sum_{j=0}^{n-1} a(j)$. By Assumption E.2, $t(n) \to \infty$ as $n \to \infty$. Let $m(t) = \max\{n \mid t(n) \leq t\}$. Thus, $m(t) \to \infty$ as $t \to \infty$.

Assumption E.4. There exists $T > 0$ such that $\forall \varepsilon > 0$,

$$\lim_{n \to \infty} P\left(\sup_{j \geq n} \max_{t \leq T} \left| \sum_{i=m(jT)}^{m(jT+t)-1} a(i)\xi_i \right| \geq \varepsilon \right) = 0.$$

Assumption E.5. The ODE (E.2) has a compact subset K of $\mathscr{R}^N$ as its set of asymptotically stable equilibrium points.

[3, Theorem 5.3.1 (pp. 191-196)] essentially says the following:

Theorem E.1 (Kushner and Clark Theorem). *Under Assumptions E.1–E.5, almost surely, $X_n \to K$ as $n \to \infty$.*

Remark E.1. We comment here on the validity of Assumptions E.1–E.5. Note that Assumptions E.1, E.2 and E.5 are essentially standard requirements. In particular, the ODE (E.2) turns out to be well-posed as a consequence of Assumption E.1. The requirement on the sequence of step-sizes summing to infinity in Assumption E.2 ensures that the algorithm does not converge prematurely since $t(n) \to \infty$ as $n \to \infty$, even though the difference between successive time points (in the algorithm's trajectory) $t(n) - t(n-1) \to 0$. Assumption E.5 holds because C is a compact set and K being a closed subset of C is also compact.

In the type of algorithms that we consider in this book, ξ_n will typically correspond to the martingale difference term M_{n+1}. In such a case, the process $N_n, n \geq 0$ defined according to $N_0 = 0$ and $N_n = \sum_{m=0}^{n-1} \xi_m, n \geq 1$ will correspond to a martingale with respect to an appropriate filtration. If this martingale is convergent (that can perhaps be shown using say a martingale convergence theorem based argument), then Assumption E.4 can be seen to easily hold as well.

Finally, the sequence $\beta_n, n \geq 0$ in (E.1) will correspond in many cases to a measurement error term. For instance, if say $h(X_n) = -\nabla J(X_n)$, where X_n is the nth parameter update and $\nabla J(X_n)$ is being estimated, i.e., is not known precisely, then β_n could correspond to the error in the gradient estimate. As an example, consider the SPSA algorithm (with projection), see Chapter 5).

$$X_{n+1} = \Gamma\left(X_n + a(n)\left(\frac{J(X_n - \delta(n)\Delta(n)) - J(X_n + \delta(n)\Delta(n))}{2\delta(n)}(\Delta(n))^{-1}\right)\right), \tag{E.4}$$

where $\Delta(n) = (\Delta_1(n), \ldots, \Delta_N(n))^T$ with $\Delta_j(n), n \geq 0, j = 1, \ldots, N$ being independent random variables with (say) $\Delta_j(n) = \pm 1$ w. p. 1/2. Also, $(\Delta(n))^{-1} = (1/\Delta_1(n), \ldots, 1/\Delta_N(n))$. Now (E.4) can be rewritten in the form (E.1) with $h(X_n) = -\nabla J(X_n)$. Also,

$$\xi_n = \frac{J(X_n - \delta(n)\Delta(n)) - J(X_n + \delta(n)\Delta(n))}{2\delta(n)}(\Delta(n))^{-1}$$
$$-E\left[\frac{J(X_n - \delta(n)\Delta(n)) - J(X_n + \delta(n)\Delta(n))}{2\delta(n)}(\Delta(n))^{-1} \mid \mathscr{F}_n\right],$$

and

$$\beta_n = E\left[\frac{J(X_n + \delta(n)\Delta(n)) - J(X_n - \delta(n)\Delta(n))}{2\delta(n)}(\Delta(n))^{-1} \mid \mathscr{F}_n\right] - \nabla J(X_n),$$

respectively, where $\mathscr{F}_n = \sigma(X_m, m \leq n; \Delta(m), m < n), n \geq 1$. Assuming that $\delta(n) \to 0$, it can be seen that $\beta_n \to 0$ as $n \to \infty$. Further, if one assumes in addition to Assumption E.2 that $\sum_n \left(\frac{a(n)}{\delta(n)}\right)^2 < \infty$, then the martingale sequence $\sum_{m=0}^{n-1} a(m)\xi_m, n \geq 1$ can be seen to be convergent. Assumption E.4 is seen to hold in such a case.

Remark E.2. Note that stability of the iterates (E.1) is guaranteed by the fact that the operator Γ projects each iterate of (E.1) to the set C that is a compact subset of $\mathscr{R}^N$. The result as stated in Theorem 5.3.1 of [3] is in fact more general than that described in Theorem E.1. The latter however suffices for our purposes. In applications where it is usually difficult to prove that the iterates of the stochastic recursion are stable, projection is a commonly used technique to enforce stability of the iterates. By choosing the constraint region C to be large enough, one can also ensure in many cases, that C contains all the asymptotically stable attractors of the unprojected ODE $\dot{X}(t) = h(X(t))$. In such a case, it might actually be useful to apply a projection based scheme since then the algorithm would not spend its resources in searching the portion of the parameter space that is known not to contain the stable attractors. In the case when there are no stable fixed points of the unprojected ODE that lie inside the constraint set C (i.e., in C^o), the algorithm will converge to a boundary point of C that is the closest to an asymptotically stable attractor of the unprojected ODE. There could also be spurious fixed points that get introduced because of the projection operation. All such points however lie on the boundary of the constraint region C (see for instance pp. 79 of [4]).

Remark E.3. As described in Assumption E.5, the set $K \subset \mathscr{R}^N$ corresponds to the set of asymptotically stable equilibria of the ODE (E.2). The set of fixed points, say $\hat{K}$, of the ODE (E.2) would contain K in addition to other fixed points that would however be unstable. A stochastic approximation procedure would typically converge to the set $\hat{K}$. It has however been shown, for instance, in [2], [5] and [1] (Chapter 4) that with a sufficiently rich noise sequence, the stochastic update in fact converges to the stable attractor set and avoids the unstable portion of $\hat{K}$ altogether. Further, in practice, it is ususally the case that the stochastic algorithm converges to the stable set (and not the unstable portion) even when no extra conditions are imposed on the noise sequence.

References

1. Borkar, V.S.: Stochastic Approximation: A Dynamical Systems Viewpoint. Cambridge University Press and Hindustan Book Agency (Jointly Published), Cambridge and New Delhi (2008)
2. Brandiere, O.: Some pathological traps for stochastic approximation. SIAM J. Contr. and Optim. 36, 1293–1314 (1998)
3. Kushner, H.J., Clark, D.S.: Stochastic Approximation Methods for Constrained and Unconstrained Systems. Springer, New York (1978)
4. Kushner, H.J., Yin, G.G.: Stochastic Approximation Algorithms and Applications. Springer, New York (1997)
5. Pemantle, R.: Nonconvergence to unstable points in urn models and stochastic approximations. Annals of Prob. 18, 698–712 (1990)

Index

acronyms, list of, XVII
applications
  communication networks, 3, 10, 257
    dynamic multi-layered pricing, 272
    optimal retransmission, 267
    random early detection, 258
  service systems, 9
  traffic signal control, 3, 10, 243
applications, service systems, 225

convergence
  almost surely, 19, 32, 283
  deterministic, 283
  in L^p, 284
  in distribution, 284
  in probability, 32, 283
  mean-square, 19, 284
  nearly uniformly, 284
  uniform, 283

lemma, Gronwall's inequality, 22, 291
lemma, Hirsch, 35, 37, 47, 48, 50, 62, 73, 88, 98, 99, 120, 181, 182, 293
Lipschitz continuity, 19, 22, 25, 32, 34, 44, 46, 60, 66, 69, 70, 83, 84, 96, 114, 118, 119, 141, 170, 181, 208, 215–217, 269, 271, 291, 292, 295

Markov decision process, 9, 188, 245
  admissible policy, 189
  Bellman equation, 17, 190
  policy iteration, 192, 194
  value iteration, 192, 193

optimization problem, 4, 151, 167
  constrained, 5, 167
  constrained, Lagrangian, 169
  deterministic, 4
  dynamic, 4
  stochastic, 4
    average cost, 270
    infinite horizon discounted cost, 190
    long-run average cost, 5, 65, 95, 133, 153, 168, 190, 263, 276
    multi-stage, 4

perturbations, Bernoulli, 7, 65
  four measurements, 107
  one measurement, 47, 49, 111
  three measurements, 109
  two measurements, 110, 160, 172, 173, 234, 237
perturbations, Cauchy, 8
  one measurement, 94
  two measurements, 94, 236
perturbations, Gaussian, 8
  one measurement, 81, 95, 100, 134, 137
  two measurements, 89, 99, 100, 136, 138, 161, 175, 176, 235, 270
perturbations, Hadamard, 52
  four measurements, 263, 265
  one measurement, 55, 63, 198, 276
  two measurements, 55, 56, 62
projection, discrete
  deterministic, 153
  generalized, 155, 157
  randomized, 154
  regular, 157

reinforcement learning, 9, 187
actor-critic algorithm, 195, 202
actor-critic algorithms, 9
function approximation, 206, 247
Q-learning algorithm, 198, 213, 245
temporal difference learning, 206

search algorithms, 13
search algorithms, global, 13
search algorithms, local, 13
descent algorithms, 14
gradient algorithm, 15
gradient algorithms, 31, 41
Jacobi algorithm, 15
Newton algorithm, 15
stochastic approximation, 7
actor-critic algorithms, 9
gradient descent, 31
infinitesimal perturbation analysis, 31
Kiefer-Wolfowitz algorithm, 7, 51, 105
SF approach, 8, 77
SF approach, Cauchy, 94, 236
SF approach, discrete parameter, 161
SF approach, Gaussian, 81, 89, 95, 99, 100, 161, 175, 235, 270
SPSA, 7, 56, 65, 73, 105
SPSA, discrete parameter, 160
SPSA, fixed parameter, 49
SPSA, one measurement, 47, 63, 276
SPSA, one sided, 49, 62, 172, 234
multi-timescale, 7, 23, 171
Newton-based
SF approach, 8, 137
SF approach, one measurement, 137
SF approach, two measurements, 138, 176
SPSA, 8
SPSA, four measurements, 107, 263, 265
SPSA, one measurement, 111
SPSA, one sided, 173, 237
SPSA, three measurements, 109
SPSA, two measurements, 110
Woodbury's inverse, 237
Robbins-Monro algorithm, 7, 18, 23, 31, 51

theorem, Borkar and Meyn, 70, 97, 119, 120, 142, 197, 204, 208, 296
theorem, envelope theorem of mathematical economics, 183
theorem, Kushner and Clark, 74, 100, 198, 204, 206, 213, 217, 272, 298
theorem, Lasalle's invariance, 146, 293
theorem, martingale convergence, 34, 47, 50, 60, 69, 72, 97, 119, 141, 181, 217, 271, 288

Printed in the United States
By Bookmasters